Adam, Apes and Anthropology

Finding the Soul of Fossil Man

By

Glenn R. Morton

DMD Publishing Co.
16075 Longvista Dr.
Dallas, Texas, 75248

This is Edition 1.0 Nov. 23, 1997

ISBN 0-9648227-2-5
Copyright ©1997, Glenn R. Morton

All rights reserved. No part of this publication may be reproduced, stored in a retrieval system, or transmitted, in any form or by any means, electronic, mechanical, photocopying, recording or otherwise without the prior permission of the copyright owner.

Scripture quotations taken from the HOLY BIBLE, NEW INTERNATIONAL VERSION
Copyright © 1973, 1976, 1984 by International Bible Society

Table of Contents

THEOLOGICAL PROBLEM OF FOSSIL MAN	1
THE PLAYERS	4
THE RESPONSE	29
TO LEAVE A LEGACY: THE NATURE OF THE ARCHEOLOGICAL RECORD	43
WHY KANGAROOS CAN'T TALK	49
A WORLD FULL OF RELIGION	62
A CURSE UPON YOUR HOUSE	88
THE HARMONY OF EYE AND EAR	97
THE PHILOSOPHER'S STONE: THE TECHNOLOGY OF FOSSIL MAN	119
INTELLIGENCE AND SOUL	156
ADAM AND EVE WERE NOT MARRIED AND OTHER ODDITIES	175
APPENDIX A: CHRONOLOGY OF HUMAN TECHNOLOGY	188

List of Tables

Table 1	Chronology of Hominid Species	4
Table 2	Tangible Evidences for Religion	63
Table 3	Improvements in Stone Cutting	104
Table 4	Early Sites with Man and Canid	147
Table 5	Randomly Diverging Sequences	175
Table 6	Mandibular Foramen Ancient and Modern Europeans	180
Table 7	Nasion Projections in Fossil Hominids	181
Table 8	Meric Index for Neanderthal and Modern Men	181

List of Illustrations

Figure 1	Taung Child	6
Figure 2	Homo erectus	14
Figure 3	Homo erectus Skull Anatomy	18
Figure 4	archaic Homo sapiens Skull Anatomy	19
Figure 5	Shanidar Neanderthal	21
Figure 6	Neanderthal Skull Anatomy	23
Figure 7	Homo sapiens Skull Anatomy	24
Figure 8	West Natron Hand Axe 1.4 Million Years Old	58
Figure 9	Large Hand Axe Middle Paleolithic	58
Figure 10	Pseudo-Venus from Wildenmannlisloch Cave	64
Figure 11	Side 1: Mas d'Azil Plaque	68
Figure 12	Side 2: Mas d'Azil Plaque	69
Figure 13	Brown Bear and Cave Bear	70
Figure 14	Drachenloch Cave	71
Figure 15	1921 Drachenloch Cross Section	73
Figure 16	Fibulas from Drachenloch Have Polished Ends	74
Figure 17	Total Energy Use for Various Organs	88
Figure 18	Nariokotome Boy (Homo erectus)	92
Figure 19	First Tallensi Two-Dimensional Artwork	99
Figure 20	Dating Rock Art	101
Figure 21	The Burin	104
Figure 22	Barbed Point Manufacture	105
Figure 23	Neanderthal Necklace from Arcy-sur-Cure	106
Figure 24	Acheulian Handaxe with Fossil Blazon	106
Figure 25	Isturitz Vulture Bone Flute	110
Figure 26	43,000 year-old Neanderthal Flute from Slovenia	112
Figure 27	Neanderthal Wall from Wildkirchli	121
Figure 28	Charcoal Particle Density, Lake George, Austral.	134
Figure 29	Oldest Hand Axe 1.4 myr	170
Figure 30	Late Hand Axe circa 300 KYR	170
Figure 31	Chromosomal Crossing Over	177
Figure 32	Adam and Eve were Not Married	177
Figure 33	Other Genetic Adams and Eves	178
Figure 34	The Construction of a Pseudogene	184

ACKNOWLEDGEMENTS

I would like to thank Jim Foley who graciously corrected numerous errors in the original manuscript and corrected some of my French translations. I would also like to thank Dr. James McIntosh, John Burgeson, Daryl Wilson, J. B. Featherstone, and James Lippard for their numerous corrections and suggestions. However, the remaining errors are mine and mine alone.

THEOLOGICAL PROBLEM OF FOSSIL MAN

Praise to the Lord, the Almighty, the King of creation!
O my soul, praise him, for he is thy health and salvation!
All ye who hear, Now to his temple draw near;
Join me in glad adoration!

It is amazing to realize that this hymn plays a role in the controversies surrounding the origin and evolution of mankind, but it does. Joachim Neumann, born in 1650, was the organist and vicar at St. Martin's in Dusseldorf, Germany, in the late 1600's. A talented composer, his congregation loved his music. "Praise to the Lord the Almighty" was his most popular. Seeking the musical inspiration, Joachim would take long walks through the countryside, normally going through a valley which was frequented by him so often that people started calling it by his name. But the name they called the valley was not Neumann.

Being a Christian, Joachim knew that he was a new man in Christ. Like many believers before him, he wanted a new name. Saul had become Paul; Simon became Peter. It was an old Christian tradition. Many contemporary composers did the same. Thus, by taking a new name, Joachim could serve two purposes at once: his Christian purpose and his musical purpose. He decided on a name which would emphasize the message in his original surname. Joachim chose the Greek rendering of Neumann, "New Man," or Neander. The valley he walked through became known as Neander Valley, or, in German, "Neander Tal." While Joachim's music is still being sung, his name is more famous for what would happen in that valley 176 years after his death—the 1856 discovery of a fossil man of unusual proportions.

As the western world became aware of the existence of an ancient man one who did not look like modern men, the intellectual battle was on. The traditional view of our origin was that God had created us just a few thousand years ago. Evolution presented a tremendous challenge to that view especially when fossils of archaic forms of humanity were found in caves. These discoveries had the appearance of supporting evolution, an atheistic form of evolution. The church immediately recognized the theological threat.

The theological difficulties the evolutionary origin of man raises are many. The issues include: the veracity of the Scriptures, the Fall, the redemption, the place of man as the highest of God's creation, the relationship of man to God, death before the Fall, and man made in the image and likeness of God. If man is a product of Nature only, then all of the above are at risk.

More than 140 years have passed since the discovery of Neanderthal, and the issue of what to do with fossil man is even more difficult to deal with. Christian writers have taken a variety of approaches, but when examined in light of modern anthropological data, all of these approaches have major problems. For example, young-earth creationists, like Lubenow, say that *Homo erectus* and Neanderthal are human and should be *Homo sapiens*.[1] This implies that classically defined *Homo erectus*, Neanderthal and *Homo sapiens* are interfertile. When members of two different ethnic groups today interbreed, they usually have children who are morphologically a mixture of their parent's traits. Therefore, if *Homo erectus* and *Homo sapiens* are interfertile we should see transitional forms between the two morphologies.

But then Lubenow says,

> "Second, *Homo erectus* demonstrates a morphological consistency throughout its two-million-year history. The fossil record does not show *erectus* evolving from something else or evolving into something else.
>
> "Third, anatomically modern *Homo sapiens*, Neanderthal, archaic *Homo sapiens*, and *Homo erectus* all lived as contemporaries at one time or another. None of them evolved from a more robust to a more gracile condition."[2]

If this is true, that there are no transitional forms (and, as we shall see, there *are* transitional forms), then *lack* of intermediates is evidence for the *lack* of interbreeding, and thus is evidence that the two forms were not the same species. Lubenow's position is logically self-contradictory.

Lubenow's position forces one of two conclusions. First the various forms may have been capable of interbreeding, but it never happened. Since *H. erectus* was all over the Old World, as was *H. sapiens*, then it means that nowhere did amorous passions arise between the two types. If they were human, then this is highly unlikely. Humans throughout history have long used rape against those they see as different. They have not paid much attention to looks. So are we to believe that NO children were produced during this time? Lubenow's rejection of the transitional forms undermines his own assumption that they were the same species. If they can't interbreed, then they aren't the same species. Secondly, if Lubenow is suggesting that there is a huge amount of morphological variation among the human species, then why would he reject Australopithecus from the human family? There is only a slightly greater variability to include him.

The typical old-earth anti-evolutionist position is no better. Holding that Adam was the first anatomically modern man and was the first man made in the image of God forces one to also hold that such activities as making spears, houses, mining, playing flutes, counting, carving statues and making clothing are not things unique to humans but are things of which animals are capable. The acceptance of the typical old-earth creationist position strips mankind of everything that makes him human and gives it to the animals. It leaves only the "image of God" as the defining feature of mankind. While theologically this is true, it is not scientifically detectable. If a creature can behave like a human, yet still be an animal, then how can I be certain that members of other ethnic groups are really human? The logical deductions from this position are too horrible to consider.

This book is being written because I find neither of the two options above appealing. This book begins with the assumption that the early part of Genesis may be teaching us about historical events. While this is an assumption, one can not ignore the very obvious observational data which science has discovered. With this being the case, it is incumbent upon anyone holding to a historical/concordistic view of early Genesis to explain how the observational data meshes with the Biblical account. A previous work, Foundation, Fall and Flood, examined a wide variety of issues relating to this type of harmonization. This book will concentrate upon the implications the behavior of fossil man have for a Christian world view. While most books touching on fossil man give only perfunctory mentions of behavior and principally examine the morphology of fossil man, this book focus on examining behavior, giving only a cursory look at morphology. There are several reasons for this. One could cite Scripture: "Man looks at the outward appearance, but the Lord looks at the heart." (1 Sam. 16:7). Behavior is as close as we humans can get to looking at the heart. To support this assertion, consider history. The United States had a terrible racial problem in which European Americans made judgements concerning the abilities and humanness of Black Americans, Native Americans and other groups, based solely upon their outward appearance. As we have moved from this myopic attitude, we have found that human abilities are universal. This painful part of our history has shown us that we are unable to determine anything about other humans from a person's looks--their outward appearance. Yet this is precisely the factor on which many Christian apologists base their decision. If Neanderthal, who looks different from modern humans, were still alive and behaved like we do, would we be able to deny him his humanity without becoming racists?

Secondly, behavior is the criterion by which we determine that someone is human in our everyday life. It is a practical Turing test.[3] The Turing test is the test applied to computer programs to determine if an artificial intelligence has been created. It is the rule that if a human cannot tell the difference between the program and a human typing on a computer terminal, then the program is an artificial intelligence. It implies that if you behave like a human, then you must be human. While this is easy to say, human behavior is difficult to define. For our purposes, we will hold to two standards. A fossil man is human *if* he does things only humans do. Thus, if a fossil man makes spears, or controls fire, something no other creature on earth does, then he is human. Secondly, if one member of a species engaged in a particular behavior, then one must conclude that all members of the species were capable of that behavior. The rule is that there are no isolated super-geniuses. While Einstein was the discoverer of General Relativity, there is no reason to believe that Plato, if given the proper education, would not also have been capable of this feat. The fact that relativity was not discovered until the 20th century does not mean that ancient Greeks, Egyptians and Sumerians were constitutionally incapable of discovering relativity. They simply hadn't thought of it; the conditions which led to the theory of relativity never occurred during Roman or Sumerian times. Thus, if one *Homo erectus* was able to make spears, then all were capable of that behavior, whether or not they had thought of it.

Finally, many Christian writers have misconstrued the anthropological data, possibly to support a preconceived theological position. This can be done in two ways: by ignoring information which is known, or by not doing enough research to know that what they are saying is incorrect. Christians deserve to have the best and most thorough research possible. Judge this book based upon that criteria. Christians also have a right, even a need, to know the problems which the popular harmonizations have. After all, as we send our children off to college, if we have not prepared them for what they will find, we have not done them any favors. With this, we will now introduce the players.

References

1. Marvin L. Lubenow, <u>Bones Of Contention</u>, (Grand Rapids: Baker Book House, 1992), p. 120
2. Marvin L. Lubenow, <u>Bones Of Contention</u>, (Grand Rapids: Baker Book House, 1992), p. 178
3. The Turing test was proposed by Alan Turing as a means to determine whether a computer program had artificial intelligence. The test was, that if one could interact with the program and not be able to tell the difference between the program and a real human, the program has become an artificial intelligence.

THE PLAYERS

The players, both living and fossil, in the drama of paleoanthropology interact with each other in fascinating ways. The amazingly interlocked relationship among those involved in the discoveries of the first two species of australopithecines reads like a Dickens novel. The fossil players, *Australopithecus*, *Homo habilis*, *Homo erectus* and *Homo sapiens*, all have traits and behaviors far beyond what most people believe. This chapter will examine the initial finds of the main divisions of fossil man. A knowledge of who is who among the fossils is essential to understanding when Adam existed. This chapter is organized in a taxonomical flow: we will look at the history of the various fossils in the order of their place in human evolution.

Anthropologists have divided the world of fossil man into two main categories, called genera: *Ardipithecus*, *Australopithecus* and *Homo*. All fossil men fit into one of these three genera. *Australopithecus* is the older genus and it contains a variable number of species depending on the views of the individual researcher. The named species in this genus are *Ardipithecus ramidus*, *Australopithecus africanus*, *A. bahrelghazali*, *A. anamensis*, *A. robustus*, *A. boisei*, and *A. afarensis*. The genus *Homo* is broken into three to six species depending upon the researcher. There are *Homo ergaster*, *H. habilis*, *H. rudolfensis*, *H. erectus*, and *H. sapiens*. *Homo sapiens* is

```
                            Millions of years ago
    Species          5.0  4.5  4.0  3.5  3.0  2.5  2.0  1.5  1.0   .5    0
                     +----+----+----+----+----+----+----+----+----+----+
Ardipithecus ramidus +   +X    +    +    +    +    +    +    +    +    +
Australopithecus anamensis +  + X+   +    +    +    +    +    +    +    +
A. afarensis         +    +   +XXXXXXXXXXX  +    +    +    +    +    +
A. africanus         +    +    +    +    XXXXXXXXXXX  +    +    +    +
A. aethiopicus       +    +    +    +    +   XXXXXXXXXXX +    +    +    +
A. robustus          +    +    +    +    +    XXXX +    +    +    +    +
A. boisei            +    +    +    +    +    XXXXXXXXXXX +    +    +    +
Homo habilis         +    +    +    +    +    + XXXXXXXXXXX+   +    +
H. rudolfensis       +    +    +    +    +    XXXXXXXXXXX  +    +    +
H. erectus           +    +    +    +    +    X    +    +    +    +    +
archaic H.sapiens    +    +    +    +    +    +   XXXXXXXXXXXXXXXXXX  +
H.sapiens neandertalensis + +  +    +    +    +    +    +    +   XXXXX+
H.sapiens            +    +    +    +    +    +    +    +    +    +  XX+
                     +----+----+----+----+----+----+----+----+----+----+
                     5.0  4.5  4.0  3.5  3.0  2.5  2.0  1.5  1.0   .5    0
```

Figure 1 Chronology of Hominid Species

further divided into varieties. The oldest fossils of *H. sapiens* are called archaic *Homo sapiens* (sometimes called *Homo heidelbergensis*). There are also *Homo sapiens neanderthalensis* and *Homo sapiens sapiens*. The last variety contains all living human beings.

Unfortunately these are not the only names that have been used in the past. A plethora of names formerly were used for these fossils. *Pithecanthropus*, *Sinanthropus*, *Meganthropus*, *Paranthropus*, *Zinjanthropus*, *Pleisianthropus*, and *Telanthropus* are some of the genus names given to the fossils in the past. Dobzhansky states,

> "Students of fossil man have a habit of giving resounding Latin specific and generic names to almost every bone fragment that they discover, which conveys the mistaken impression that there existed in the past many different manlike species and genera. Weidenreich, however, has pointed out that when the remains of human or pre-human forms which lived more or less simultaneously, are compared, the differences between them are only of the order of those found between the now living human races. There is no fossil evidence of the existence at any one time of more than a single human or human-like species, except possibly, in the case of the Australopithecine."[1]

4

Despite the age of this citation (1955) nothing has really changed the fact that, for most of the past five million years, only one type of human has existed at any one time. If a researcher discovers an example of an already existing form of fossil man, then his discovery is not big scientific news. If he finds a new species, it is better. But to discover a new genus (which implies major morphological differences) draws attention immediately to the find and to the discoverer. In the early part of this century, nearly every new fossil find was given a unique genus name. It was almost as if each generation of fossil man was a different genus from its parents! The fossils widely known as Peking Man were given the generic name *Sinanthropus*. But these fossils were very similar to the fossils found earlier by Dubois in Java, which he had named *Pithecanthropus*. Eventually anthropology put its house in order and made a more compact classification scheme.

This consolidation took place at a conference in 1950 at Cold Spring Harbor, Long Island, New York. Ernst Mayr, a German-trained taxonomist, presented a paper that began the process of consolidation. Mayr, using the same techniques he used on non-human fossils, argued that there were really only three human species, and they were all to be classified in the genus *Homo*. He classified *Australopithecus africanus* as *Homo transvaalensis*. Thus, there were three species of Homo: *Homo transvaalensis*, *Homo erectus* and *Homo sapiens*. He wrote:

> "After due consideration of the many differences between Modern man, Java man, and the South African ape-man, I did not find any morphological characters that would necessitate separating them into several genera."[2]

Mayr's criticism went too far. Anthropologists were not agreeable to doing away with *Australopithecus* as a genus. But even without the demise of this genus, Mayr had accomplished a huge task by simplifying the plethora of names into two genera and a few species.

Australopithecus

During the early part of the 20th century, colonials with their strange accents were not favored or given much chance of advancement by the British scientific establishment. G. Elliot Smith was an Australian who eventually became a world famous anatomist. Unlike native-born Englishmen, he had had a difficult struggle working up the English scientific ladder. Smith, who had worked at the University of Cairo and the University of Manchester prior to being appointed to University College, London, had earned his way up the ladder by developing a way to study the brains of fossil animals by making endocranial casts. He would coat the interior of the skull with liquid latex rubber; after it had set, he would pull the rubber out. This made a mold of the brain that could then be filled with plaster to make a cast. Such casts showed some amazing things about fossil brains. Smith discovered a small crease in the rear of the brain and called it the lunate sulcus because it was shaped like the crescent moon. This feature was visible on the brains and casts of apes, but it was rarely visible on human brains. In human brains the expansion of the cortex hides this feature.

Another colonial who had a difficult time with the British scientific establishment was Raymond Dart. Dart was an intelligent young man who had won a scholarship to the University of Queensland. There, his quick intellect gained him entrance to the University of Sydney where he studied medicine. In 1914 the British Association for the Advancement of Science came to Australia for some presentations. Dart was given a job assisting Arthur Smith, a famous anatomist, in the preparation of his talk. This became an important connection for Dart because Arthur Smith had an even more famous brother, G. Elliot Smith.

After obtaining his medical degree, Dart went to London as a member of the World War I Australian Army Medical Corps. When he was demobilized at the end of the war, he heard that G. Elliot Smith needed an assistant. Dart, through his connection with Elliot's brother, was given the job. At first it was a dream come true; Dart was working for the best man in the anatomical field. But something went wrong. Less than a year after Dart gained the assistantship, Elliot Smith sent him to the United States to work with a lesser-known anatomist, R. J. Terry. Dart was not able to return to England for a year. When he did return to England, he was there just over a year before Smith shipped him off to the ends of the earth to take the anatomy position at the three-year-old University of Witswatersrand in Johannesburg. Some have said that Smith decided on this treatment because he considered Dart flighty and unorthodox in his views. Dart had gained a reputation for scorning accepted opinion, and this was something that the staid British establishment would not tolerate.[3]

Johannesburg in 1922 was still a rough boom town founded on the wealth of gold mining when Raymond Dart and his wife, Dora, moved there (or were exiled there, depending upon the viewpoint). The story is told that when Dart and his wife arrived at the medical school she burst into tears at seeing the facilities. The walls of the dissecting lab were supposedly pockmarked where the students had been practicing tennis and soccer. The truth of the matter was that the pockmarks were not from tennis balls but from other round objects obtainable from corpses and the game was baseball not tennis. The "bats" were human femurs.[4] Two students caught playing this macabre game were removed from classes by Prof. Gerrit Schepers but readmitted by Dart when their parents strenuously objected to such harsh punishment.

After two years, Dart had raised the standards of the school. Since the school had no collection of reference skeletal material, he paid his students for donating to the school the most unusual set of fossil bones. It was by this process that Dart stumbled across the first Australopithecus. Josephine Salmons was a student of Dart and a family friend of the director of the Northern Lime Company at Taung, South Africa. While at dinner at her friend's house she spotted a fossilized skull of a baboon displayed on the mantel. When she reported it to Dart, he dismissed it because there were only two fossil primate species known south of the Sahara. When Miss Salmons brought the specimen to Dart, he realized that she was correct: it was a baboon. Dart arranged with the mine manager, G.W. Barlow, to send him any fossils that might turn up from the quarrying operation.[5] This arrangement soon paid off.

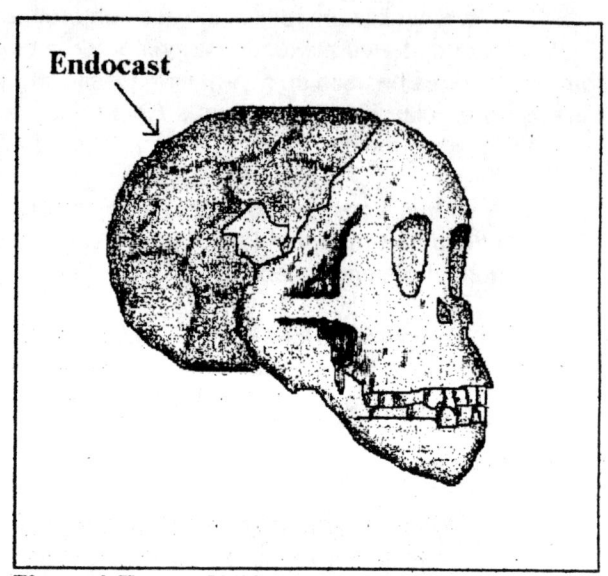

Figure 1 Taung Child

A miner named de Bruyn had been collecting fossils for a long time as a hobby. He was persuaded to send a couple of boxes of fossils to Dart. They arrived in November of 1924 while Dart and his wife were dressing in preparation for hosting the wedding of a friend. Dart took a break from dressing and examined the two boxes. The first box showed nothing unusual, but in the second he immediately recognized a natural endocast of a brain. Because of his work with G. Elliot Smith, Dart immediately recognized that the brain did not belong to an ape. Two fissures at the back of the brain were three times as far apart as they were in chimpanzee brains. This brain had a cortex that was more expanded than a chimp's, but not as expanded as a human. Seeing only the brain with no skull to identify the owner of the brain, Dart began a frantic search of the box to find the skull. At the bottom of the box, he found a rock into which the brain fit. As this was going on, the groom appeared, urging Dart to clean up and come to the wedding. Dart locked the fossil in his wardrobe and went to the wedding.

It took a month for Dart to remove enough rock matrix from the fossil to expose its face. Two days before Christmas he was finally able to see the jaw, with its full set of deciduous teeth, the permanent molars just beginning to erupt. On this basis he concluded that the creature was not an adult but a juvenile. The morphology of the creature was more human-like than any living ape. The brain was higher and more rounded than the flat brains of apes. There was no sign of the brow ridges that appear on the apes and the jaw and teeth were much more similar to those of modern humans. The foramen magnum, the hole through which the spinal cord enters the skull, was situated under the skull as it is in humans rather than nearer the back as is the case with apes. The position of the foramen magnum was indicative of an upright locomotion, like that of humans. Dart named the creature *Australopithecus africanus*.

Dart spent the first two weeks of 1925 writing the report which he mailed off to the British publication *Nature*. When the editor received the package containing the article and photo, he sought out advice from experts like Elliot Smith, Arthur Keith and W. L.H. Duckworth. They gave permission to publish the article, but things quickly got out of hand. The news media in Johannesburg broke the story of a new ape-man on Feb 4, 1925, before the *Nature* article was published. This did not improve matters.

While the initial responses by the scientific community were congratulatory in nature, they quickly turned less favorable. The English establishment had always viewed Dart as a loose cannon. News accounts from the papers did not disabuse them of that notion. In *Nature*, one week after Dart's article appeared, the establishment began raining on Dart's parade. Arthur Keith called Dart's claim that *Australopithecus* was in the human lineage preposterous and sarcastically praised Dart for discovering a new form of ape. Others, noting the lack of detailed knowledge of the precise layer in which the Taung fossil had been buried, suggested that Dart had mistaken recent fossil material for old. In other words, these were merely new bones that had fallen into an old hole. Dart watched as almost the entire anthropological world rejected his opinions on the Taung fossil. His former teacher, G. Elliot Smith, offered little support for his former student.

His only real supporter of his view was Robert Broom, an eccentric Scottish medical doctor with an eclectic career. He had attended only four years of grammar school[6] due to poverty. Because of his work as a lab assistant, he was able to enroll in medical school in Glasglow University, graduating in 1889 with honors in midwifery. Broom, during his Glasglow days, had also been a medalist in geology. When he arrived in South Africa in 1897, he became fascinated with the landscape there. His knowledge of the local geology, which he picked up during his medical trips, won him, in 1903, a professorship of geology at the University of Stellenbosch, a conservative religious institution. His belief in evolution caused his dismissal from that institution in 1910 and he was unable to obtain another professorship.[7] To support himself, he practiced medicine again in the Karroo region of South Africa, where he began to study the vertebrate fossils of the Karroo Formation. His collection of these mammal-like reptiles and the scientific descriptions which he published won him wide praise and a reputation as one of the world's leading paleontologists. By 1920, his work with the fossils had won him a position as a Fellow of the Royal Society, the most prestigious society in the British scientific establishment.

Two weeks after the *Nature* article was published, Broom burst into Dart's lab, ignored Dart and fell on his knees in front of the Taung fossil. When asked what he was doing, Broom said something about "adoring our ancestor". Broom would be the best supporter Dart would ever have. His support was much appreciated, but did not immediately help Dart.

Dart ran into further bad luck when, in 1930, he finished his monograph on *Australopithecus* and submitted it to the Royal Society of London. The ultimate establishment rebuff to his views came when the Royal Society decided not to publish his monograph but instead to publish Arthur Keith's *New Discoveries Relating to the Antiquity of Man* in which Keith laid out a devastating dismissal of Dart's find. Keith's work was perceived to be the final judgment on the creature.

The next year Dart and his wife took the Taung fossil to London to try to win support for his views. He was attempting to win Elliot Smith's support and to persuade the Royal Society to change its mind and publish his monograph, but he was upstaged by the arrival in London of Davidson Black, who was bringing Peking man to town for a visit.[8] The discovery of *Homo erectus* at Zhoukoudian near Beijing had thoroughly excited the anthropological world. Elliott Smith gave an impressive speech in support of Peking man, outlining its significance. His presentations were filled with slides, maps, cross-sections of the cave and casts of the skull for the audience to see. Dart had no fancy professionally-prepared displays for his talk and he was doomed to follow Smith. Dart took the lectern with only the Taung skull in his hand. He said nothing new about the fossil which had not been said years earlier. Halfway through the talk, it was obvious from the faces of his audience that they didn't really want to hear what Dart had to say.

After the speech the ultimate disaster was barely avoided. Some of Dart's friends took him out to dinner to cheer him up, sending Dora, his wife, back to the hotel with the fossil. Unfortunately, Dora left the package containing the fossil in the cab when she got out at the hotel. The package traveled about London all night, sitting on the seat with numerous fares, and was not discovered until the next morning. The taxi driver turned it over to the police, who thought they might have a new murder case. Meanwhile, Dart had returned from dinner, discovered the missing fossil, and notified the police. Happily, the fossil was reunited with Dart, who took it back to South Africa. For the next twenty years, he did little work in the realm of paleoanthropology.

By 1934, at the age of 68, Robert Broom retired from practicing medicine, but poverty forced him to take a position at the Transvaal Museum in Praetoria. In 1933, as the president of the South African Association for the Advancement of Science, Broom was unable to afford train fare for a trip to give a speech. Dart wrote a letter to General Jan Smuts, the newly elected Prime Minister, who arranged for the appointment. However, the Transvaal Museum was not happy to accept Broom into its organization. Years earlier, when Broom was studying the vertebrate fossils of the Karroo, there were persistent rumors (some believe proof) that Broom had been helping to finance himself by selling museum fossils to the American Museum of Natural History. Needless to say, the museum was not eager to accept someone they viewed as a thief back into their midst. However, since the Prime Minister himself had pushed the issue, they had little choice.

The first two years of Broom's stay at the Museum were spent studying mammal-like reptiles which were common in the Karroo. But before long, the challenge of finding an adult *Australopithecus* was taken up. Dart had done nothing in the field for several years. Broom did not want to have to travel to Taung which was quite far away, and looked around for another, closer cave. Two people brought Broom a fossil mandible from Sterkfontein limeworks. Broom dismissed their fossil. A few months later, they returned with another jaw, only this one contained some teeth. Broom was interested and asked to be taken to Sterkfontein on the first free weekend. Broom quickly assessed the situation at Sterkfontein and returned the following day to begin excavations with an entire museum crew. G.W. Barlow, the same man who had been in charge at Taung when Dart was sent his package containing the Taung child, was now in charge of Sterkfontein. He was quickly turned into an ally. Broom told Barlow what he was looking for and nine days later, Barlow walked up to Broom, handed him an endocast of an australopithecine brain and asked "Is this what you're after?"

Broom and the students on his crew used dynamite to remove the fossils from the hard limestone matrix. This procedure would eventually bring the establishment down upon Broom years later. After each blast, They carefully examined the rubble. Eventually this revealed up to eight australopithecine skulls. These skulls proved what Dart had been saying, that *Australopithecus* was hominid. But again, the British and European scientific establishments ignored these discoveries.

In 1938, Broom made what was probably his greatest discovery, thanks to the efforts of a schoolboy named Gert Terblanche. The boy lived near Sterkfontein and worked as a guide in the caves, which were advertised by the owner of the Sterkfontein Limeworks. Gert found a palate with teeth at Kromdraai, a hill about a kilometer from Sterkfontein. The jaw was very large and the creature had massive teeth. Eventually this creature was named *Australopithecus robustus*. It was the second of the currently recognized species of *Australopithecus* to be found.

World War II interrupted the excavations. Broom was asked to restart the search for the missing link in 1946, and so he returned to Sterkfontein. His methods brought quick condemnation from members of the Historical Monuments Commission, who still harbored resentments against Broom's past behavior with regard to the selling of fossils, and his present arrogance. The use of dynamite and pickaxes was criticized for losing valuable information on the stratigraphic levels. This was a valid concern. The Commission granted Broom a permit to dig at Kromdraai. He ignored this and dug at Sterkfontein. The Commission demanded that a "competent geologist" should accompany Broom. He ignored them, and dug anyway. After all, he had been a professor of geology for seven years during his career and was not about to be deterred. The Commission eventually gave up trying to control the great Robert Broom.

In 1947, the ostracism of *Australopithecus* began to end. During the 1940s, Wilfred Le Gros Clark had replaced Arthur Keith as the preeminent English anatomist. Broom had bombarded Le Gros Clark with letters supporting the hominid status of the australopithecines. In one exchange, Le Gros Clark had indicated that he needed to see photographs of the original, as well as careful drawings and casts before science could make a judgment on the fossils. Broom shot back,

> "You say anatomists in England will have to suspend judgment until casts etc. etc.... English judgment may be of a high order, but when *Australopithecus* was discovered in 1924 England did not suspend judgment. Four English scientists at once expressed their opinion that it was a chimpanzee."[9].

Wilfred Le Gros Clark was thus persuaded to come to South Africa to examine the hominid material himself. He had come prepared with a list of traits that apes have that humans do not and a list of traits that humans have which apes do not. The traits Le Gros Clark observed first hand in the fossils fit those of a human. Le Gros Clark's reputation was impeccable. When he left South Africa to attend a Louis Leakey-convened conference in Nairobi, Kenya, he was fully converted to the view that *Australopithecus* belonged in the human lineage. This was the beginning of the end of the establishment's resistance. When Arthur Keith read in the *Nature* report of the Nairobi conference confirming that Le Gros Clark had thrown his support in favor of *Australopithecus*, he wrote an admission to the same journal stating that Dart had been correct and that he had been wrong.

At the end of his life, Robert Broom was a driven man, trying to finish his monograph on the South African Australopithecines. This was to be the crown for his long career. On April 6, 1951, he completed the final corrections to this work. Reportedly, he said, "Now that's finished...and so am I." He died that night.[10]

Since that time several other australopithecine species have been found. In November, 1974, Donald Johanson discovered the fourth of the presently accepted australopithecine species. This one was found in Hadar, Ethiopia and was classified as *Australopithecus afarensis*. This is the famous fossil which was named "Lucy." Subsequent studies have convinced many anthropologists that this species was the ancestor of man rather than *A. africanus*. In 1995, Meave Leakey, daughter-in-law to the man who organized the Nairobi conference which was the turning point in Australopithecus's fortunes, reported finding another species of australopithecine, *A. anamensis*, in rocks dating four million years old.

The Traits of *Australopithecus*

What are the traits that convince anthropologists that australopithecines deserve to be classed with man rather than with the apes? There are several. The jaw is like that of a modern man rather than that of an ape. Christopher Wills relates,

> "...when the jaw was viewed from above, the teeth were evenly spaced along it in a smooth parabolic curve, with the lines of molars on each side of the jaw diverging rather than running parallel as they do in apes. And one prominent feature of the dentition in the great apes, a gap in the row of teeth of the lower jaw called a diastema, was not present. Because the upper canines project so markedly in the apes, the diastema is needed to allow the jaws to close properly. But the Taung child's upper canines did not project, and there was no diastema. In short, the teeth and jaw were remarkably humanlike, indeed far more humanlike than the rest of the skull with its mild prognathism and small cranial vault."[11]

The Taung child had back teeth that were large in proportion to the front teeth. This is a human characteristic, not an apelike one. *Australopithecus* was also lacking the apelike characteristic of having large canines. Humans have small canines. When sufficient teeth were recovered from the australopithecines it was shown that the shape of the various teeth did not match those of the apes but were more manlike.

Another trait which is human-like in the australopithecines is the position of the foramen magnum. This is the hole in the skull through which the spinal cord enters the brain. In adult apes this hole is toward the back of the skull. This is because apes do not walk upright. In order for apes to see as they knuckle walk, the skull must be situated so that the foramen magnum is on the back of the skull rather than under the skull as in humans. In order for humans to see as we walk, the skull must be supported from below. If our foramen magnum were like that of apes, we would always look at the ceiling. If theirs was like ours, apes would always be looking at the ground, unable to see approaching predators. *Australopithecus* had a foramen magnum like ours, *under the skull*.[12] He walked upright. Dart was criticized when he first suggested this possibility because all he had was a juvenile, and it is difficult to prove bipedal locomotion from a juvenile. Discoveries by Broom in the late 30's and 40's showed that adult australopithecines also had the foramen magnum under the skull.[13] In 1947 Broom and Robinson uncovered an australopithecine pelvis at Sterkfontein which proved that *Australopithecus* was bipedal.[14] B In 1978 Andrew Hill discovered the 3.7-million-year-old footprints an australopithecine had left in a volcanic ash at Laetoli, Tanzania.[15] These prints prove that someone was walking bipedally at that time. Additionally, these prints show that their feet were shaped like ours.[16]

The most often cited trait of *Australopithecus* is the brain size: his cranial capacity. The cranial capacity of a fossil hominid is estimated either by filling the skull with mustard seed and then measuring the volume of the seed or by making a latex mold of the interior of the skull and measuring its volume. Such measurements of the brain sizes of australopithecines yield values from 350-650 cubic centimeters.[17] Their average brain size is 464 cc. The older australopithecines, like *A. afarensis*, have smaller brain sizes. To put this in perspective, chimpanzee brains range from 282-500 cc with an average value of 383 cc and modern humans have an average cranial capacity of 1370 cc.

The australopithecines range in height from three and a half feet[18] to five feet. They range in weight from 66 to 200 pounds.[19] Their canines are smaller than the canines of a chimpanzee but are bigger than those of modern men.

Geographically, australopithecines are the most limited in extent of all the hominids. They have been found in South Africa and throughout the Rift Valley of East Africa. Just within the past couple of years *Australopithecus* has been reported in Chad.[20] This extends the range of *Australopithecus* 2,500 miles to the west of the Rift Valley. Nevertheless, all discoveries have been from Africa alone.

Homo habilis

The next important fossil species recognized today is *Homo habilis*. This species was defined in 1964 by Louis Leakey, Phillip Tobias and John Napier. The fossil *habilis* was first found in Olduvai Gorge, in Tanzania. The players in this drama are quite intriguing as well.

Olduvai Gorge is named for a plant that grows abundantly in the region. This plant is called *oltupai* by the local Masai tribesmen.[21] The German, Hans Reck, who first studied this canyon in 1913, westernized the name and called it Olduvai (pronounced Oldoway). The story is told that a butterfly expert, named Kattwinkel, accidentally discovered Olduvai Gorge when chasing a specimen. Not paying attention to where he was running, he nearly plunged to his death when his intended specimen flew to the edge of a 300 foot precipice.[22] In any event, Kattwinkel did bring back some mammalian fossils from this remote location. Hans Reck, seeing what Kattwinkel had, wanted more and organized an expedition to the gorge.

The 1913 Reck expedition was very productive. Reck returned to Germany with over 1700 fossils, a knowledge of the geology of the area and one human fossil. Reck dubbed this fossil *Oldoway man*. The skull and skeleton were entirely modern but had been found in layers which contained mammalian fossils believed to be at least a million years old. Further, the skeleton had been "flexed."[23] This is a burial procedure in which the body is tied up with the knees drawn to the chest as if the body was in a crouch. This position indicated intentional burial, not a chance burial. If modern man were intentionally burying his comrades one million years ago, that would be an astounding discovery. Reck claimed that this was the case. Needless to say, this claim was viewed with suspicion. Most authorities believed that Oldoway man was a recent burial in the older sediments. They cited the fact that Reck had not found a single stone tool to go with the modern man.[24] Since modern man made tools, Reck must be wrong. Reck organized a second expedition to attempt to prove his position, but this was 1914. World War I had broken out, making it impossible for the expedition to proceed.

Louis Leakey was the next person to have an impact on Olduvai Gorge. Leakey was born on August 7, 1903 in Kabete, Kenya, to missionary parents. His childhood was spent with the Kikuyu people. At the age of 13, he was initiated into the Kikuyu tribe. His experiences gave him a fluency in Kiswahili and an intimate knowledge of the countryside. Like many colonials, Louis was sent to England to obtain his education. He studied anthropology at St. John's College at Cambridge under Arthur Keith, the man who gave Dart such trouble over the australopithecines. In the mid-1920's Louis was struck in the head during a rugby match. He was advised by a doctor to take a year off from college to recuperate so he returned to Kenya and began looking for fossils, an event which changed his life. The first thing Louis did was to contact Hans Reck about Olduvai Gorge. While Leakey was unable to go to the gorge, though this contact would prove fruitful five years later.

While Leakey was not a person many would include in the ranks of Christendom, the faith of his parents did leave a lasting impression on his life and the direction of his research. Johanson and Shreeve write:

> "Keith is often credited for planting in Leakey his unwavering conviction in the antiquity of man—a belief that Leakey himself would carry to inspiring heights. But the seed may already have been sown. Deeply religious by nature and training, Louis was still considering following his father's path into the ministry as late as 1925. Leakey was no creationist. But Christianity, unlike many other religions, teaches that man alone is endowed with a soul, and that his dominance over the rest of creation is part of the divine order. It is hard to imagine how Leakey's thinking could *not* have been influenced by a belief in this special relationship between man and the Creator. Such a faith does not easily accommodate the notion of a recent derivation of humanity from a chimpanzee look-alike."[25]

By 1931 Leakey had led two expeditions into Africa, but his connection with Hans Reck led to a 1931 joint expedition to Olduvai Gorge with Reck. Leakey had bet Reck that within 24 hours Leakey would find the stone tools that Reck had not found in 1913. Reck took the bet. The expedition reached the Gorge at 10 A. M., September 26, 1931. By the next morning Leakey handed Reck a perfectly made Acheulean hand axe made of volcanic rock. Hans Reck paid the bet happily, thinking that he had just proven his critics wrong. As it turned out, hand axes were lying all over the place.

How had Reck's 1913 expedition failed to find these ubiquitous tools? Reck had fallen victim to something every fossil hunter has experienced. If you don't know what to expect or are expecting something different, it is occasionally impossible to find what is there. Reck had been expecting to find *flint* tools as all good European anthropologists are taught. The problem was that there is no flint in that part of Africa; the ancient men could not comply with European expectations. Leakey, because he had grown up in the area, knew what to look for.

I have had this experience. Once my employer told me to take a large check as a donation to a college. The geology dean asked me if there was anything they could do for me while I was in town. Smelling the perfect opportunity, I asked them if they could show me a good place to find trilobites, as I had never excavated one before. They took me to a cliff, called Lusters Gate, along a road several miles out of town. They told me that this was their absolutely top secret location for getting prized trilobites. They left me out there on my own for an entire morning. I climbed the cliff, searching for trilobites. Growing up in northeastern Oklahoma, I was trained to see most fossils in limestone. Finding fossils there requires looking for subtle difference of color and shading on the limestone. I used my childhood experience. I fell from the cliff, landing head first on the ground, collected five ticks and numerous cuts from local thorn bushes, but I found *no* trilobites. When I got back to town, the professor showed me what I should have been looking for. The trilobites were the same color as the rest of the rock and were mere bumps in bas-relief which eroded out of the cliff face. I had not known what to look for. The professor gave me two, which I still have, as a consolation prize for my efforts.

Leakey had gone to Olduvai with Reck almost convinced that Reck was wrong about the age of Oldoway man. Leakey had generally accepted the criticism that it was an intrusive burial. Quickly Leakey was converted to Reck's position and indeed claimed in *The Times* that Reck had been correct. Oldoway man was the most ancient known man. Unfortunately, Leakey was quickly proven wrong, by, among others, a geologist named Percy Boswell. The skeleton was no older than 20,000 years.[26] Leakey was developing a reputation for rashness.

Leakey's next two discoveries didn't help. He discovered two very modern skulls with tools and extinct elephant bones at Kanjera and he found a human-like jaw at Kanam. Leakey showed more impetuousness and poor judgment when he proclaimed that Kanam man was older than Reck's Oldoway man. People didn't believe Leakey immediately because his past track record was less than sterling. Arthur Smith Woodward organized a conference to examine these new claims. Leakey presented his best case and the next day the others discussed the data. The convocation decided that Leakey was correct and they agreed with the proposition that Kanam and Kanjera men were proof of an ancient lineage for Homo. Everybody was convinced and congratulated Leakey except Percy Boswell, who supported Piltdown as the ancestor of humanity and began to pick apart the evidence Leakey had assembled. Boswell continued to attack Leakey's position, and finally Leakey made a disastrous suggestion: he suggested that the Royal Society should pay for Boswell's trip to Kenya to examine the Kanam site for himself. The Royal Society complied.

Leakey and Boswell went to Kanjera, where things began to unravel. Leakey could not pinpoint exactly where he had recovered the jaw. The best he could do was to say that it was within 10 meters of a given spot. Boswell and Leakey went to Kanam next, where the local fishermen had removed the iron stakes he had used to mark the location. But much worse, the photo Leakey had used in his London talk (of the site where the Kanam mandible was found) was not a picture of the location where he had found the mandible. The photo was of a site a few hundred meters away. When Boswell got back to London, he wrote a highly critical letter to *Nature* describing how Leakey's data was misleading at the very least. Since no one likes to be misled, least of all the scientific community, Leakey's reputation plummeted.

To make matters worse, the time in the field had placed a great strain on Leakey's marriage. By 1934 Louis and his wife, Frida, were separated. By 1936, Leakey was living with Mary Nichol, an archeological illustrator, while awaiting the final divorce decree from Frida.[27] Since divorce was a social taboo, this was the final straw for Leakey's hopes of an academic career; his ability to obtain grants for his work was equally in shambles. Louis, with his new wife Mary, left for Kenya where he could glean only the tiniest grants to support his dreams of discovery. These events did teach him to cover his bases if another discovery were to come his way.

Throughout the rest of the 1930s and 1940s Louis and Mary worked at an incredible pace. They opened up one archeological site after another. In 1948 they discovered a skull of *Proconsul*, a Miocene ape believed to be ancestral to *Australopithecus*. The discoveries they made during that time made the trespasses of Kanam fade. They worked at Olduvai Gorge off and on over the years. They would travel across the trackless plain to the Gorge, work as long as their supplies would last, and then return again after they could scrape up enough money to make another trip.

In the 1950s the Leakeys returned to Olduvai more often because Louis now had a museum salary to help support them. They found fossil mammals, such as a baboon as large as a gorilla, pigs the size of a hippopotamus, and buffalo with horns extending for 7 feet. They also found many primitive stone tools and bones butchered by the stone tools, but they had no clue as to who was making and using the tools. During this time they worked out the geology of the Gorge in great detail. Bed I was the lowest and Bed IV was the uppermost. Each bed was separated from the others by volcanic ash. On a July morning in 1959, Mary Leakey took her dogs for a walk in FLK, a gully (korongo) named for Frida Leakey, Louis' first wife. Louis was sick in camp with a malaria attack. It was that morning that Mary found the skull that was named Zinjanthropus or Nutcracker man after the huge teeth that it had. Zinjanthropus was a robust australopithecine, and is today classified as *Australopithecus boisei*. The skull, was in a layer with primitive stone tools dated to 1.75 million years. This made big news at the time because it was the earliest evidence of tool-making man. Overnight, Leakey's work had tripled the time fossil man had lived on the earth.

When the discovery was announced, it brought fame and funding to Leakey. Prior to the discovery of *Zinjanthropus*, the National Geographic Society had never given a penny to Leakey, but now they funded him lavishly and wrote articles about him and his work in their magazine. It was this funding along with lots of digging, that brought about an increased pace of discoveries, eventually leading to the realization that there was another species in the beds with *Zinjanthropus*. It would be called *Homo habilis*.

In November, 1960, once again at Olduvai, Leakey struck pay dirt. This time it was his son, Johnny, who found the fossil. Jonathon Leakey had been allowed to excavate a sabertooth tiger. During his excavation, he discovered a mandible and two cranial fragments. Labeled Olduvai Hominid 7 (OH 7), the cranial fragments allowed an estimate of the cranial capacity of 650 cc.[28] This brain size was considerably larger than any that had been found in an australopithecine. That same year a very human-like set of foot bones and hand bones were also found in Bed I. These showed a fully upright means of locomotion.

Further discoveries indicated that there was a big-brained hominid in the same strata as *Zinjanthopus*. As a result, Leakey called in Phillip Tobias, a former student of Raymond Dart, and John Napier to help analyze these fossils. Napier was brought in to examine the anatomy. Walker and Shipman relate:

> "John's [Napier] analysis of the anatomy of the new finds was that the species from which they were drawn--like modern humans--had both power and precision grips, a perfect combination for the earliest toolmaker. The finding delighted Louis to no end, since he had never been comfortable with the hypothesis that the small brained, heavy-skulled Zinj was the toolmaker at Olduvai."[29]

In 1964, Leakey, Tobias and Napier published a paper suggesting that the real tool maker at Olduvai was *Homo habilis*, not *Zinjanthropus*. This was the first definition of *habilis*, a name suggested by Raymond Dart. From the start, *H. habilis* has been a controversial species. The fossils to which the label *H. habilis* has been attached are widely varied in morphology. Walker and Shipman write:

> "These discrepancies have made for a diverse group of *habilis* specimens that have different sized and shaped teeth, different palates, markedly different faces (for example, 1470's face is about twice as long as the face of 1813 from between the eyes to the bony sockets for the front teeth), and tremendously variable cranial capacities, from just over 500 cc to about 800 cc. The fundamental question is whether all of these specimens are too variable to be grouped into one species. Differences in shape and dimensions within a single species is a phenomenon often attributed to sexual dimorphism, meaning that size and shape varies with sex. Sexual dimorphism is common among primates, with male gorillas weighing twice as much as females, or male baboons having greatly elongated canine teeth and long, doggy faces while females have small canines and much shorter faces. The typical pattern in primates is for sexual dimorphism to make the males bigger and more robust than females, but some species, like chimps, have only a mild degree of sexual dimorphism and gibbons have none."[30]

The Traits of *Homo habilis*

The one trait that has defined *habilis* is its "intermediate position" between *Australopithecus* and *erectus*.[31] The problem is that not all the fossils that are attributed to *habilis* are intermediate. When put on a time chart, the traits do not show a simple pattern going from *Australopithecus* to *Homo*. It is possible that there are two or more species involved which have become lumped together taxonomically.

The most important traits found in *habilis* concern the brain. As noted above, the brain size varies from

500 to 800 cc, but even more important than that is the fact that certain brain structures, which are unique to *Homo sapiens* today, are first found in *Homo habilis*. When the inside of a skull is examined, one finds certain bumps and ridges hardened into the bone. These shapes reflect the shape of the brain that once occupied that skull. While the brain may be softer than bone at death, in the juvenile state the brain's growth impresses the brain's surface onto the soft, juvenile bone. In humans and their immediate predecessors, Neanderthal and *Homo erectus*, the interior of the skull preserves a ridge that corresponds to areas known as Broca's area and Wernicke's area. These two areas are involved in the production and comprehension of speech. An enlarged Broca's area is unique to humans. Dean Falk reports,

> "The one area of the cortex that best distinguishes an ape brain from a human brain is the frontal lobes. A triangular fold of gray matter known as Broca's area appears in left frontal lobes of humans and is associated with speech. This area does not appear in ape brains. Not only is the lower frontal lobe the best area for distinguishing an ape brain from a human brain, but it is also the area which is most likely to leave a good impression on the inside of the skull and therefore appear on an endocast --a happy accident."[32]

An enlarged Broca's area first appears in the skulls of *Homo habilis*. KNM-ER 1470, is a *habilis* skull from Kenya dating to around two million years ago. It clearly has this enlarged area impressed upon the interior of the skull.[33] No other animal on earth possesses an enlarged Broca's area—only man. This feature is nearly a defining trait of humanity.

This simple fact has tremendous implications for Christian apologetics. Language was given to Adam by God through the act of Adam naming all the animals. Yet here we have a being who does not look a lot like us, has a smaller brain, and yet possesses the brain area involved in speech. What are Christian apologists to do with *Homo habilis*? Even a skeptic of the early appearance of language, Terrence Deacon admits that *Homo habilis* exceeded the language ability of modern apes more than 2 million years ago.[34]

Another manlike feature of *habilis* is a pronounced asymmetry between the two hemispheres of the brain. Shreeve notes,

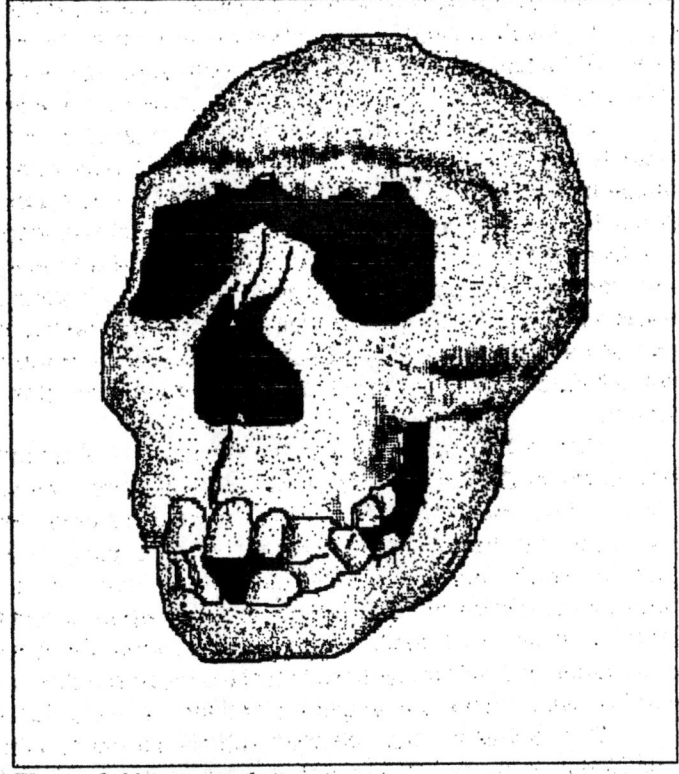

Figure 2 *Homo erectus*

> "According to Ralph Holloway of Columbia University, the leading authority on ancient hominid brain structure, the markings revealing Broca's and Wernicke's areas appear millions of years before the Creative Explosion was allegedly triggered by the emergence of language, certainly by the time of *Homo habilis*. Holloway has also shown that *habilis* skulls reveal cerebral asymmetry: a left-hemisphere lopsidedness, which is associated in our species with language. More recently, Terry Deacon of Harvard University has pointed to language-related structures in the prefrontal cortex of the brain that also began to swell beginning with *Homo habilis*."[35]

Whatever the status of the postcranial (body) bones, the brain of *Homo habilis* appears quite human.

Homo erectus

When Darwin published The Origin of Species in 1859, there was only one known fossil man and that was Neanderthal, who had been found merely three years earlier in Germany. Darwin did not deal with the origin of man in his first book because he knew that it would be entirely too controversial. But this did not stop others from engaging in discussions of the issue. Ernst Haeckel, a biologist, was the most vocal of Darwin's supporters in Germany. He had been a student of the most famous biologist of that time, Rudolf Virchow, who we will see involved in the aftermath of the discovery of Neanderthal. Haeckel is most famous for proposing the "law" that animals replay their evolutionary pathway when an embryo develops into an adult. This was phrased in the English world by the saying, "Ontogeny recapitulates phylogeny." In 1868, Haeckel proposed that a new species of fossil man would be found, which would be the "missing link" between man and the apes. He even gave this creature a name, *Pithecanthropus alalus*, meaning "ape-man without speech."

The term "missing link", coined by Haeckel, has remained with us to this day. But it was occasionally an appellation of mirth in earlier days. In 1877, Darwin was to give a lecture to the students of Cambridge. Darwin's wife, Emma, noticed some cords that had been hung from the ceiling and strung across the gallery. She wondered what they were for. She would soon find out. A monkey puppet appeared on the first of the cords. This caused a lot of commotion until it was removed by an administrator. A ring with ribbons tied on it came next. This, of course, was the missing link!

Virchow, a man who demanded evidence, opposed Haeckel's methods of science which seemed to make up evidence and to name creatures which didn't exist. However, in spite of Virchow's opposition, Haeckel's writings became quite influential across the European continent. Haeckel was an enthusiastic speaker and was given many opportunities to speak. At one such engagement at the University of Amsterdam, Haeckel outlined his views on the origin of man. He felt that the missing link would be a stooped, half ape, half manlike animal. The missing link would have to walk in a fashion intermediate between humans and apes. He believed that mankind was most closely related to the gibbon rather than to the African apes. Because of this, he felt that the missing link would be found, not in Africa, but in Asia--southeast Asia to be specific. Haeckel believed that the most likely place was Pakistan, but he encouraged the students at the lecture to examine the caves of the Malaysian islands.

One young man who heard this talk, a lecturer at the medical school named Eugene Dubois, became fascinated by the problem of human origins. Dubois was appointed to work with Max Furbringer, a leading comparative anatomist, but their relationship was not a good one. Dubois was easily offended and did not want his mentor attending his lectures out of fear that others would think that Dubois could not handle the job on his own. Dubois did not like criticism or questions and was also angry about the authorship of a paper that he and Furbringer had worked on together but on which only Furbringer's name appeared. Due to the increasing tensions between him and Furbringer, as well as Dubois' increasing interest in human origins, Dubois decided to quit the university and go to the Dutch East Indies in search of the missing link. Dubois was so angry that he did not even tell Furbringer when the ship was leaving in spite of the fact that Furbringer had specifically asked to see them off.

Dubois had signed on with the Royal Dutch East Indies Army as a surgeon. It was going to be a very different life for Dubois, his wife, Anna, and their child. The ship arrived at Padang, Sumatra, in December 1887. At first, Dubois' medical duties occupied most of his time. He spent his time off duty searching for fossils in the local caves, but he could see that this was not going to work. There was too little time for fossils and the distances which could be traveled during his hours off duty were too short. What Dubois did find was an impressive collection of fossil vertebrates. He began to badger the authorities for time and funds with which to search for fossils. Eventually, after two years, when the East Indies Committee for Scientific Research granted Dubois funds, the military relieved him of his medical duties, and he became a full time fossil hunter. He was also given the use of some civil engineers and fifty convicts for his excavations.

Dubois was no more pleased with his current situation than he had been with his life in academia back in the Netherlands. He complained about everything. He called the Malays "indolent" and complained that the natives were hiding caves from him. He complained about the living conditions in the field (something he should have expected), about the bouts of fever he had, and the uselessness of one engineer assigned to his staff and the death of the other. Seven of his convict laborers, seeing a chance for a better life, ran away; others were sick with fever.

In 1890, Dubois moved to Java due to the discovery of a fossil human skull at Wadjak. This skull is today classified as *Homo sapiens*. On Java he found an enigmatic jaw fragment with one premolar tooth. Neither Dubois nor anyone else has been able to make much of this fossil. In August, 1891, reports surfaced that animal fossils were seen eroding out of the banks of the Solo River near Trinil, which lay at the foot of the volcano Lawu. Dubois moved his operations there. This proved to be a fortuitous decision, and he ordered his convicts to dig where the fossils were. This was a different procedure than was normally followed. Usually anthropologists excavate only when they find a human fossil of interest eroding out of the soil. In this case, Dubois had not seen anything of anthropological interest but had dug anyway. The convicts dug a huge pit that eventually measured 40 feet across and 50 feet deep.

Dubois returned to his base camp at Tulungagang and awaited the results of the dig. In September 1891, the crew sent him a hominoid molar tooth. Dubois named it *Anthropopithecus erectus* after an extinct ape which had recently been found in the Siwalik Hills in India. Dubois had never seen that fossil, so the comparison was made from a written description. In October, Dubois received a skullcap which was similar in morphological form to that of the first Neanderthal but was a much smaller skull.

In May, 1892, after the rainy season had ended, the convicts continued their work. In August, 1892 they found a femur that was reported to have come from the same stratigraphic level as the skull cap and tooth. In reality, it was found some distance away but no one is sure exactly how far; the reports have said that it was found either 10, 12 or 15 meters upstream from the skull cap. The femur was like that of modern humans. Up to this point, Dubois had thought the skull belonged to a chimpanzee or gibbon, and he added the name *erectus* to his fossil. Not having any comparative material, Dubois wrote to Max Weber, an anatomist back in the Netherlands, asking for a chimpanzee's skull. While he was awaiting the arrival of the chimp skull, he continued to call the fossil *Anthropopithecus*. Within nine days of the December 18th arrival of the chimp's skull, Dubois changed his mind and renamed the fossil *Pithecanthropus*. The chimpanzee skull clearly showed that the cranial vault of Dubois' fossil was too small to be a human and too large to be an ape. Dubois decided that he had found Haeckel's *Pithecanthropus*. Dubois named it *Pithecanthropus erectus*. Eventually Haeckel would write Dubois, "congratulations to the discoverer of *Pithecanthropus* from its inventor."[36]

While the find was certainly supportive of Haeckel's views that the missing link would be found in Asia, it was not quite what Haeckel had in mind. The femur of this creature was straight, like a human's; Haeckel had expected a stooped-walking creature. Pithecanthropus walked upright. Dubois estimated the brain to have been 1000 cc, which is at the lower end of the normal human brain size range.[37] Modern estimates place it between 900-940 cc.

It is interesting that this fossil would eventually be called *Homo erectus* and yet the name itself was first used as a joke. Some anonymous writer played off the name *Anthropopithecus erectus* to construct his own name which referred to the writer not to a fossil. Walker and Shipman relate,

> "On February 3, 1893, a local newspaper, the *Bataviaasch Niewsblad*, carried a semisatirical account of his discovery, taken from his November 1892 quarterly report that used the name *Anthropithecus erectus*. Dubois clipped the article and saved it for the rest of his life. Ironically, this article is signed *Homo erectus*; it is by many years the earliest known publication that uses the name in conjunction with this fossil. But science moves slowly. In the scientific literature the suggestion to revise *Pithecanthropus erectus* to *Homo erectus* was made in the mid-1940's, but it was not until 1960 that the change was formally accepted."[38]

Throughout the rest of 1893, Dubois worked to write a monograph on his fossil. It was a hard period in his life. News arrived that his father had died; his wife miscarried, and he was often ill from fever. The fact that no stone tools were found with the fossil was a difficulty that needed addressing. There was no one around trained in science with whom he could discuss the meaning and implications of the find. There was also no one around to tell Dubois how important the geological description of the deposit would be. He spent many hours in conversations with Adam Prentice, a manager of a coffee plantation, for want of a more suitable conversationalist. By the end of the year, Dubois finished his monograph (which lacked a clear geological description of the site). He published it in a special paper in Batavia (now Jakarta) rather than submitting it to a reputable journal. Cheerfully, he mailed it out to the scholars in Europe, confident that fame awaited him now.

In early 1894, Dubois and his family set out for Holland, stopping first in India to see *Anthropopithecus* for himself. Dubois was upbeat for this first leg of the trip until he began to hear of the reaction to his monograph. The loose standards for archaeological digs in the nineteenth century raised many problems for Dubois and they were now coming home to roost. Scholars questioned whether the femur had been from the same individual as the skull cap. The engineers in charge of the dig had not made maps nor had they made any measurements of the exact stratigraphic positions of the finds. All they had provided was a written description of the layer in which the material had been found. Dubois himself had not been at the site to verify anything. Virchow, in Germany, believed that the skull was that of an ape but the femur was clearly human, leading to the conclusion that Dubois was putting two unrelated species together into one being. If Dubois thought that his old mentor, Furbringer, attending his lectures was a slap at his competence, this blast from Virchow must have really stung.

The response in England and Ireland was better. Daniel Cunningham, an Irish anatomist, was open to the idea that *Pithecanthropus* had given rise to Neanderthal and through them, modern humans. Dubois took his box of bones to England to be examined by a newly emerging anatomical expert named Arthur Keith, the same man who would later oppose Dart's view of *Australopithecus*. Keith supported Dubois' interpretation that these bones were all from the same individual but surprised Dubois by suggesting a third alternative. The skull was not an ape's nor was it a transitional form. It was the skull of a primitive human.

This conclusion and other events were beginning to convince Dubois that he should withdraw from the scientific scene. As bad as things were, his one real supporter, Gustav Schwalbe, a German anatomist, committed an even worse sin. In 1899, Schwalbe had obtained a cast of the skull from Dubois, wrote an even more impressive monograph on the fossil, and went on the lecture circuit. While Schwalbe agreed with Dubois about the transitional nature of the fossil, Dubois felt that Schwalbe had stolen the fossil and the credit. This was the last straw for Dubois, who removed his fossils from the scientific stage and refused to let researchers examine them. Some reports relate that Dubois buried his bones under the floor boards of his dining room. Others say he kept them in a cabinet in his dining room. During the many years in which Dubois was out of circulation, rumors floated that he was insane and had destroyed the fossils.

The next big find of *Homo erectus* took place in China. Davidson Black was a Canadian medical doctor who had worked with G. Elliot Smith on the Piltdown finds. The exposure to this ultimately fraudulent fossil gave Black an intense interest in fossil man. When he received an opportunity in 1919 to teach anatomy, neurology and embryology at the Peking Union Medical College, he accepted it. The position was paid for by the Rockefeller Foundation and fossils were not in their plans. Black was chided several times by his sponsors to keep his attention on anatomy not paleontology.

China was a different place. When Black and his wife arrived they found that their assigned quarters had no doors and no glass in the windows. A colleague quietly talked to the Chinese staff and then took the Blacks on an all-day tour of Peking. When they returned, their quarters were fixed up. But that was not the end of the cultural differences. When Black needed cadavers he was told to contact the police. Walker and Shipman tell us,

> "When Black attempted to obtain cadavers for gross anatomy dissections, he was told to contact the police, who obligingly sent over a cartload of executed prisoners, all headless. This would not do, Black advised the chief of police, explaining his needs more fully. The next time, the prisoners were sent over intact, and alive, with cheerful instructions that Black was to execute them any way he liked. He didn't like, and worked out other arrangements."[39]

Life in China was interesting.

Black, a Canadian, entered a country where, by law, the University of Uppsala, Sweden had a monopoly on all paleontological work. This monopoly had been obtained by Johan Gunnar Andersson, a mining engineer and avid fossil hunter. It was because of this monopoly that the American Museum of Natural History was forced to work in the Gobi desert, where Roy Chapman made major dinosaur discoveries. Because of this law, Black had to work under the umbrella of the Swedes.

In 1921 the Swedish team examined the cave at Zhoukoudian. They found a few stone tools that Andersson correctly identified, and an Austrian named Otto Zdansky found two human teeth. He kept the existence of these two teeth to himself until 1926 when he reported them in a meeting at Beijing which Black attended. Black was enthusiastic about this evidence. He also had connections with the Rockefeller Foundation to obtain funding. Black was a good salesman. The Rockefeller Foundation was an American organization and, as such, would be partial to the views of one of the leading American paleontologists, Henry Fairfield Osborn. Osborn was the president of the American Museum of Natural History. He believed that in order for an ape to evolve into man, he must leave the forest and live on the open plain. Because of this, Osborn felt that Asia was the most likely locale of the transition between apes and man. Thus Black sold the meager two teeth as evidence of a "striking confirmation"[40] of Osborn's prediction that Asia would be the cradle of mankind. Black wrote, "The

Chou Kou Tien discovery therefore furnishes one more link in the already strong chain of evidence supporting the hypothesis of the Central Asiatic Origin of the *Hominidae*."[41] Nothing like playing to one's audience.

The excavation of Zhoukoudian began on March 27, 1927. The field director was a Swede named Birger Bohlin. By the autumn, only one additional tooth had been discovered. A few more human teeth were found over the next two years, along with thousands of mammal bones. Black had published a report on the new fossil man and called him *Sinanthropus pekinensis*, "Chinese man from Peking." With only three teeth to go on, it could certainly be argued that he had published too early with too little data.

But 1929 brought a change in fortunes. On December 2, W. C. Pei, a Chinese paleontologist who had replaced Bohlin as the field director was working in an extremely cramped area of the cave. The only light he had was produced by candles. It was there that he discovered a skull that resembled the one found by Dubois. The fossil was brought out of the earth in a wet and soft state. Pei and his colleagues slowly dried the fossil by the fire. Next they wrapped it in gauze, covered it with plaster and quilts and took it on a bus ride to Beijing. On December 6, the fossil reached Black at the medical school. Black presented the find at a meeting of the Geological Survey of China on December 28, 1929. The team talked about the geology of the site, the extinct mammals found with the skull and the skull itself. There was also evidence from burnt bones that this being had used fire and collected quartz crystals.[42]

The fossils made worldwide headlines and propelled Black into the limelight. Black took a trip in 1931 to England to show off his fossils and gain the support of his former teacher, G. Elliot Smith. Black arrived in town just in time to ruin the visit of Smith's other student, Raymond Dart, who was bringing the Taung child to town. One student went away happy, the other discouraged.

The fact that this creature looked similar to those found by Dubois did not prevent the assignment of a new name. This was the practice of anthropologists at that time. This constantly proliferating set of names was not brought under control until 1950 at the Cold Spring Harbor Symposium. Noel Boaz writes,

> "The biggest bombshell dropped on the Old Guard, however, came from Ernst Mayr, a German-trained ornithologist and specialist in the naming (taxonomy) of species in nature. Using the new yardstick of variability within populations, he stated that 'after due consideration of the many differences between Modern man, Java man, and the South African ape-man, I did not find any morphological characters that would necessitate separating them into several genera.' He suggested that *all* the fossil human-like specimens that anthropologists had discovered after so much laborious effort over the preceding century be simply ascribed to *one* genus, our own—*Homo*. In other words, the entire 'Age of Description,' from before Darwin to Cold Spring Harbor, was a waste of time. His opinion was that the differences were not as great as between genera of other animals. This assertion meant that the wonderfully diverse lexicon of human paleontology, a virtual linguistic playground for the classically educated, with melliferous names such as *Plesianthropus transvaalensis*, *Meganthropus palaeojavanicus*, *Africanthropus njarensis*, *Sinanthropus pekinensis*, *Pithecanthropus erectus*, and so on, were to be replaced. Everything was now to be simply *Homo*, with three species: *Homo transvaalensis*, *Homo erectus*, and *Homo sapiens*."
>
> "Mayr's proposal went so far that even Washburn argued that at least the South African *Australopithecus* be retained (instead of *Homo transvaalensis*) because it showed such significantly more primitive anatomy than members of the genus *Homo*. Mayr simply countered that the population is what the species designates. How one determines a genus is arbitrary. The definition is gauged by the relative amount of difference that one sees between the genera of other animals and, in Mayr's opinion, hominid fossils don't show very much difference. To anthropologists, this statement was a bit like telling a new mother that her baby looks like every other baby. It did not go over well."[43]

One final piece of information needs to be mentioned in regard to *Homo erectus*. There are some minor morphological differences between *H. erectus* found in Asia and the *H. erectus* found in Africa. Some taxonomists prefer to place the African variety of *erectus* into another species. The African *H. erectus* is given the name *H. ergaster*. For this book, to make things simple, I have chosen to include all into the classification of *H. erectus*.[44]

The Traits of *Homo erectus*

H. erectus has a low vaulted skull as all fossil men do until the advent of *Homo sapiens*. By this it is meant that the forehead did not extend above the eyes like it does with modern humans; it sloped back from the eyes. The eyes are surrounded by large brow ridges. The jaw is more massive than *Homo sapiens*, but the teeth cusp patterns are similar to those of *Homo sapiens* and *Australopithecus*. Like *Homo sapiens*, *Homo erectus* often lacked the third molars, the wisdom teeth.[45] Looking at the skull from the front, it is widest just below the eyes, while modern human skulls are more straight-sided. Looked at from the top, these skulls have the characteristic shape of a flask.[46]

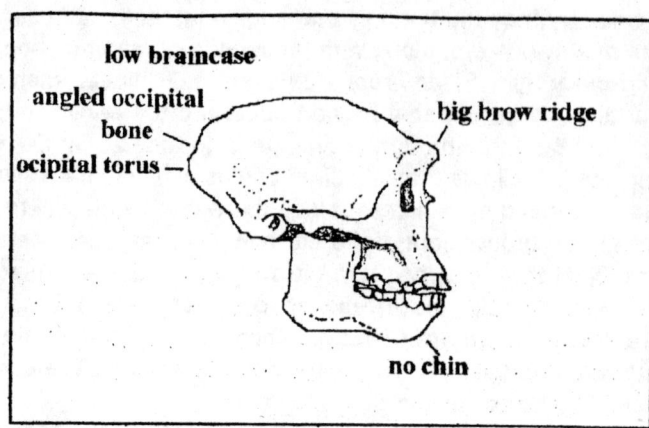

Figure 3 Homo erectus skull anatomy

The height of an adult *erectus* was quite similar to that of modern humans. In 1984, a nearly complete *erectus* skeleton was found at Nariokotome, Kenya. It was of a youth around 12 years of age. He already stood at a height of 5 feet 3 inches, and it has been estimated that as an adult he would have stood over six feet tall. The limb bones are very human-like. *Erectus* was fully bipedal. His leg bones were straight like ours.

The cranial capacity of *erectus* varies from 775 cc to 1225 cc.[47] The older ones have smaller brains than the younger ones. The 1.5 million-year-old Nariokotome boy would have had an adult cranial capacity of 909 cc; the half-million-year-old *Homo erectus* fossils from Zhou Kou Dian, China, are above 1000 cc. There are some exceptions to this general rule such as Olduvai Hominid 9 which dates to 1.2 million years and has a cranial capacity of 1067.[48]

The interior of the *Homo erectus* skulls showed the same impressions of Broca's and Wernicke's areas which are associated with speech in modern humans. Brain lateralization is apparent in the *erectus* skulls. This is the phenomenon in which one hemisphere of the brain is larger than the other and it also is correlated with speech abilities.[49] The evidence for a complex culture is consistent with the concept that these beings had speech. There is evidence that *Homo erectus* worked wood, tanned hides, mastered fires, lived in huts and left at least one possible religious artifact, the Berekhat Ram figurine—a statue of a naked female.[50] Objects like these were used in religious worship as late as Roman times.

Homo erectus is found in strata from nearly two million years old to possibly as recent as 30,000 years.[51] This youngest *erectus* skull is from Ngandong in Java. The owner walked the earth after the first anatomically modern humans appeared. The geographical range of *erectus* is equally broad. They are found from Africa to China to Java to southeastern Europe. What is quite interesting today is that they appear in three of these regions almost simultaneously. In Georgia, the Dmanisi jaw dates between 1.6 and 1.8 MYR (MYR=million years). In Java the earliest dates for *erectus* are 1.8 MYR.[52] The oldest *H. erectus* is from Africa and dates to 1.9 MYR.[53] This widespread occurrence of the same species over such a vast area strongly implies that *Homo erectus* first appeared much earlier. A species would need a certain amount of time to spread out over the landscape.

Archaic *Homo sapiens*

At the turn of the 20th century, Otto Schoetensack was a paleontologist at the University of Heidelberg. He had become interested in the origin of man. He knew of Neanderthal and *Pithecanthropus*

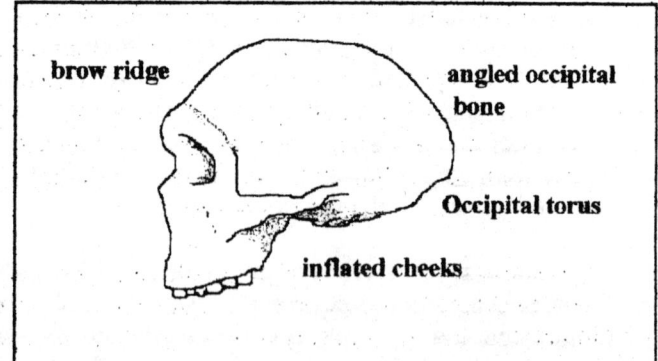

Figure 4 archaic *Homo sapiens* skull anatomy

erectus. But he had a deeply held belief that a transitional form was to be found in strata older than those in which Neanderthal had been found. Schoetensack had convinced the owner of a Mauer, Germany quarry, Joseph Rosch, to donate fossils to the university. Rosch was cooperative and was instructed to be on the look out for any human or human-like fossils. Twenty years passed. On October 21, 1907, Rosch informed Schoetensack that a human fossil had been found. Schoetensack immediately left for Mauer.[54] The jaw had been found in a layer buried by 24 meters (78 feet) of sediment. This was an old fossil.[55] The jaw was massive lacking a chin but the teeth were quite human. It was given the name *Homo heidelbergensis*,

The species name has had a variable success. Forms like the Mauer mandible have been labeled archaic *Homo sapiens* at some times and by some authorities. Sometimes these transitional forms between *Homo erectus* and Neanderthal are called by the original name *H. Heidelbergensis*. These forms are found in Europe, Africa and China. The Asian hominids of the same time frame, while sometimes being called by the same name, are much more erectus-like and, as noted above, erectus-like hominids are found as late as 30,000 years in the east.[56] These fossil men were found from early times and at first anthropologists did not know what to do with them. The problem they faced was that there is a continuum in morphological form between the *erectus* and *sapiens*. Brian Fagan notes,

> "The entire evolutionary process--the transition from developed *Homo erectus* to early archaic *Homo sapiens*, then, finally, into *Homo sapiens sapiens*--does not appear to have occurred rapidly. Rather, it was a slow and continuous development that took as much as half a million years. Consequently, it is very hard to draw a clear taxonomic boundary between *Homo erectus* and archaic *Homo sapiens* on the one hand, and between archaic and anatomically modern *Homo sapiens* on the other. Both *Homo erectus* and archaic *Homo sapiens* display considerable anatomical variation."[57]

The time frame over which archaic *Homo sapiens* lived is from somewhere around 500 thousand years ago until around 30 thousand years ago.[58] The oldest is the Petralona skull, found in northeast Greece,[59] dated to between 250 thousand and 600 thousand years ago, with most experts thinking it is around 400 thousand. It is clearly neither a modern man nor is it an *erectus*. It has a cranial capacity of 1200 cc, fully within the modern range. It still has a sloped forehead but not as much as exists in older hominids. The jaw is still very massive with a projecting face. The brow ridges are still prominent.

Similar forms have been discovered in the French Pyrenees at the cave of Arago. Henri de Lumley began excavations, in 1964, and by 1971 found a face and a parietal bone. This material dates to around 400,000 years. Other fossils which fall into the classification of archaic *Homo sapiens* are the Steinheim skull dated to around 300,000 years and the Mauer mandible with a similar or slightly earlier date, The latest possible archaic *Homo sapiens*, dated at 33,000 years, is from Egypt.[60]

The most recent find of what probably will eventually be classified as archaic *Homo sapiens* is that of the fossil men from the Atapuerca hills of northern Spain. There, between 1995 and 1996, Jose de Castro and Antonio Rosas uncovered the fragmented remains of six individuals. They are currently claiming a new species for these finds since they believe that the Atapuerca men show both modern and Neanderthal-like traits. However, it is unlikely that this species name will be accepted by the anthropological community. In such a case, it is likely that they will be classed into archaic *Homo sapiens*.

The cranial capacities of archaic *Homo sapiens* range from 1070 to 1470 cc. with an average of 1271 cc. This is a fully modern cranial capacity. As in the case of *Homo habilis* and *Homo erectus*, their brains show all the signs of language capabilities. Other than the details of the morphology, archaic *Homo sapiens* possessed all of the important traits modern humans have.

The Neanderthals

Neanderthal tried very hard to be discovered but people refused to recognize that he existed. The earliest Neanderthal was found in late 1829 or early 1830 by Charles Schmerling. Schmerling, a physician from Belgium, had excavated a cave along the Meuse River. From the Engis Cave, he had removed the skulls of three individuals associated with the bones of extinct animals. The first skull has since disintegrated so it is unknown what type if individual it was. The second skull is a fully modern male which turned out to be only 8,000 years old. The final skull was that of a two and a half year old Neanderthal child. The fact that it was a child's skull

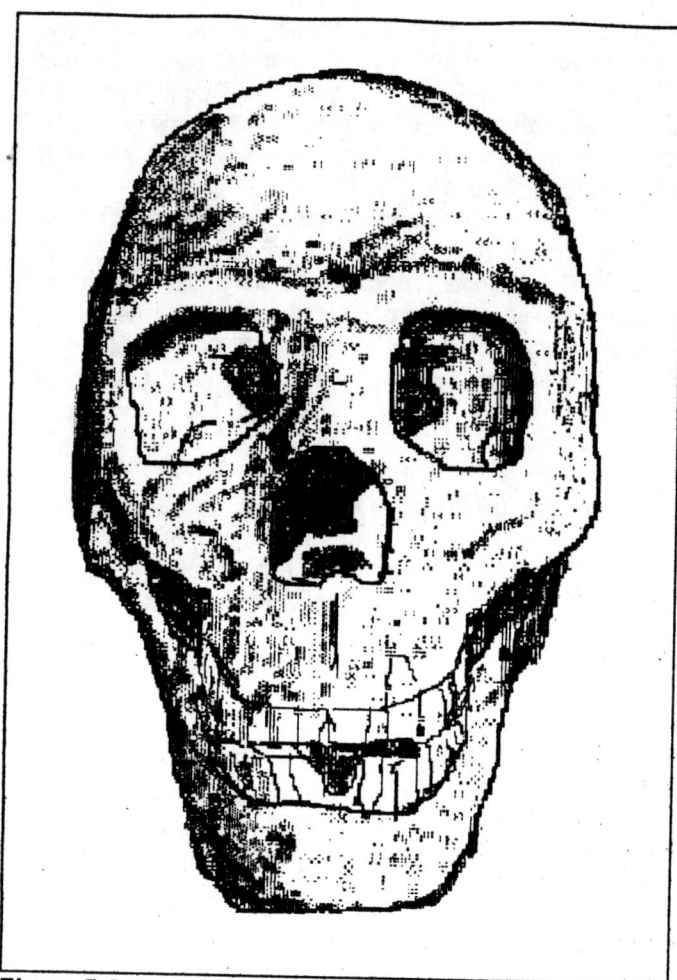

Figure 5 Shanidar Neanderthal

disguised its true nature. Schmerling's work at the nearby cave of Engihoul produced a human limb bone, also associated with extinct animals. Schmerling knew that he had proof of the antiquity of man, but when he published in 1834, the scientific world ignored him.[61] They were not ready for Neanderthal.

The second time Neanderthal tried to be discovered was far to the south. In 1848, during the construction of military fortifications, a human skull was unearthed from the north face of the Forbes quarry at Gibraltar. No one knew what to do with it. The local scientific society, the Gibraltar Scientific Society, makes only the briefest mention of it in the minutes for March 3, 1848.[62] This skull was then placed on the shelf of a local museum where it lay forgotten for the next 16 years

The third time is the charm, they say. In 1856 quarrymen laid charges in a cave 60 feet above the Dussel River in Germany. When the blast had done its damage, the foremen sent workers to clear the rubble. Among limestone, flint, and mud, the workers found bones. They believed that the bones belonged to a cave bear. The foreman had the bones set aside for a local school teacher, Johann Karl Fuhlrott, who was widely known in that region as a collector of fossils, plants and rocks. A few weeks later, Fuhlrott was told to come get the bones.

Fuhlrott knew immediately what the bones were. They were human, but not quite human. He raced to the quarry looking for more pieces of bone. He searched in vain among the debris at the quarry. There were no more pieces to be had. However, what Fuhlrott had in his hand was enough to start an intellectual firestorm, but not before three years had passed and Darwin had published The Origin of Species. Neanderthal man had to wait for his time in the sun. The immediate scientific reception was chilly indeed.

Fuhlrott had taken the bones to Hermann Schaaffhausen, an anatomist at the University of Bonn. Schaaffhausen believed that the bones represented an ancient form of man who had inhabited Europe prior to the Celts. On Feb. 4, 1857, Schaaffhausen presented a preliminary report to the Lower Rhine Medical and Natural History Society at Bonn. On June 2, 1857, Fuhlrott and Schaaffhausen presented two papers to the Natural History Society of Prussian Rhineland and Westphalia. Fuhlrott presented the data he possessed about the locale of the find; the fossils had been covered by four to five feet of sediment. Fuhlrott reported that the bones had belonged to a single specimen, according to the testimony of the workmen he had interrogated. He also reported the fact that dendritic mineralization had been observed on the bones. At that time, dendritic mineralization was taken as proof of antiquity as had been advocated by H. von Meyer, a colleague of Schaaffhausen's at Bonn. However, von Meyer was on the verge of changing his opinion about that detail and would soon criticize Neanderthal.

After Fuhlrott sat down, Schaaffhausen rose to give his paper. He made the points that the morphology of the skull was unique and unknown among modern peoples, that the bones came from a time prior to the occupation of Europe by the Celts and Germans and that these bones were associated with extinct animals. After this, he began the description and display of the skull itself. The skull was long and narrow with large brow ridges. The rear of the skull had an occipital bun. This is a protrusion or bump in the center of the back of the skull. Some modern Europeans have such a feature albeit reduced in size from that of the Neanderthal. The skull was not domed like modern man's but had a short forehead like that of an ape. A lower right arm had been recovered along with the left humerus and ulna. The ulna had a healed fracture. This injury early in life had left the left arm useless, so it was smaller than the right, and made it impossible for this person to straighten his arm. Schaaffhausen noted this but no one seemed to hear. The femora were extremely robust with huge muscle attachments. This fellow was strong--much stronger than modern humans.[63] Schaaffhausen made the case that the bones were not pathological, because he knew this was likely to become the focus of the argument. The only two men who left the meeting believing the bones were the bones of an ancient man were the two who had entered the room believing that fact, i.e., Fuhlrott and Schaaffhausen. The rest awaited the pronouncement of the German scientific establishment.

The first attack came from Professor A.F. Mayer. He argued that the bowed legs proved that this wretched creature had been a horseman from his youth. Further, Mayer noted that the Neanderthal man had been aflicted with rickets as a child. He concluded with the ludicrous charge that the skeleton belonged to a Mongolian Cossack who had deserted from the Tchernitcheff's Russian army in 1814, which had come through Germany to attack Napoleon in France.[64] This, of course, played into the racial sentiments of the day and, in addition, was quite patriotic.

Virchow, the great man of German medicine, accepted Mayer's conclusions and proclaimed the bones those of a "recently deceased pathological idiot suffering from rickets."[65] Virchow's opinion carried lots of weight and is still cited today by those like the young-earth creationists who are opposed to the concept of evolution and must explain the morphological differences by other means.[66] Virchow ignored several facts that were contrary to his theory. The bones of Neanderthal are very robust by comparison with anatomically modern man. This fact indicates that Neanderthal was extremely muscular, and is not what one expects from the bones of an individual suffering from the diseases Virchow and Meyer were claiming. Even the broken ulna showed evidence of having healed in a normal rather than diseased manner. The problem was that Virchow did not believe in the mutability of species and thus any deviation from the human norm was to be explained on the basis of disease. Virchow even used the fracture as evidence that the Neanderthal individual could not have belonged to a primitive hunter-gatherer society. Virchow firmly believed that no member of such a society could possibly survive such a serious injury because primitive peoples would not be compassionate!

The controversy continued throughout the 1860s, 1870s and into the 1880s. Neither those who believed that Neanderthal was diseased nor those who believed in evolution were able to extract from the earth sufficient evidence to destroy the opposing argument. But in July 1886, two Belgians, an amateur fossil hunter named Marcel de Puydt and a geologist, Marie Joseph Maximin Lohest, were working the cave at Spy d'Orneau. In the lowest levels of the cave, they found two nearly complete skeletons of Neanderthals associated with Mousterian stone tools. Since Lohest was an expert in Paleozoic fish and de Puydt was a lawyer, they needed the help of an expert on fossil men. Lohest called in an anatomist, Julien Fraipont, to help describe the finds. These three men wrote what became one of the best 19th century monographs on a fossil find. They documented everything, thus allowing the presentation of a conclusive case for the antiquity of the skeletons. Their similarity to the bones of the Neander valley and Gibraltar proved that the Neanderthals formed a consistent morphological group separate from modern men.

The skeleton showed some characteristics that were intermediate between ape and man although these differences were exaggerated at the time. The tibial plateau in modern men is nearly horizontal.[67] In Neanderthal it was tilted at 18 degrees, and in apes the value is even greater. Fraipont argued that this meant that Neanderthal walked with bent knees. It was unfortunate that this was an erroneous conclusion because it contributed to the view that Neanderthal was a brutish ape-man. This conclusion was not overturned for more than 70 years when the true meaning of the tibial plateau was found.

In spite of that, the finds at Spy convinced the academic community that Neanderthal was an ancestor of modern man. This conclusion has had a long history of being accepted and being rejected ever since. Over the past 20 years, as the replacement theory of human origins has been in the ascendancy, the amount of Neanderthal influence on human makeup has been minimized. The replacement theory postulates that anatomically modern men arose in Africa around 100-200 thousand years ago and pushed all other forms of man to extinction. This school of thought minimizes the genetic intermingling between the moderns and the ancients. The other school of thought holds that mankind in each region evolved into modern man without a huge migration out of Africa. This school of thought emphasizes the amount of Neanderthal ancestry in modern Europeans. Whatever view a researcher takes, Neanderthal has never been out of the limelight since the Spy discovery.

The Traits of Neanderthal

Neanderthal has a variety of traits that connect him to modern man. The most important characteristic is the size of the brain-case. Their brains were quite large, the average size exceeding that of modern man. Females had brain sizes averaging around 1300 cc while the males averaged about 1600 cc. The largest was the Amud man with 1740 cc.[68] Based upon cranial evidence, the Neanderthal was mostly right-handed as we are.[69] The inside of their skulls show the characteristic impressions of Broca's areas which in man has been associated with speech. Because of this, while early excavators were quick to claim that the Neanderthal brain was quite primitive, modern authorities differ. Stringer and Gamble state,

> "As far as Neanderthal brains are concerned, the pendulum has swung to the other extreme, partly in reaction to previous errors of interpretation and partly in recognition of our present ignorance. Thus many palaeoanthropologists are now reluctant to judge the quality of Neanderthal brains at all, or are content to assume that there was little intellectual difference from modern humans. Opinions tend to fluctuate."[70]

Like all previous hominids[71], Neanderthal lacked a chin, although in a relatively few specimens, there is some slight chin development.[72] There are two ways to mechanically strengthen the front of the jaw so that it can bear the forces that it encounters. The bone of the mandible can be thickened inside, a development which is called the 'simian shelf.' This is a shelf of bone which stretches between the two sides of the jaw and helps hold the two sides together. In anatomically modern man, the thickening of the mandible occurs on the outside, producing our characteristic jutting chin.

Neanderthals probably would not have won a modern beauty contest. Their noses were large and at the bridge of the nose, it jutted out from the face at nearly a right angle. They had double arched brow ridges over their eyes and their forehead was flatter than modern man's. They had a pronounced occipital bun, a bulge at the back of the skull which is a trait some anthropologists believe was given to modern Europeans by Neanderthal ancestors. Their front teeth were shovel-shaped and their back teeth had strange roots that were called taurodont. Shreeve cites other features which may have come down to us from the Neanderthal,

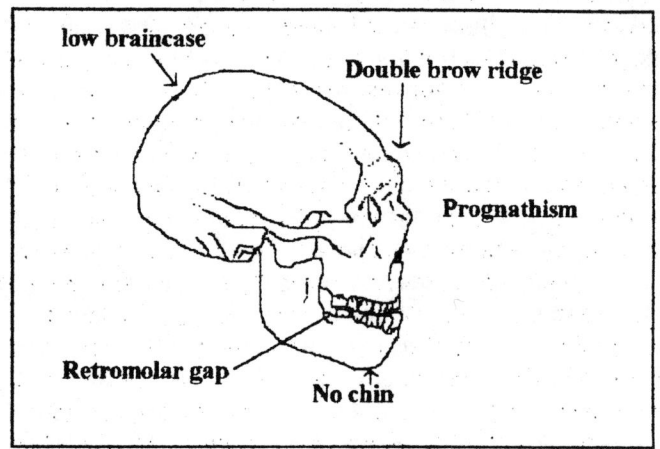

Figure 6 Neanderthal skull anatomy

> "Frayer's own reading of the record reveals a number of overlooked traits that clearly and specifically link the Neandertals to the Cro-Magnons. One such trait is the shape of the opening of the nerve canal in the lower jaw, a spot where dentists often give a pain-blocking injection. In many Neandertals, the upper portion of the opening is covered by a broad bony ridge, a curious feature also carried by a significant number of Cro-Magnons. But none of the alleged 'ancestors of us all' fossils from Africa have it, and it is extremely rare in modern people outside Europe. By far the simplest explanation is that the Cro-Magnons inherited the trait directly from their Neandertal forebears. Frayer knows, of course, that it takes more than a single, rather obscure feature like this, or even an assortment of features, to convince people that Neandertals deserve a place in our ancestry. But he isn't about to stop trying. 'If there is a Neandertal heaven, maybe they are up there smiling down on me right now,' he concluded. 'Nobody else is.'"[73]

As noted above, the skeleton of Neanderthal was quite robust with large joints. Their muscles left huge attachment marks which implies that these people lived a life of extreme exertion. The skeleton also shows some evidence of the predominant posture. Neanderthals squatted on their haunches a lot.[74] Their leg bones display squatting facets that are similar to those on modern peoples in cultures where squatting is a way of life.

In 1996 Jeffrey Schwartz and Ian Tattersall discovered two jutting protrusions of bones in the nasal passages of Neanderthal.[75] They also lacked tear ducts.[76] Because of this, Schwartz and Tattersall suggest that Neanderthal is an entirely different species. Some Christian apologists have grabbed hold of this and jumped to the conclusion that anyone with huge nasal bones can't be *human*.[77] Of course, as Phillip V. Tobias of the University of the Witwatersrand in Johannesburg, South Africa, points out, "Comparable bony protrusions appear in the nasal cavities of the modern southern African people known as the San."[78] So, if this trait excludes Neanderthal from the human race, then the primitive San are excluded also. Of course, this is ridiculous; the San are quite human, as were the Neanderthal.

Homo sapiens sapiens

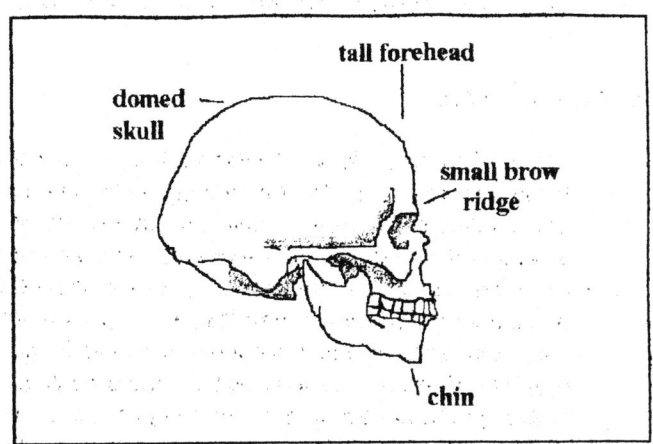

Figure 7 *Homo sapiens* skull anatomy

Homo sapiens sapiens is the form of mankind into which all modern peoples are placed. There is no uniformly accepted definition of anatomically modern man,[79] but the fossils that have generally been identified as modern are judged so based upon the skull.[80] The earliest *Homo sapiens sapiens* judged to have a modern cranium still show some archaic postcranial features[81] and archaic cranial features such as the lack of a well-developed chin.[82] The brutal fact is, however, that there are no fossils which are suddenly totally modern. Modernity is gradually acquired with more and more skeletal features attaining the shape of our skeletons. Even Cro-Magnon man carries a few archaic features although he is called fully modern. Neanderthals had a bony covering over the opening of the nerve canal on the mandible as we saw above. This trait is extremely rare in populations outside of Europe and in the African ancestors of modern man. Yet Cro-Magnon (who is considered a fully modern *Homo sapiens*) carries this archaic feature as do many modern Europeans.[83]

Modern humans are first found at Klasies River Mouth, South Africa, 120,000 years or so ago.[84] They spread over the old world at various rates. By 100,000 years ago, fossils of modern men are found buried at Qafzeh in the Near East.[85] It was not until 67,000 years ago that evidence of modern man, the Liujiang skull, is found in China.[86] However, even this skull is quite robust compared with the more delicate men of this age.[87] Modern man is found in Malaysia and New Guinea by 40,000 years ago, Europe by 35,000 years ago,[88] Japan by 32,000 years ago,[89] Australia by 30,000 years ago,[90] and the New World most certainly by 13,000 years ago, but maybe as early as 30,000 years ago.[91]

Due to the vagaries of archaeological history, most laymen believe that when modern men appeared on earth, they brought modern behavior. Christian apologists have often repeated this concept. Hugh Ross writes:

> "While bipedal, tool-using, large brained hominids roamed the earth at least as long ago as one million years, evidence for religious relics and altars dates back only 8,000 to 24,000 years. Thus the secular anthropological date for the first spirit creatures is in complete agreement with the biblical date."[92]

And Wilcox writes:

> "Both cultural and physical evidence suggests an abrupt establishment of the image about 100,000 years ago."[93]

This is not the case. Modern man did not behave in modern ways during the first half of its existence. By this I mean that they did not make modern tools and their social life was somewhat different. However, different culture does not mean different species. The mystery is why modern bodies didn't start behaving in modern ways for so long. Stringer and Gamble write:

"The fossil evidence we have discussed so far in this chapter builds a case for a morphological transition right across Africa between Ancients and early Moderns via the borderline Moderns some time before 100,000 years ago, and certainly well before the Upper Palaeolithic (or Late Stone Age) which began c. 40,000 years ago. We have seen throughout the last few chapters that we can no longer expect (as others before us did) that anatomy and behaviour -- as indicated in the archaeological record -- go hand in hand. The earliest anatomically modern specimens of Africa and the Levant may have looked like us in many ways, but they were not us. They lack the 100,000 years of change which have made us what we are today. This leaves us with one of the big questions concerning our origins, why then *did* these populations look so much like us?"[94]

and Tattersall notes,

"What I find particularly interesting, though, is that at just about the time of the latest Neanderthal fossil (not, of course, the same thing as the last Neanderthal) known from the Levant, we find the first evidence of Upper Paleolithic industries in the region. In Europe such industries are associated with undisputed moderns; and although the very earliest Levantine Upper Paleolithic (from the 45 kyr [kyr=thousands of years ago--GRM] site of Boker Tachtit, in the Negev desert) is not associated with human fossils, it's reasonable to believe that anatomically modern humans (who had, after all, been around in the area for 50 kyr) were responsible for it. Can we then say that Neanderthals disappeared from the area when *Homo sapiens* stopped simply looking modern and began to behave in a modern fashion, too? I'd personally bet that we can; but until we have more and more precisely dated archaeological sites and hominid fossils from this time and place, we won't be sure."

"The subject of behavior is complicated by the fact that whereas in Europe Upper Paleolithic stone and bone tools were associated from the beginning with evidence of 'creativity' in the form of engravings, sculpture, notation, musical instruments, and so forth, this was not the case in the Levant. What's more, the earliest Upper Paleolithic tools from Boker Tachtit, while fully Upper Paleolithic in concept, were made using techniques that had been current in the Middle Paleolithic. However, since anatomically modern humans had made Middle Paleolithic tools for the first 50 kyr of their existence, we probably shouldn't find this too surprising."[95]

To further complicate matters, the latest Neanderthals behaved in a fashion which we would regard as modern. Schepartz notes,

"Mellars admits that 'there is now unmistakable evidence that the final Neanderthal populations in Western Europe were behaving in a way which-by most conventional archeological criteria--was entirely Upper Paleolithic not only in a basic technological sense but also in at least some spheres of cognitive or symbolic expression.'"[96]

Cranial capacity of the vast majority of normal modern humans range from 900 to 2000 cc with an average of 1345 cc.[97] There are documented cases of people with normal intelligence having smaller brain sizes.[98] The lower part of this range overlaps that of *Homo erectus* which obviously leads to the conclusion that if one wishes to reject *Homo erectus* from humanity because of his brain size, then one must also reject many modern humans for the same reason.

Time: The Final Player

The division of time over the period of the hominids needs to be outlined so that various terms used by anthropologists can be understood. The earliest hominid fossil appears just after the start of the Pliocene at 5.5 million years ago. The beginning of the Pliocene is demarcated by the filling of the Mediterranean basin, an event which altered the climate of Africa and Europe and brought major environmental changes to the region. The Pliocene ends 1.8 million years ago. The period from 1.8 million years ago until 12,000 years ago constitutes the Pleistocene. This is the time of glaciations. Depending upon the researcher, there are between five and eight glacial periods during this time. The sea level fell and rose in sync with the glaciers. The final geologic period, the Recent, begins when the last glaciers melted around 12,000 years ago. All of recorded human history has occurred during the most recent half of this period.

Other terms that are used are based upon stone tools. The Lower Paleolithic refers to stone tool industries that are dominated by hand axes. The Middle Paleolithic is characterized by industries where flake tools predominate. In the Upper Paleolithic, blade tools take precedence. Blade tools are defined as tools that are more than twice as long as they are wide. These terms are not, strictly speaking, chronological since they refer to technology and technology has changed at various times in different parts of the world. Generally speaking, the Lower Paleolithic is associated with tools made by *Homo erectus* and archaic *Homo sapiens*. Middle Paleolithic tools were made by archaic *Homo sapiens*, Neanderthals and anatomically modern *Homo sapiens*. Upper Paleolithic tools were made by both Neanderthal and Homo sapiens.

In spite of what some anthropologists say, over the past century and a half, other anthropologists have uncovered much evidence that men in the fossil record were very similar to us. The players in our past are real beings, who were once flesh and blood. Christian apologetics must incorporate these findings into a picture consistent with the Scripture which does not ignore or misrepresent the data that has been found. We will examine how various writers have dealt with these issues in the next chapter.

References

1. Theodosius Dobzhansky, Evolution, Genetics and Man, (New York: John Wiley and Sons, Inc., 1955), p. 333
2. Ernst Mayr cited by Noel T. Boaz, Quarry, (New York: Macmillan, 1993), p. 10.
3. Erik Trinkaus and Pat Shipman, The Neandertals (New York: Vintage Books, 1992), p.228-229.
4. Alan Walker and Pat Shipman, The Wisdom of the Bones, (New York: Alfred Knopf, 1996), p. 88-89
5. Alan R. Walker and Pat Shipman, Wisdom of the Bones, (New York: Adolph Knopf, 1996), p. 95.
6. Donald Johanson and Maitland Edey, Lucy, (New York: Simon and Schuster, 1981), p. 47.
7. Alan R. Walker and Pat Shipman, Wisdom of the Bones, (New York: Adolph Knopf, 1996), p. 93.
8. Alan R. Walker and Pat Shipman, Wisdom of the Bones, (New York: Adolph Knopf, 1996), p. 93.
9. Donald C. Johanson and Maitland Edey, Lucy, (New York: Simon and Schuster, 1981), p. 75.
10. Donald Johanson and Blake Edgar, From Lucy to Language, (New York: Simon & Schuster, 1997), p. 136
11. Christopher Wills, The Runaway Brain, (New York: Harper Collins, 1993), p. 86-87
12. Donald C. Johanson and Maitland Edey, Lucy, (New York: Simon and Schuster, 1981), p. 51
13. Michael Day, Guide to Fossil Man (Cleveland: World Publishing Co., 1965), p. 171
14. C. Loring Brace et al, Atlas of Human Evolution, (New York: Holt Rinehart and Winston, 1979), p. 32,34
15. Ian Tattersall, The Fossil Trail (New York: Oxford University Press, 1995), p. 148
16. Niles Eldredge and Ian Tattersall, The Myths of Human Evolution, (New York: Columbia University Press, 1982), p. 7
17. Dean Falk, Braindance, (New York: Henry Holt, 1992), p. 36 see also Ralph L. Holloway, "The Cast of Fossil Hominid Brains," Scientific American, July 1974, p. 111. The *Australopithecus* STW 505 has a cranial capacity of 650 cc. Jim Foley, Personal Communication. Oct. 3, 1997.
18. Donald Johanson and James Shreeve, Lucy's Child (New York: William Morrow, 1989), p. 86-87. See also Clive Gamble, Timewalkers, (Cambridge: Harvard University Press, 1994), p.59.
19. Clive Gamble, Timewalkers, (Cambridge: Harvard University Press, 1994), p.59.
20. Michel Brunet et al, "The First Australopithecine 2,500 kilometres west of the Rift Valley (Chad)", Nature, 378, Nov. 16, 1995, p. 273-274 esp 274.
21. Alan R Walker and Pat Shipman, The Wisdom of the Bones, (New York: Alfred Knopf, 1996), p.99.
22. Donald Johanson and James Shreeve, Lucy's Child (New York: William Morrow, 1989), p. 59
23. Alan Walker and Pat Shipman, The Wisdom of the Bones, (New York: Alfred Knopf, 1996), p.100.
24. Ian Tattersall, The Fossil Trail (New York: Oxford University Press, 1995), p. 81-82
25. Donald Johanson and James Shreeve, Lucy's Child (New York: William Morrow, 1989), p. 60.
26. Donald Johanson and James Shreeve, Lucy's Child (New York: William Morrow, 1989), p. 61.
27. Alan R. Walker and Pat Shipman, The Wisdom of the Bones, (New York: Alfred A. Knopf, 1996), p. 101 see also Donald Johanson and James Shreeve, Lucy's Child (New York: William Morrow, 1989), p. 62
28. Donald Johanson and James Shreeve, Lucy's Child, (New York: William Morrow and Co., Inc., 1989), p. 69-70.
29. Alan Walker and Pat Shipman, The Wisdom of the Bones, (New York: Alfred Knopf, 1996), p. 107-108.

30. Alan Walker and Pat Shipman, The Wisdom of the Bones, (New York: Alfred Knopf, 1996), p. 123-124.
31. Alan Walker and Pat Shipman, The Wisdom of the Bones, (New York: Alfred Knopf, 1996), p. 133.
32. Dean Falk, Braindance, (New York: Henry Holt and Co., 1992), p. 48-49.
33. Dean Falk, Comments, Current Anthropology, 30:2, April, 1989, p. 141-142.
34. Terrence W. Deacon, The Symbolic Species, (New York: W. W. Norton and Co., 1997), p. 251
35. James R. Shreeve, The Neandertal Enigma, (New York: William Morrow and Co., 1995), p. 274-275
36. Ian Tattersall, The Fossil Trail, (Oxford: Oxford University Press, 1995), p. 35.
37. Ian Tattersall, The Fossil Trail, (Oxford: Oxford University Press, 1995), p. 35
38. Alan Walker and Pat Shipman, The Wisdom of the Bones, (New York: Alfred Knopf, 1996), p. 45.
39. Alan Walker and Pat Shipman, The Wisdom of the Bones, (New York: Alfred Knopf, 1996), p. 56
40. Alan Walker and Pat Shipman, The Wisdom of the Bones, (New York: Alfred Knopf, 1996), p. 57-58
41. Alan Walker and Pat Shipman, The Wisdom of the Bones, (New York: Alfred Knopf, 1996), p. 58
42. W.C. Pei, "Notice of the Discovery of Quartz and Other Stone artifacts in the Lower Pleistocene Hominid-Bearing Sediments of the Choukoutien Cave Deposit," Bulletin of the Geological Society of China, 11:2:1931:109-146, p.120
43. Noel Boaz, Quarry, (New York: The Free Press, 1993), p. 10.
44. Ian Tattersall, The Fossil Trail, (Oxford: Oxford University Press, 1995), p. 35.
45. Christopher Dean, "Jaws and Teeth," Cambridge Encyclopedia of Human Evolution, (Cambridge: Cambridge University Press, 1992), p. 59
46. Christopher Wills, The Runaway Brain, (New York: Basicbooks, 1993), p. 153.
47. Bernard Campbell, Human Evolution, (Chicago: Aldine Publishing Co., 1974), p. 272
48. Ian Tattersall, The Fossil Trail, (Oxford: Oxford University Press, 1995), p. 108
49. Clive Gamble, Timewalkers, (Cambridge: Harvard University Press, 1994), p. 172
50. Desmond Morris, The Human Animal, (New York: Crown Publishing, 1994), p. 186-188.
51. C. C. Swisher III, W. J. Rink, S. C. Antón, H. P. Schwarcz, G. H. Curtis, A. Suprijo, Widiasmoro, "Latest Homo erectus of Java: Potential Contemporaneity with Homo sapiens in Southeast Asia" Science Volume 274, Number 5294, Issue of 13 December 1996, pp. 1870-1874
52. Vadim A. Ranov, Eudald Carbonell, and Xose Pedro Rodriguez, "Kuldara: Earliest Human Occupation in Central Asia in Its Afro-Asian Context," Current Anthropology, 36:2, April 1995, p. 337-346, p. 342.
53. Roy Larick and Russell L. Ciochon, "The African Emergence and Early Asian Dispersal of the Genus Homo,", American Scientist, 84(Nov/Dec. 1996), p. 538-551, p. 540.
54. Erik Trinkaus and Pat Shipman, The Neandertals, (New York: Random House, 1992), p. 173
55. Donald Johanson and Blake Edgar, From Lucy to Language, (New York: Simon & Schuster, 1997), p. 196
56. C. C. Swisher III, W. J. Rink, S. C. Antón, H. P. Schwarcz, G. H. Curtis, A. Suprijo, Widiasmoro, "Latest Homo erectus of Java: Potential Contemporaneity with Homo sapiens in Southeast Asia" Science Volume 274, Number 5294, Issue of 13 December 1996, pp. 1870-1874
57. Brian M. Fagan, The Journey From Eden, (London: Thames and Hudson, 1990), p.60
58. Robert G. Bednarik, "Early Subterranean Chert Mining," The Artefact, 15:(1992), pp 11-24, p. 13; see also, P. M. Vermeersch, et al., "33,000-yr Old Chert Mining Site and Related Homo in the Egyptian Nile Valley," Nature, 309(May 24, 1984): 342-344, p. 342
59. Ian Tattersall, The Fossil Trail, (Oxford: Oxford University Press, 1995), p.174
60. P. M. Vermeersch, et al., "33,000-yr Old Chert Mining Site and Related Homo in the Egyptian Nile Valley," Nature, 309(May 24, 1984): pp. 342-344.
61. Erik Trinkhaus and Pat Shipman, The Neanderthals, (New York: Vintage Books, 1992), p. 38-39.
62. Erik Trinkaus and Pat Shipman, The Neanderthals, (New York: Vintage Books, 1992), p. 45.
63. Erik Trinkaus and Pat Shipman, The Neanderthals, (New York: Vintage Books, 1992), p. 50-52.
64. Eric Trinkaus and Pat Shipman, The Neandertals, (New York: Vintage Books, 1992), p.58.
65. Brian M. Fagan, The Journey From Eden, (New York: Thames and Hudson, 1990), p. 76.
66. Marvin L. Lubenow, Bones of Contention, (Grand Rapids: Baker Books, 1992), p. 76-77.
67. Eric Trinkaus and Pat Shipman, The Neandertals, (New York: Vintage Books, 1992), p. 130.
68. Christopher Stringer and Clive Gamble, In Search of the Neanderthals, (New York: Thames and Hudson, 1993), p. 82.
69. Dean Falk, "Hominid Paleoneurology", Ann Rev. Anthropology, 1987, 16:13-30, p. 16.
70. Christopher Stringer and Clive Gamble, In Search of the Neanderthals, (New York: Thames and Hudson,

1993), p. 82.

71. For this trait in Homo erectus (Peking Man) see, Michael H. Day, Guide to Fossil Man, (Cleveland: The World Publishing Co., 1965), p. 253

72. Christopher Stringer and Clive Gamble, In Search of the Neanderthals, (New York: Thames and Hudson, 1993), p. 80-81.

73. James R. Shreeve, The Neandertal Enigma, (New York: William Morrow and Co., 1995), p. 126-127.

74. Christopher Stringer and Clive Gamble, In Search of the Neanderthals, (New York: Thames and Hudson, 1993), p. 93.

75. Jeffrey H. Schwartz and Ian Tattersall, "Significance of Some Previously Unrecognized Apomorphies in the Nasal Region of Homo neanderthalensis," Proceedings of the National Academy of Sciences USA, 93(1996):10852-10854.

76. "Nosing into Neandertal Anatomy," Science News, Oct. 12, 1996, p. 234.

77. Hugh Ross, "No Tears for Neanderthal", Facts & Faith, 10(1996):4, p. 11.

78. "Nosing into Neandertal Anatomy," Science News, Oct. 12, 1996, p. 234.

79. Milford Wolpoff and Rachel Caspari, "The Modernity Mess," Journal of Human Evolution, (1996), 30:167-171, p. 169

80. S. E. Churchill et al, "Morphological Affinities of the Proximal Ulna from Klasies River Main Site: Archaic or Modern?" Journal of Human Evolution, (1996) 31:213-237, p 231

81. S. E. Churchill et al, "Morphological Affinities of the Proximal Ulna from Klasies River Main Site: Archaic or Modern?" Journal of Human Evolution, (1996) 31:213-237, p 233.

82. Milford Wolpoff and Rachel Caspari, "The Modernity Mess," Journal of Human Evolution, (1996), 30:167-171, p. 168

83. James R. Shreeve, The Neandertal Enigma, (New York: William Morrow and Co., 1995), p. 126-127; Milford Wolpoff and Rachel Caspari, Race and Evolution, (New York: Simon & Schuster, 1997), pp 329-346

84. Richard G. Klein, The Human Career, (Chicago: The University of Chicago Press, 1989), p. 353 and James R. Shreeve, The Neandertal Enigma, (New York: William Morrow and Co., 1995), p. 214-215.

85. Chris Stringer and Clive Gamble, In Search of the Neanderthals, (New York: Thames and Hudson, 1993), p. 218.

86. C. C. Swisher III, W. J. Rink, S. C. Antón, H. P. Schwarcz, G. H. Curtis, A. Suprijo, Widiasmoro, "Latest Homo erectus of Java: Potential Contemporaneity with Homo sapiens in Southeast Asia" Science Volume 274, Number 5294, Issue of 13 December 1996, pp. 1870-1874

87. C. C. Swisher III, W. J. Rink, S. C. Antón, H. P. Schwarcz, G. H. Curtis, A. Suprijo, Widiasmoro, "Latest Homo erectus of Java: Potential Contemporaneity with Homo sapiens in Southeast Asia" Science Volume 274, Number 5294, Issue of 13 December 1996, pp. 1870-1874; Gunter Brauer, "The Evolution of Modern Humans: a Comparison of the African and non-African Evidence," in Paul C. Mellars and Chris B. Stringer ed. The Human Revolution. (Princeton: Princeton University Press, 1989), pp. 123-153, p. 143

88. L. A. Schepartz, "Language and Modern Human Origins," Yearbook of Physical Anthropology, 36:91-126(1993), p. 91.

89. E. Trinkaus, et al. "Early Modern Human Remains from Eastern Asia: the Yamashita-cho 1 Immature Postcrania," Journal of Human Evolution(1996), 30:299-314, p. 300.

90. C. C. Swisher III, W. J. Rink, S. C. Antón, H. P. Schwarcz, G. H. Curtis, A. Suprijo, Widiasmoro, "Latest Homo erectus of Java: Potential Contemporaneity with Homo sapiens in Southeast Asia" Science Volume 274, Number 5294, Issue of 13 December 1996, pp. 1870-1874

91. L. Luca Cavalli-Sforza, Paoli Menozzi and Alberto Piazzi, The History and Geography of Human Genes, (Princeton: Princeton University Press, 1994), p. 306

92. Hugh Ross, The Fingerprint of God, (Orange: Promise Publishing, 1991), p. 159-160.

93. David L. Wilcox, "Adam, Where Are You? Changing Paradigms in Paleoanthropology," Perspectives on Science and Christian Faith, 48:2(June 1996), p. 94

94. Chris Stringer and Clive Gamble, In Search of the Neanderthals, (New York: Thames and Hudson, 1993), p.131.

95. Ian Tattersall, The Fossil Trail (New York: Oxford University Press, 1995), p.225.

96. L. A. Schepartz, "Language and Modern Human Origins," Yearbook of Physical Anthropology, 36:91-126(1993), p. 120. See also Paul Mellars, "Major Issues in the Emergence of Modern Humans," Current Anthropology, 30(1989):3:349-385.

97. John H. Relethford, <u>Fundamentals of Biological Anthropology</u>, (Toronto: Mayfield Publishing Co., 1994), p. 192

98. There is a report that Daniel Lyon, an Irishman who died in 1907 at age 46 had a 650 cc cranial capacity. Reports are that he was "normal". This is the size of *Homo habilis* and the gorilla. <u>Guiness Book of Records 1996</u>, (New York: Facts on File, 1995), p. 14

THE RESPONSE

Christians have responded to the concept of mankind's evolution in a variety of ways depending upon the theology which is being defended. Responses from young-earth creationists have varied much in their understanding of the humanity of the various fossil men discussed in the last chapter. Generally *Australopithecus*, if mentioned at all, is viewed as an ape. *Homo erectus* (e.g., Pithecanthropus which is Java man, Peking Man etc) is viewed variably. Neanderthal tends to be seen as a human being, although degenerate. However, it is interesting that the young-earth creationists as a whole are far more willing to accept the humanity of these various fossil men, even *Homo erectus*, than are the old-earth creationists. The young-earth view requires that there be victims of the Noahic flood and the fossil men would fill that role marvelously, but Henry Morris, the leading young-earth creationist, has specifically stated that he believes nearly all fossil men are post-diluvial.[1] Old-earth creationists generally do not believe in a global flood and thus cannot easily handle an ancient, widespread humanity. They usually prefer to have Adam created within the past 100,000 years, and identify Adam as the first anatomically modern man. Anatomically modern men are spiritual; all others are not. This view requires, then, that at least some of the fossil men be pre-Adamite and raises severe problems when we consider the behavior of these fossil men.

Early Reactions to Evolution

Most early reactions to man's evolution were based upon a young-earth interpretation of the Bible. As Andrew White says, "Darwin's Origin of Species had come into the theological world like a plough into an ant-hill. Everywhere those thus rudely awakened from their old comfort and repose had swarmed forth angry and confused."[2] The implications to the origin of man were clear to everyone and the way had been prepared by the publication in England in 1860 of a German theology book that questioned literalism.[3] The earliest responses to the concept of man's evolution did not take the threat seriously. Leading clergymen, such as Samuel Wilberforce, did not attack Darwin because of his facts, but on the basis of the inconsistency Darwin's views presented to Christians. Wilberforce is reported to have written that Darwin was guilty of "a tendency to limit God's glory in creation," that "the principle of natural selection is absolutely incompatible with the word of God," and that evolution "contradicts the revealed relations of creation to its Creator."[4] While these statements may or may not be true, they show that Wilberforce really did not understand the issue before him. The issue was not the incompatibility of his view of Scripture with evolution. The important issue ignored by Wilberforce, was the historicity of the Scriptural account. Was man created in the fashion described by the Bible? Was man a creation of God at all? These issues would be decided based upon whether or not the observational data supported Darwin. If it did, to simply say that evolution was incompatible with Christianity forced a terrible choice upon a scholar. He could either believe what his eyes told him and disbelieve the Bible, or he could ignore what he saw and believe the Bible. History has shown that most scholars have chosen to believe what they see. The abandonment of the intellectual high ground, caused by reactions like Wilberforce's, have had the effect of removing Christianity from the intellectual dialogue over the past century. And over the past century, as educational levels have increased, more and more lay people have become acquainted with precisely this dilemma.

One other item to notice in Wilberforce's response is that he gives no framework into which the data cited by Darwin could be fit. Human beings seem designed to need explanatory frameworks, but all Wilberforce offered were reasons why evolution couldn't be true. It was a negative response rather than a positive one. Christian scholars have rarely proposed frameworks into which to place the data acquired by science, and indeed, have viewed scientists and science with suspicion. This retreat left the intellectual playing field wide open to those who had an explanation for the observational data. And this explanation could handle the data down to extremely detailed levels. Christians did not put a team on the field but sat on the sidelines yelling that the opponents' team on the field could not win, when in fact, that team had won by default.

Others, like George Mivart, attempted to harmonize evolution with the Scriptural origin of man. Mivart took the approach that it was all right for man's body to evolve but his spirit had to have been an act of divine creation. Mivart wrote to his friend, Edmund Bishop, "As to the account of Creation, God need not have written such misleading words as 'and the evening and the morning were the first (second,third,etc.) day."[5] This type of response was equally ineffective at solving the problem or achieving the respect of scientists like Huxley, who wrote:

> "If we are to listen to many expositors of no mean authority, we must believe that what seems so clearly defined in Genesis—as if very great pains had been taken that there should be no possibility of mistake—is not the meaning of the text at all. The account is divided into periods that we may make just as long or as short as convenience requires. We are also to understand that it is consistent with the original text to believe that the most complex plants and animals may have been evolved by natural processes, lasting for millions of years, out of structureless rudiments. A person who is not a Hebrew scholar can only stand aside and admire the marvelous flexibility of a language which admits of such diverse interpretations."[6]

This approach also left much to be desired because it, too, missed the real issue, which is the Bible's connection to history. Agnostics like Huxley knew this truth intuitively. Without a connection to history, the Biblical account becomes something that cannot be trusted. Mivart's view only left the most tenuous connection between God, the events of the Bible and the origin of man.

In the 1890s Sir William Dawson, an influential Christian writer, dealt with the human fossil material of his day by including all fossil men into *Homo sapiens*.[7] These included Cro-magnon and Neandertal. He based his conclusion upon the brain size and morphology of the fossils. His book would be the last one which could so easily handle the issue of fossil man. There were no other fossil men known to him at the time he wrote the book, probably in late 1893. As Dawson's book went to press, Eugene Dubois was publishing his full account of Java man. This fossil would be much harder to deal with than the large brained Cro-magnons or Neanderthals.[8] From this moment on, Christians would have to deal with a being which was considerably different from modern men.

The earliest "modern young-earth creationist," George McCready Price, set the tone for much of what was to follow. Early in this century, Price began publishing a sequence of books challenging evolution. Eventually, Price's books would become quite influential as the seed for Henry Morris's views on geology. Price viewed fossil man as unconnected with *Homo sapiens*, but the result of lies or carelessness of scientists. Price's contempt for modern science set the tone which today so pervades young-earth creationism. Of *Homo erectus* (Java Man) he wrote:

> "When these remains were first brought to Europe, Virchow, the foremost authority of that time, declared this skull to be that of a gibbon. The same opinion has been expressed by many others. Richard Hertwig says: 'The opinion that is most probably correct is that the fragments belonged to an anthropomorphic ape of extraordinary size and abnormal cranial capacity, and with a relatively large brain.'
> "Nothing but its large size would hinder anyone from calling it an ape."[9]

Price's view of Neanderthal and Cro-Magnon was different. These specimens he seemed to view as perfectly human. He would even agree that these fossil men were associated with extinct animals. But then he would use those facts to conclude that modern man had witnessed the formation of the entire geologic column. He said,

> "But in this fact, if it be a fact, that Man lived under the wholly strange and different conditions of 'Pliocene' or perhaps 'Miocene time,' is THE VERY STRONGEST POSSIBLE ARGUMENT that I can conceive of for the necessity of a complete reconstruction of geological theory—I mean, of course, apart altogether from the preposterous way in which the life succession was assumed and built up and then treated as an actual fact."[10]

and

> "Man must have seen the entire elevation or at least the completion of practically all the great mountains of the world, such as the Rockies, the Andes, the Alps, the Carpathians, the Caucasus, the Himalayas, etc."[11]

He has committed the logical error of assuming that if man lived in association with one extinct animal then he must have lived in association with all extinct animals. In spite of this type of logical error, George McCready Price was extremely influential. Henry Morris states of Price's influence on his life,

> "I first encountered his name in one of Harry Rimmer's books (see the discussion of Rimmer later in this chapter) and thereupon looked up his book The New Geology in the library at Rice Institute, where I was teaching at the time. This was in early 1943, and it was a life changing experience for me. I eventually acquired and read most of his other books as well."[12]

Harry Rimmer, of whom Morris speaks above, did not discuss fossil man at all in the two books of his that I have been able to obtain. His book Modern Science and the Genesis Record, first published in 1937, makes no mention of the scientific evidence about fossil man at all. This is one form of Christian response to the possibility that man evolved--ignore the data entirely.

Byron C. Nelson, an influential early old-earth creationist, discussed the place of fossil men in his 1948 book Before Abraham. Unlike most modern old-earth creationists, Nelson included all fossil men in humanity. This would include both Java man and Peking man (now both considered Homo erectus). He erroneously claimed that all but one of the fossils found at Zhoukoudian, China, were of a "modern" type.[13] This was not true, but it avoided theological problems. But he had caught some of Prices' distrust of science. He accused the evolutionist of picking and choosing the fossils they used. This charge is one often leveled by Christian apologists today. His book would be the last one which did not have to deal with Australopithecus although he should have. It was in 1947, during the time Nelson prepared his manuscript, that Arthur Keith accepted Australopithecus as part of the human lineage.[14]

An even more influential book by an old-earth creationist, Bernard Ramm, makes no mention whatsoever of the various forms of fossil men.[15] Ramm a theologian chose to examine the theological issues ignoring the practical questions of what one is to do with the various fossils identified in the human lineages. Ramm left his readers no framework for understanding fossil man other than an approval for an ancient humanity.

Young-Earth Creationist Views of Fossil Man

For the purpose of the following discussion, I will define the young-earth creationists as those closely associated with either the Institute for Creation Research or the Creation Research Society. We will examine the statements that have been made by members of this movement about the various forms of fossil man.

Australopithecus

The typical young-earth response to Australopithecus seems to be that he cannot be an evolutionary ancestor. Duane Gish wrote:

> "If Australopithecus, Homo habilis, and Homo erectus existed contemporaneously, how could one have been ancestral to another? And how could any of these creatures be ancestral to Man, when Man's artifacts are found at a lower stratigraphic level, directly underneath, and thus earlier in time to these supposed ancestors of Man? If the facts are correct as Leakey has reported them, then obviously none of these creatures could have been ancestral to Man, and that leaves Man's ancestral tree absolutely bare."[16]

and

> "We conclude that the australopithecines (A. africanus, H. africanus, H. habilis, A. boisei, A. robustus, A. afarensis) were apes, with no genetic relationship either to Man or to any of the extant apes."[17]

Gish rejects the australopithecines as human ancestors and relegates them to an entirely different and separate group from man or the apes. But his reasoning, used by many creationists, is highly flawed. The reasoning he uses would be the same as saying that if I and my great-grandmother lived simultaneously, she could not be my ancestor. But in fact, she and I did live simultaneously for two years and she is my ancestor. Himalayan cats were derived from a cross between the Siamese and Persian varieties, yet all three live today on earth quite nicely. The existence of the Himalayan cat did not require the destruction of all Siamese and Persian kitties.

John Klotz, in one of the earlier creationist discussions of *Australopithecus*, as well as Henry Morris and Gary Parker, used similar reasoning. Klotz wrote,

> "Yet all recognize that this indicates *Australopithecus* was contemporaneous with some form very similar to, if not identical with man. Hence he could not be man's ancestor."[18]

and Morris and Parker argue,

> "The australopithecines could not have been our ancestors, of course, if people were walking around *before* Lucy and her kin were fossilized -- and there is some evidence to suggest just that."[19]

Gish's earliest comment on *Australopithecus* was more logical than the one presented above. Gish wrote:

> "Combining all of the above considerations with the undoubted fact that the australopithecines possessed ape-sized brains, strongly indicates, I believe, that they were nothing more than aberrant apes, ecologically similar, perhaps, to galada baboons."[20] [Gish misspelled "gelada--GRM]

While the australopithecines may or may not have been human in any sense we would recognize, the logic used by the young-earth creationists in their rejection seems lacking. The creationist argument here ignores the fact that australopithecines are found in the geologic column 3 million years *earlier* than *Homo habilis*; *habilis* is found in strata *earlier* by 600 thousand years than strata containing *Homo erectus*. While there was a period of overlap, the ancestor did precede the descendent in all cases.

Admittedly, there is much doubt about the 'humanity' of the australopithecines and there is much doubt about whether any australopithecine manufactured tools, but at one time it was believed that they did. The australopithecines do not show Broca's area and Wernicke's area, brain structures that are associated with speech. There is no evidence of clothing.

Homo habilis

The young-earth creationist view on *Homo habilis* is even less consistent. Caledonio Garcia-Pozuelo-Ramos, a young-earth creationist, rejects *habilis* stating that he presents no human characteristics.[21] The quotation by Gish above clearly dismisses *habilis* from the human lineage, but again for the spurious reason that he may have lived at the same time as *Homo erectus*. Gish further writes of the notion that *habilis* should be in the genus *Homo*:

> "Today, most anthropologists reject this notion and maintain that *H. habilis* was a variety of *Australopithecus*, albeit perhaps somewhat more 'advanced.'"[22]

This simply is not true. Anthropologists almost universally include *habilis* in the genus *Homo*. A minority of authorities, including Richard Klein, have expressed reservations about the *habilis* taxon.[23] Klein has expressed the concern that *habilis* contains more than one species. Even he does not doubt that some of the fossils currently attributed to *Homo habilis* belong in the genus *Homo*. Statements of this nature, which display a lack of knowledge about anthropology, abound in Christian apologetical literature.

Marvin Lubenow, the young-earth creationist most knowledgeable about anthropology, writes:

> "However, even if *Homo habilis* were a legitimate taxon, and 1470, 1481, and 1590 were proper members of their taxon, *Homo habilis* could not be the evolutionary ancestor of *Homo erectus* because the two groups lived at the same time as contemporaries.
>
> "In chapter 5 we discussed the process by which one species allegedly evolves into

another species (regardless of whether the evolutionist believes in phyletic gradualism or punctuated equilibria). For habilis to evolve into erectus, habilis must precede erectus in time."[24]

This approach, in addition to being logically flawed as noted above, ignores the fact that the earliest *Homo habilis* does precede *Homo erectus*. The earliest *Homo habilis* is from rocks dated at 2.4 million years old.[25] This was found at Lake Baringo in 1967 by John Martyn. The earliest *Homo erectus* is dated at 1.9 million years old.[26] There is no requirement for *Homo habilis* to become extinct for it to give rise to *Homo erectus*, any more than your mother had to die to give birth to you. This having been said, it may come about that erectus lived much longer ago than is currently recognized. But this will have to be established by future discoveries.

By making statements like the above, apologists give no credit to the intelligence of their opponents. Anthropologists are smart enough to know that if they believe *habilis* gave rise to *erectus*, then *habilis* must precede *erectus* in time. Underestimating one's opponent is not a good strategy.

Henry Morris presents a confusing response to these various fossils. He does not seem to be aware of the difference between *habilis* and *erectus*. He wrote:

> "There is no doubt that *Homo erectus* ('erect man') had an upright posture. That he was truly human, rather than an erect ape, has recently been confirmed by studies of the brain endocast from the skull known as '1470,' discovered a number of years ago by Richard Leakey.
>
> 'An endocast from the Kenya National Museum, a *Homo habilis* specimen known as ER 1470, reproduces a humanlike frontal lobe, including what appears to be Broca's area.'"[27]

Broca's area is a part of the brain that controls speech. While Morris may think he is establishing the humanity of *erectus*, he has just proclaimed *Homo habilis* human.

One issue which must dealt with is the apparently widespread belief that the small size of the habilis brain places them absolutely among the apes. Davis and Kenyon state,

> "Some design proponents point out that its brain case, while larger than most australopithecines, is really too small for it to be classified as human (650 cc compared to about 1400 cc for modern man). They interpret it instead as an extinct primate."[28]

Young-earth creationists also follow this line of reasoning. John Klotz writes in a section defining what man is,

> "Man's brain is two-and-a-half to three times as big as the brian of the largest ape, the gorilla, and his brain is also relatively the largest."[29]

Klotz then goes on to point out the brainsize of the various fossil men, implying that brainsize is important. But it is not. There has been one normally intelligent human, alive in this century, having a 650 cc brain.[30] The man was Daniel Lyon, an Irishman who died in 1907 at age 46. Regardless that this brainsize is smaller than that of *Homo erectus* and a one year old *Homo sapiens* infant,[31] it seems quite presumptive to place him among the apes merely upon the basis of his brain size. Considering that *habilis* also had an enlarged Broca's area, which is indicative of language and a brainsize equal to that of Daniel Lyon, one cannot rule him out of the human family.

Homo erectus

Young-earth creationists have generally accepted *Homo erectus* as being fully human in a theological sense. The statement by Morris above is a good illustration of that acceptance, although there are a few exceptions.[32] However, as noted above, Morris has presented a very confusing response to the problem presented by fossil man. While he accepts their humanity above, he rejects them as seen here:

> "These discoveries, along with many evidences from other parts of the world, show that modern man lived at the same time as *Australopithecus*, *Homo erectus*, and other such supposed ancestors. Whatever these latter extinct creatures may have been, they could not have been

ancestors of man."[33]

If *Homo erectus* is human and lived in the past, then he has every chance to be the ancestor of modern men. This is especially true if the earth is only 6000 years old. While accepting the humanity of *erectus* he now denies the possibility that *erectus* is an ancestor. Yet if *erectus* is human, he could have interbred with our ancestors. However, things get even muddier when one looks back a few pages in the same book. He remembers:

> "When I was in school, I was taught that the three conclusive proofs of human evolution were Piltdown man, Peking man and Java man. These famous discoveries, however, are no longer taken seriously. Piltdown man was a hoax, Peking man has been lost for forty years and Java man was later admitted by its discoverer to be an artificial construct of a human thighbone and the skull of a gibbon. Other former 'stars' in the ape-man extravaganza were Nebraska man (an extinct pig) and Neanderthal man (now universally acknowledged to be modern man).
> "The current 'star' in this long-running show is a supposed hominid (ape-man) named *Australopithecus* (meaning 'ape of the south'), associated with a varied collection of fossil evidence, including Leakey's Skull 1470 and Carl Johanson's Lucy, as well as Mary Leakey's Laetoli fossil footprints."[34]

His lack of anthropological knowledge comes shining through. First, Peking man and Java man are both *Homo erectus*. Above, he said they were and were not our ancestors and now he tells us that no one pays attention to them. This is simply not the case. No anthropologist questions the place of *Homo erectus* in our ancestry. Casts were made of the Peking Man skulls and, while the original bones were lost during World War II, the casts are still studied. Secondly, Dr. Morris does not know the name of the discoverer of "Lucy". It was Donald Johanson, not Carl Johanson. A simple look at the original literature would reveal this. And this is not an isolated mistake. He makes the same mistake a year earlier in his book <u>Biblical Basis for Modern Science</u>.[35] Furthermore, in the above quotation, Morris calls 1470 an australopithecine but the quotation at the top of this section has him calling it an *erectus*, but then quoting Leakey calling it an *habilis*. In point of fact, 1470 is considered a *habilis*, Morris' confusion notwithstanding.

Part of the reason Morris takes this approach to Java man concerns the statements Eugene Dubois made late in his life. He revised his opinion of Java man. Morris and Parker write:

> "Eugene Dubois later dismissed his own find as the unrelated parts of a human and a giant gibbon, and since he had found—but kept secret for thirty years—a human skull discovered at the same level, he knew his other finds should not be called the ancestors of human beings. Although I'm completely unable to offer any reason why, Donald Johanson (1981) refers to Java man as if it were still considered a valid fossil, and Richard Lewontin (1981) wants (*Pithecanthropus*) taught as one of five 'facts of evolution' he cites."[36]

The reason Johanson and Lewontin refer to *Homo erectus* as a valid fossil is that no credible anthropologist believes that Java Man is a gibbon. Since Dubois' day many partial skeletons, and one nearly complete skeleton, of *Homo erectus* have been found. The characteristic *erectus* skulls, with cranial capacities many times that of any gibbon, have been found elsewhere in the world in association with femurs similar to the one found in Java. Furthermore, a gibbon has a 100 cc brain; *Homo erectus* has a 1000 cc brain. One can certainly tell the difference between a gibbon and a creature with the cranial capacity of a modern human. Finally, the set of morphological characteristics seen in Java Man are seen in similar fossils from elsewhere.

Morris' views on *Homo erectus* are quite inconsistent. Since three of these citations are contained in the same book within ten pages of each other, it is difficult to even talk about an opinion *du jour*, rather than an opinion *du page*.

Other young-earth creationists are uncertain of the status of *Homo erectus*. Frair and Davis contradict Morris' view of Java Man and at least show that they are aware that Java Man *is Homo erectus*, and they acknowledge that *Homo erectus* is not an ape,

> "Around the turn of the century a number of fossil hominids thought to be ancestral to humans were known to paleontology. Essentially those were what we today call Neanderthal man

(now generally acknowledged to be a race of *Homo sapiens* and clearly human) and *Homo erectus*, an example of which is Java man. Those forms surely were not apes or anything like apes, so they cannot be considered transitional forms between people and apes, or between people and the apelike, hypothetical common ancestor of creationists and other skeptics of the day."[37]

They further suggest a way to determine the status of *erectus*. They write:

"In our opinion it is possible that these creatures were human, although there is no clearcut evidence of any religious life among them. It is possible that some were specialized or degenerate branches of humanity."[38]

We will see that there is indeed such evidence for a religious life.

Gish believes that some *Homo erectus* have no link to modern man and at least some of them should be classified with Neanderthal. He advocates,

"At this time, while the evidence in most instances is still very fragmentary, and published reports in some instances have been strongly influenced by preconceived ideas, it is our opinion that some specimens attributed to *Homo erectus*, such as Java Man and Peking Man, are definitely from the ape family with no link of any kind to Man. In other cases (some of which have not been described here) specimens have been attributed to *Homo erectus* which otherwise would have been attributed to Neanderthal Man if the authorities making this decision had not believed that the fossil creature was too old to have been Neanderthal Man."[39]

Lubenow takes a very direct approach to dealing with erectus. He suggests,

"Surprisingly, *Homo erectus* furnishes us with powerful evidence that falsifies the concept of human evolution. Three questions are crucial. First, is *Homo erectus* morphologically distinct enough to warrant its being classified as a species separate from *Homo sapiens*? The evidence clearly says no. By every legitimate standard applicable, the fossil and cultural evidence indicate that it should be included in the *Homo sapiens* taxon."[40]

Homo erectus may very well be the same biological species as *H. sapiens*. Many anthropologists like Wolpof, Jelinek and others have suggested that *Homo erectus* should be included into our species *Homo sapiens*. However, that does not mean that *Homo erectus* was exactly like *us* morphologically. *Homo erectus*, as noted earlier, has no chin with which to buttress the front end of the jaw, but has a simian shelf which is on the interior of the jaws. Apes also have this feature but modern man does not. *Homo erectus* had huge brow ridges, which are not a modern human feature. Their skulls were smaller and the occipital and temporal bones possessed differences from those of modern man. Overall, his bones were thicker and more robust than ours. While I agree that he was human in the theological sense, he was not a modern human in the morphological sense in spite of what young-earth creationists wish to imply.

As has been seen, the young-earth positions on *Homo erectus* are quite different, varying by the author and even page by page in the same book. Part of the problem is that they don't know whether it is better to accept the evolutionary implications of the major morphological change between *sapiens* and *erectus* or to have human-like, non-spiritual beings walking the earth. This dilemma seems to draw them in two directions at once.

One Dallas area creationist deserves special mention in the rejection of *Homo erectus*. Jobe Martin wrote,

"So there is no hard evidence that Peking Man is an ancestor of Homo sapiens. Some photographs of Peking skulls remain. The skulls were broken into from the rear and most probably, the brains served as food for true Homo sapiens. It would hardly be likely that the ancient ancestors of man lived concurrently with man and that his brains would be considered a delicacy of his great-grandchildren, homo sapiens. As early as 1957, French paleontologist, Dr. Marcellin Boule, proposed that the people who made the tools that killed Peking Man were true Homo sapiens."[41]

Martin's conclusion is suspect because his data is erroneous. Marcellin Boule could hardly have been making such suggestions in 1957 since he had been dead for 15 years, having died on July 4, 1942! Furthermore, photographs are not the only things remaining of Peking Man. Franz Weidenreich made some excellent casts of the Peking Man fossils which exist to this day.[42]

Neanderthal

Among the young-earth creationists it is almost unanimous that Neanderthal is included in humanity. Dennis Peterson writes:

"This is appropriate since both Cro-Magnon Man and Neanderthal Man are both true humans indeed."[43]

Frair and Davis agree,

"The evidence is more than suggestive therefore, that Neanderthal man is a member in very good standing of the human race."[44]

Morris and Parker:

"Nowadays, evolutionists agree completely with creationists: Neanderthals were just plain people, no more different from people living today than one living nation is different from another."[45]

Other young-earth creationists like Klotz,[46] Huse[47], and Wysong[48] and Lubenow concur that Neanderthal must be classified as a true *Homo sapiens*.

The Place of Fossil Man

One of the most curious features of the young-earth view of fossil man is the general rejection of fossil man as being pre-diluvial. Henry Morris writes:

"The other difficulty is the rarity of human fossils deposited by the Flood. As discussed in Chapter 14, the antediluvian population could well have been at least as large as the present world population. Why, then, do we find so few human fossils and remains of the pre-Flood civilizations? As we shall see in Chapter 15, most of the human fossils that have been found (Neanderthal, etc.) are probably post-Flood.
"The answer could well be that man is the most mobile of all creatures, and thus would be able to survive the Flood waters by swimming, climbing, rafting and other means, much longer than other creatures. When finally overtaken and drowned, the bodies of the antediluvian men and women would finally merely decay and be dispersed, never being caught and buried in sediments at all."[49]

and

"A few fossils possibly of antediluvian men have been found and others may be unearthed in the future, but these are bound to be very rare."[50]

The response of the young-earth creationists has lacked consistency. This reflects the absence of a well thought out defendable theoretical position. There is no consensus on whether the various forms are apes, men, preflood or postflood. The result is total confusion.

Old-Earth Creationist Views of Fossil Man

Old-earth creationists have an entirely different problem than the young-earthers. Since they believe in the scientifically accepted age for the earth, they are forced to choose some point at which man became man. In order to retain any touchstone with the Biblical account there must be a point at which some hominid became a man in a theological sense. The problem devolves into one of choosing the time of Adam's creation. The choices made by various authors raise several disturbing questions. I will present several of the viewpoints but this is hardly meant to be a comprehensive list. Since this section is meant to be a discussion of creationist reactions, I will not discuss any Christian views in which mankind simply evolved with no divine intervention. That view does not fit into a creationist paradigm.

Most old-earth anti-evolutionists cite the origin of anatomically modern men 100,000 years ago as the demarcation between animal and man. David Wilcox in a recent article wrote,

> "Both cultural and physical evidence suggests an abrupt establishment of the image about 100,000 years ago."[51]

But as we saw, the earliest anatomically modern men were doing the same things, creating the same tools and thus were culturally identical with archaic *Homo sapiens*. By this we mean that they made the same tool types in the same manner. This lack of a cultural divide with the appearance of anatomically modern men invalidates the view Wilcox is advocating. However, Wilcox at least is aware that anatomically modern men appeared on earth as long ago as 100,000 years ago. Some are not.

Hugh Ross sets the date much younger than this. He claims that if Adam was created before 60,000 years ago, the Bible is untrue. He states,

> "However, the dates for these finds are well within the biblically acceptable range for the appearance of Adam and Eve — somewhere between 10,000 and 60,000 years ago according to Bible scholars who have carefully analyzed the genealogies."[52]

Of all the hominid fossils, Phillip Johnson writes,

> "The hominids, like the mammal-like reptiles, provide, at most, some plausible candidates for identification as ancestors, if we assume in advance that ancestors must have existed. That 130 years of very determined efforts to confirm Darwinism have done no better than to find a few ambiguous supporting examples is significant negative evidence."[53]

As we shall see, this limitation on the creation of Adam creates terrible problems for Ross' and Johnson's old-earth position. Like Ross, Johnson clearly rejects any ancestors for modern man. But as we shall see in the next few chapters, these 'rejects' behaved very much like us in the manner in which they used their very primitive technology.

Bernard Ramm is equally uncertain of the date of Adam and Eve, whom to call human and who to exclude from humanity. Ramm writes:

> "The chief problem with an origin of man at 500,000 B.C. is the connexion of Gen. 3 with Gen. 4. We might stretch the tables of ancestors a few thousand years, but can we stretch them 200,000 years? In the fourth and fifth chapters of Genesis we have lists of names, ages of people, towns, agriculture, metallurgy, and music. This implies the ability to write, to count, to build, to farm, to smelt, and to compose. Further, this is done by the immediate descendants of Adam. Civilization does not reveal any evidence of its existence till about 8000 B.C. or, to some 16,000 B.C. We can hardly push it back to 500,000 B.C. It is problematic to interpret Adam as having been created at 200,000 B.C. or earlier, with civilization not coming into existence till say 8000 B.C."[54]

Yet Ramm acknowledges that there is no date given for the creation of mankind. He tells us,

> "The Bible itself offers no dates for the creation of man. We mean by this that there is no such statement in the text of the Bible at any place. We may feel that 4000 B.C. or 15,000 B.C. is more consonant with the Bible than a date of 500,000 B.C. But we must admit that any date

of the antiquity of man is an inference from Scripture, not a plain declaration of Scripture.

"If the anthropologists are generally correct in their dating of man (and we believe they are), and if the Bible contains no specific data as to the origin of man, we are then free to try to work out a theory of the relationship between the two, respecting both the inspiration of Scripture and the facts of science."[55]

Like Ramm, John Wiester is uncertain of the time of origin for mankind. He writes;

"It is therefore not possible to determine whether Adam and Eve were the ancestors of Neanderthal man or Cro-Magnon man. Evidence of characteristics associated with human beings appear as early as Neanderthal man. To date it is impossible to identify such characteristics any earlier."[56]

While including the possibility of Neanderthal and Cro-magnon being human, Wiester specifically excludes *habilis* and *erectus*, stating,

"I believe we can dismiss *Homo habilis* and *Homo erectus* as likely candidates for Adam and Eve. For one thing science is not certain whether they led to *Homo sapiens* at all. They may have become extinct. Furthermore, the present fossil evidence does not indicate they possessed those traits that we consider uniquely human."[57]

While there has been some discussion about the taxonomic status of *Homo habilis*, there has not been much recent (last 30 years) questioning of the ancestral status of *Homo erectus*. Wiester is factually wrong in this. And subsequent discoveries show clearly that *Homo erectus* did have some significant traits in common with us.

David Wilcox disagrees with Wiester on even the possibility of Neanderthal being human. He notes:

"Despite their large brains, they showed cultural stasis--no indications of representational art or record keeping, and no sign that language was part of their 'adaptive complex'--things which the Cro-Magnon sites following them showed from the start."[58]

He concludes that only modern humans are able to be included in what is theologically considered humanity, e.g., having the image of God. He clearly states,

"Based on this, I judge the anatomically modern *Homo sapiens* of Africa and the Levant (which appeared about 100,000 years ago) as a unified species, differentiated from the archaic groups of '*H. sapiens*' which preceded and paralleled him."[59]

Hugh Ross takes a similar view of Neanderthal man. He writes,

"Bipedal, tool-using, large-brained primates (called hominids by anthropologists) may have roamed the earth as long ago as one million years, but religious relics and altars date back only 8,000 to 24,000 years. Thus, the secular archaeological date for the first spirit creatures is in complete agreement with the biblical date.

"Some differences, however, between the Bible and secular anthropology remain. By the biblical definition, these hominids may have been intelligent mammals, but they were not humans. Nor did Adam and Eve physically descend from them. (According to Genesis 1:26-28 the human species was created complete and brand-new by God through His own personal miraculous intervention.) Even here, though, support from anthropology is emerging. New evidence indicates that the various hominid species may have gone extinct before, or as a result of, the appearance of modern humans."[60]

and

"From a Biblical standpoint, I see Neanderthals as one of the *nephesh*, soulish (not spiritual) creatures God made before he made humans. In other words, the Neanderthals must have been a bipedal mammalian species created a few tens of thousands of years before Adam and Eve. Neanderthals became extinct, possibly as the result of some climactic upheaval, at least several thousand years before the creation of Adam and Eve."[61]

There are several factual problems with Ross' position here. As we saw, modern men were on earth for the past 120,000 years. This means that modern man, *Homo erectus* and Neanderthal lived on earth together for 70,000 years. The ancient forms of man did not go extinct before or even as a result of the appearance of modern men. Secondly, when we see in future chapters what Neanderthal did, including inventing the oldest musical instruments, one must wonder what advantage this approach has. It is rare, these days, that I find myself in agreement with young-earth creationists, but here I agree with John Morris,

> "But this leads to a host of theological problems some of which are discussed in the pages to follow. For example, to the consistent old-earth 'creationist,' Neanderthal Man, who lived long ago according to evolution theory and radioisotope dating methods, was only a human-like animal, who talked, painted pictures, buried his dead, etc., but who had no eternal spirit, and was not related to modern man."[62]

Robert Gange rejected all hominids except modern man when he wrote in 1986,

> "But if one ignores the opinion of the dominant personalities and asks what the actual data teach regarding human fossils, the answer is that they show man existing as man for about 40 thousand years or so. The skulls and bones found in layers that are dated much older than this are *not* human, but rather belong to prehistoric creatures from whom most of the investigators presume man to have come."[63]

Unfortunately for Gange, it was in the late 1980s that the anatomically modern human remains at Qafzeh and Klasies River Mouth Cave were dated to around 90,000 and 120,000 years respectively, effectively tripling the time that modern man has been on the earth.

Percival Davis and Dean Kenyon reject *Homo erectus* as human based upon the supposed idea he didn't behave like us. They state,

> "It had significant anatomical differences from modern man that have prevented its classification as *Homo sapiens*. It also left no evidence that it buried its dead, no signs of art, or other recognizably human culture."[64]

Five years prior to the publication of their book, evidence for art among *Homo erectus* was being found and while still controversial, is accepted by some portions of the anthropological community.[65] The site of Bilzingsleben, Germany, shows evidence of huts in a village having been made by *Homo erectus*. We will hear more of this important site which has been known for over a decade but never mentioned in Christian apologetical writings.

The Place of Fossil Man

Old-earth creationists have an equally difficult time as the young-earth creationists figuring out where fossil man fits into the biblical scheme. Rarely do they tell you whether fossil man was pre- or post-Flood. And when they do venture a guess their suggestions run into a multitude of anthropological problems. Hugh Ross suggests,

> "Let's compare this chronology with the present archaeological and anthropological data. Some time before about 35,000 years ago, humans and civilization sprang up in the Mesopotamian flood plain, centered in Babel. Roughly 33,000 years ago, humans began to spread out over Africa, Asia, Australia, and Europe. About 12,000 years ago, large numbers of people began to settle in North and South America. About 11,000 years ago, migration from Siberia to the Americas ceased.
>
> "My guess is that Peleg lived about 11,000 years ago. This scenario and its dates remain

tentative, of course."[66]

Dr. Ross chooses these dates because of his belief that mankind can be no older than 60,000 years. However he ignores the fact that there is no evidence of the type of civilization he is referring to dated 33,000 years ago in Mesopotamia. He also ignores the fact that anatomically modern humans are not found first in Mesopotamia but at Klasies River Mouth Cave, South Africa, dated at 120,000 years ago.[67] And they spread from there, not Mesopotamia, to the other parts of the world.

Bernard Ramm relates the problem of placing fossil man into a biblical framework,

"If the evidence is certain that the American Indian was in America around 8000 B.C. to 10,000 B.C., then a universal flood or a universal destruction of man, must be before that time, and due to Genesis and Babylonian parallels there is hardly an evangelical scholar who wishes to put the flood as early as 8000 B.C. to 10,000 B.C."[68]

Regardless of what evangelical scholars wish, if the flood account and the creation account cannot be adequately harmonized with the scientific data, either the Bible is false or one must search for a new viewpoint.

Other old-earthers like Davis and Kenyon, Wiester, Gange and Wilcox do not suggest ways in which to place fossil man into a biblical framework. Thus, while old-earth creationists deal satisfactorily with the data for an old earth, they do not deal well with how fossil man fits into a biblical framework.

Conclusion

The response that the various schools of thought within Christendom have made to the concept of human evolution has been contradictory, confusing and invalidated by data now available. This is not a good situation. We will now examine the nature of the anthropological record, what one can find, what one cannot find, and how the record is deposited.

References

1. Henry M. Morris, The Biblical Basis for Modern Science, (Grand Rapids: Baker Bookhouse, 1984), p. 363, 422.
2. Andrew D. White, A History of the Warfare of Science with Theology, (New York: George Braziller, 1955), p. 70.
3. Martin J. S. Rudwick, The Meaning of Fossils, (Chicago: Chicago University Press, 1985), p. 241.
4. Andrew D. White, A History of the Warfare of Science with Theology, (New York: George Braziller, 1955), p. 70.
5. George Mivart, cited by Bolton Davidheiser, Evolution and Christian Faith, (Grand Rapids: Baker Book House, 1969), p. 85.
6. Thomas H. Huxley, "Lectures on Evolution" in Agnosticism and Christianity, Buffalo: Prometheus Books, 1992), p. 14
7. Sir J. William Dawson, The Meeting Place of Geology and History, (New York: Fleming H. Revell and Co. 1989), p. 61
8. Erik Trinkaus and Pat Shipman, The Neandertals, (New York: Random House, 1992), p. 139
9. George McCready Price, Evolutionary Geology and the New Catastrophism, (Mountain View, California: Pacific Press Publishing Assoc., 1926), p. 296.
10. George McCready Price, The Fundamentals of Geology, (Kansas City:Pacific Press Publishing Assoc., 1913), p. 218-219.
11. George McCready Price, Evolutionary Geology and the New Catastrophism, (Mountain View, California: Pacific Press Publishing Assoc., 1926), p. 318.
12. Henry M. Morris, A History of Modern Creationism, (San Diego: Master Book Publishers, 1984), p. 80
13. Byron C. Nelson, Before Abraham, (Minneapolis: Augsburg Press, 1948), p. 89
14. Arthur Keith, "Australopithecus or Dartians," Nature, 159(1947):377
15. Bernard Ramm, The Christian View of Science and Scripture, (Grand Rapids: Wm B. Eerdmans, 1954)
16. Duane T. Gish, Evolution: The Challenge of the Fossil Record,(El Cajon: Creation-Life Publishers, 1985), P. 171.

17. Duane T. Gish, <u>Evolution: The Challenge of the Fossil Record</u>, (El Cajon: Creation-Life Publishers, 1985), P. 179.
18. John W. Klotz, "The Case for Evolution," in Paul A. Zimmerman, <u>Darwin, Evolution, and Creation</u>, (St. Louis: Concordia Publishing House, 1959), p. 129.
19. Henry M. Morris and Gary E. Parker, <u>What is Creation Science?</u> (San Diego: Creation-Life Publishers, 1982), p. 124-125
20. Duane Gish, <u>Evolution: The Fossils Say No!</u>, (San Diego: Creation-Life Publishers, 1978), p. 123.
21. Celedonio Garcia-Pozuelo-Ramos, "Taxonomy of Primates," <u>Creation Research Society Quarterly</u> 34(Sept. 1997):93-, p. 102.
22. Duane T. Gish, <u>Creation Scientists Answer their Critics</u>, (El Cajon: Institute for Creation Research, 1993), p. 133
23. Richard G. Klein, <u>The Human Career</u>, (Chicago: The University of Chicago Press, 1989), p. 100
24. Marvin L. Lubenow, <u>Bones of Contention,</u> (Grand Rapids: Baker Books, 1992), p. 127.
25. Christopher Wills, <u>The Runaway Brain</u>, (New York: Harper Collins, 1993), p.132
26. Roy Larick and Russell L. Ciochon, "The African Emergence and Early Asian Dispersals of the Genus *Homo*," <u>American Scientist</u>, 84, (Nov/Dec., 1996), p. 540.
27. Dean Falk "The Petrified Brain," <u>Natural History</u>, vol 93, September 1984, p. 38 cited by Henry M. Morris, <u>Creation and the Modern Christian</u>, (El Cajon, California: Master Book Publishers, 1985), p.183-184.
28. Percival Davis and Dean H. Kenyon, <u>Of Pandas and People</u>, 2nd edition (Dallas: Haughton Publishing Co., 1993), p. 109
29. John W. Klotz, <u>Studies in Creation</u>, (St. Louis: Concordia Publishing Co., 1985), p. 151
30. <u>Guiness Book of Records 1996</u>, (New York: Facts on File, 1995), p. 14
31. Alan Walker and Pat Shipman, <u>Wisdom of the Bones</u>, (New York: Alfred Knopf, 1996), p. 212
32. Malcolm Bowden states: "The Java man fossils consisted of a large ape type skull cap and a human leg bone. The Pekin man fossils were all apes." see Malcolm Bowden, "*Homo erectus*–A Fabricated Class?" <u>Creation Research Society Quarterly</u>, 34(June 1997), p 22-23.
33. Henry M. Morris, <u>Creation and the Modern Christian</u>, (El Cajon, California: Master Book Publishers, 1985), p.189.
34. Henry M. Morris, <u>Creation and the Modern Christian</u>, (El Cajon, California: Master Book Publishers, 1985), p.181.
35. Henry M. Morris, <u>biblical Basis for Modern Science</u>, (Grand Rapids: Baker Bookhouse, 1984), p. 398
36. Henry Morris and Gary E. Parker, <u>What is Creation Science?</u>, (San Diego: Creation-Life Publishers, 1982), p. 118.
37. Wayne Frair and Percival Davis, <u>A Case For Creation</u>, 3rd ed., (Chicago: Moody Press, 1983), p. 57.
38. Wayne Frair and Percival Davis, <u>A Case For Creation</u>, 3rd ed., (Chicago: Moody Press, 1983), p. 123.
39. Duane T. Gish, <u>Evolution: The Challenge of the Fossil Record</u>, (El Cajon: Creation-Life Publishers, 1985), P. 203-204
40. Marvin L. Lubenow, <u>Bones of Contention,</u> (Grand Rapids: Baker Books, 1992), p. 120.
41. Jobe Martin, <u>The Evolution of a Creationist</u>, (Rockwall, Texas: biblical Discipleship Publishers, 1994), p. 94.
42. Ian Tattersall, <u>The Fossil Trail</u>, (New York: Oxford University Press, 1995), p. 67.
43. Dennis R. Petersen, <u>Unlocking the Mysteries of Creation</u>, (South Lake Tahoe: Creation Resource Foundation, 1990), p. 120.
44. Wayne Frair and Percival Davis, <u>A Case for Creation</u>, 3rd Edition, (Chicago: Moody Press, 1983), p. 125.
45. Henry M. Morris and Gary E. Parker, <u>What is Creation Science?</u> (San Diego: Creation-Life Publishers, 1982), p. 117.
46. John W. Klotz, "The Case for Evolution," in Paul A. Zimmerman, <u>Darwin, Evolution, and Creation</u>, (St. Louis: Concordia, 1959), p. 130.
47. Scott M. Huse, <u>The Collapse of Evolution</u>, (Grand Rapids: Baker Book House, 1983), p. 101.
48. Randy L. Wysong, <u>The Creation-Evolution Controversy</u>, (Midland Mich.: Inquiry Press, 1976), p. 298.
49. Henry M. Morris, <u>The Biblical Basis for Modern Science</u>, (Grand Rapids: Baker Bookhouse, 1984), p. 363.
50. Henry M. Morris, <u>The Biblical Basis for Modern Science</u>, (Grand Rapids: Baker Bookhouse, 1984), p. 422.
51. David L. Wilcox, "Adam, Where Are You? Changing Paradigms in Paleoanthropology," <u>Perspectives on Science and Christian Faith</u>, 48:2(June 1996), p. 94.
52. Hugh Ross, "Art and Fabric Shed New Light on Human History," <u>Facts & Faith</u>, 9:3 (1995)p. 2. See also, Hugh

Ross, "Chromosome Study Stuns Evolutionists," Facts & Faith, 9:3,(1995) p. 3 and Hugh Ross, "Searching for Adam," Facts & Faith, Vol 10, No. 1, p. 4

53. Phillip E. Johnson, Darwin on Trial, (Downers Grove: Intervarsity Press, 1993), p. 86.
54. Bernard Ramm, The Christian View of Science and Scripture, (Grand Rapids: Eerdmans Publishing Co., 1954), p. 228.
55. Bernard Ramm, The Christian View of Science and Scripture, (Grand Rapids: Eerdmans Publishing Co., 1954), p. 220.
56. John Wiester, The Genesis Connection, (Nashville: Thomas Nelson Publishers, 1983), p. 189.
57. John Wiester, The Genesis Connection, (Nashville: Thomas Nelson Publishers, 1983), p. 188.
58. David L. Wilcox, "Adam, Where Are You? Changing Paradigms in Paleoanthropology," Perspectives on Science and Christian Faith , 48:2(June 1996), p. 93.
59. David L. Wilcox, "Adam, Where Are You? Changing Paradigms in Paleoanthropology," Perspectives on Science and Christian Faith , 48:2(June 1996), p. 94.
60. Hugh Ross, Creation and Time, (Colorado Springs: NavPress, 1993), p. 141.
61. Hugh Ross, "Link with Neanderthals Cut by Computer," Facts & Faith, 9:3, 3rd Qtr. 1995, p. 2.
62. John Morris, The Young-earth, (Colorado Springs: Master Books, 1994), p. 27.
63. Robert Gange, Origins and Destiny, (Waco: Word Publishers, 1986), p. 119.
64. Percival Davis and Dean H. Kenyon, Of Pandas and People, 2nd edition (Dallas: Haughton Publishing Co., 1993), p. 110.
65. Paul G. Bahn, "Comments", Rock Art Research 5:2(1988): 91-107, p. 96. See also Robert G. Bednarik, "On Lower Paleolithic Cognitive Development," 23rd Chacmool Conference Calgary 1990, pp 427-435, p. 432.
66. Hugh Ross, "The Broken Tie That Binds," Facts & Faith, 10:3, Third Quarter 1996, p. 6.
67. Richard G. Klein, The Human Career, (Chicago: The University of Chicago Press, 1989), p. 353 and James R. Shreeve, The Neandertal Enigma, (New York: William Morrow and Co., 1995), p. 214-215.
68. Bernard Ramm, The Christian View of Science and Scripture, (Grand Rapids: Eerdmans Publishing Co., 1954), p. 234.

TO LEAVE A LEGACY: THE NATURE OF THE ARCHEOLOGICAL RECORD

Unless you are very famous, in 100 years, what will be left on earth as evidence that you lived? Your clothing will have been given away, worn, worn out, then discarded. Your old clothing thrown into the city dump will be unlikely to point to you as a person. In 100 years if an archeologist dug those clothes up, there would be no reason for him to connect them with you. Most of the books in your house will be thrown away, recycled, or decayed to powder due to the acid in the pages. A few might make it into the decor of a restaurant. Your records, tapes and CDs to which you enjoy listening will all be broken. Your house may survive, assuming that it is kept in good condition and no one bulldozes it for some urban renewal project. A few pictures of you may survive in the hands of your descendants; otherwise what pictures survive will be found for sale in future flea markets as nameless faces. There will be some government records: the deed to your house, the tax rolls, maybe your military service record. There will, of course, be your headstone at your grave, if it is not vandalized.

In 1000 years, very few of your personal possessions will still be intact. Unless your house is made almost entirely of stone it will not be standing in 1000 years. Termites and dry rot will destroy the timbers in that time. Most of your papers and books will be gone, rotted away. Fires, wars and lack of care will likely have destroyed the government records. What government clerk would care to preserve 900-year-old records of deed transactions? Is a dead man going to sue him for selling the property? How many of us know our genealogy back 1000 years to know that we are heirs to that man whose land has been wrongfully sold? The pottery in your house will probably survive, in broken form, for 1000 years. The name you scratched into the roadway's wet cement might survive if the road is abandoned, forgotten and buried by the accumulation of dirt covering it. During each successive year the chances increase for the name's destruction. Your tombstone may survive if it is granite. Soft limestone headstones will be worn away by the light acid in the rains, eradicating at the very least the name if not the entire stone. Several limestone markers in a downtown Dallas graveyard are illegible after only 100-150 years. If it is a durable form of limestone it can survive for several millennia if the rainfall is not too great. Each year brings more exposure to the forces of destruction. The best way for any object to be preserved in the archeological record is for it to be buried and protected from the vicissitudes of nature. Buried objects are much more likely to be preserved.

After 10,000 years, all that will remain of our civilization will be the stone monuments and other objects made of equally durable materials. One example of metallic objects that will last that long is coins that have been buried in non-acidic soils.

> "The works that have survived from ancient Egypt consist almost entirely of spectacular funerary monuments, such as the world-famous pyramids and temples. Very little is known about the nonreligious architecture--the palaces, public buildings, and private houses--of ancient Egypt. The principal reason for this is that religious buildings and tombs were constructed of such durable materials as limestone, sandstone, and granite. Moreover, their location was favorable to their enduring: they were built on the edge of the desert, beyond the reach of the Nile, which overflowed its banks every year. Palaces, public buildings, houses, and other examples of secular architecture, on the other hand, generally employed perishable materials, such as reeds, wood, and mud brick (sun-dried blocks of clay, precursors of kiln-fired bricks). Furthermore, the secular buildings of ancient Egypt were situated in the damp, fertile portion of the Nile Valley and thus were subject to annual flooding."[1]

After 10,000 years, only objects made of pottery, bone or stone are likely to survive. A few rare exceptions have been found, but their rarity only serves to accentuate the low numbers. All photographs will be destroyed, newspapers rotted, high school diplomas lost. Nothing is likely to survive from our personal existence. One exception might be the humble hotdog, which contains so many preservatives that it may never decay.

The archeological record is very incomplete and fragmentary. What is generally not understood is that each year brings the destruction of some of the artifacts associated with a culture. The severity of the destruction depends on the durability of an artifact and the location at which it resides. Items made of stone, bone and other hard and durable material can last a long time before the forces of destruction overcome them. Items made of wood, animal skins, paper, or other vegetable matter, like cotton, are easily destroyed. Year after year, the relentless parade of destruction continues. Year after year, erosion wears away fossils on rock outcrops, the stone art of past civilizations, the pottery and the remains of buildings. The buildings will collapse, erode and disappear. After 100,000 years, what little remains will leave future scientists questioning whether we were truly human or merely soulless brutes.

There is another aspect of the archeological record which suggests that what we will leave to our descendants is quite different from what we received from our ancestors. The ancient people had a poorer technology. They used more wood in their daily activities than we do. Since wood rots away very quickly, many of the manufactured objects of their culture have disappeared. Jim Hamm notes of Native American technology,

> "In the 1930's, in a cave in the Gila River area of New Mexico, a cache of several thousand reed arrows was found. Less than a dozen of the arrows had stone points, the rest were hardwood blunts."[2]

A "hardwood blunt" is an arrow with a wooden bulb on it. It has no stone tool attached to it. If it weren't for this cache of blunts, we would not know that most arrows had no preservable stone arrowhead. In another few thousand years after the wood rotted, nothing but the arrowheads would be left, leaving us only 0.1% or so of their technology. From this we would have to reconstruct their lives.

This century in Australia, aborigines lived in a way which would leave little record of their existence.

> "In western Australia today there are aborigines who make very crude-looking stone tools. But their wooden implements are very elaborate, with fancy painting on bark, and beautiful spearthrowers and shafts. They also have extremely complex social systems, cosmology, and narrative traditions. If you were to dig up one of their sites a thousand years from now, however, all you would see would be the clunky stone tools. Does this mean those aborigines were technologically inferior? Not at all. They were simply relying on perishable materials."[3]

This reliance on perishable materials is the bane of the archaeologist. Painted bark disappears rapidly; grass dolls degrade. Clothing made of vegetable matter, such as grass skirts, would be long beyond the reach of the archaeologist. Feathers used for decoration would not survive to leave their mark. So, if all the evidence of working with vegetable matter disappears, how do we know it ever existed? Some things leave indirect evidence. Wood working requires sharp stones to be used to shape the wood. The act of whittling the wood leaves a particular pattern of scratches on the stone tool's cutting edge. Sometimes microscopic examination of these tools can show what material was shaped by the tool.[4]

How does the increase in technology effect what we see in the fossil record? An artifact cannot be fossilized until sometime after it is first made. Similarly an animal cannot be fossilized until after it has evolved. This creates an interesting situation. It is unlikely that the very first example of an animal or an artifact will be fossilized. We do not know when the first scythe was made and when the harvesting of grains began. What we know is that the oldest flint sickle we know of is from 13,000 years ago.[5] Since it is unlikely that we found the very earliest flint sickle, all we can say with absolute certainty is that the harvesting of grains by means of a sickle began at least by this time and probably somewhat earlier. How much earlier can only be estimated statistically. Sometimes the first and second appearance of a given object in the fossil record are separated by thousands or millions of years. The point is that the very first example of any object in the fossil record is very, very unlikely to be the earliest example that existed.

The first occurrence of life on land is found in rocks dated to 1.2 billion years, while the second dates from 800 million.[6] This is a gap of 400 million years (MYR) with no intervening evidence in the record. The earliest occurrence of caecilians, a type of amphibian, dates to 180 million years (MYR) while the second earliest find is dated to 80 million years.[7] There is a gap of 5 million years between the first and second oldest amphibians.[8] Prior to 1986, there was no fossil record of tarsiers. At that time a discovery in Thailand proved that tarsiers lived on earth during the early Miocene. There was a 20 million-year gap between the first and second occurrence of tarsiers. Then in 1994 a new discovery was made in which a tarsier was found in Eocene rocks dated 30 million years old.[9] Here we have a gap of 10 million years followed by a gap of twenty million years in the history of the tarsiers. The earliest turtle is found 60 million years before the second oldest turtle fossil.[10] There are gaps of 8 million years, 7 million years, and 7 million years between the earliest four fossil birds.[11] There have been some recently discovered fossil birds that might close some of these gaps but the dating is still controversial. The fossil record of coprolites, fossil dung, is also gapped. The earliest coprolite showing digested land plants dates at 412 MYR, the next at 390 MYR and the third at 90 MYR[12] Until 1995, the oldest fossil evidence of mushrooms was one found in Dominican amber dated 25-30 million years old. Then a mushroom was found in 90-94 MYR old Cretaceous amber,[13] a gap of 60 million years. There is a 25 million-year gap between the first and second shark fossils.[14] There is a 50 MYR gap between the first occurrence of a land plant and the second occurrence.[15] Spiders fossils have a 45 million year gap between the first and second occurrences.[16] This list could be extended to nearly every single artifact in the fossil record.

The most important lesson the above facts teach is that the earliest find of an object in the archaeological record is *not* necessarily the first occurrence of that object on earth. When looking at the fossil record of man, the earliest known *Homo erectus* is from rocks dated nearly 2 million years; it is not necessarily the case that *Homo erectus* evolved at that time. He may have lived long prior. We can never be sure. Are we looking in the correct place? Were earlier erectus' living in areas where the soil would dissolve their bones? It is a fact that Bones will totally disappear in one year of burial in a tropical forest environment. We cannot be too dogmatic that we know when something arose. Robert Bednarik notes:

"Unless we believe that hominids lived only in regions of high-pH soils, especially on limestone karsts, we have to concede that our record of their distribution, in both space and time, is fundamentally flawed."[17]

What is the cause of this gapped nature of the fossil record? Preservation is part of the answer. As one goes further back in time toward the origin of any object, say iron nails, there is a greater chance that the nails were rusted away. Each year that a nail lies in the earth, destructive forces work on it. Secondly, as one approaches the time of origin for iron nails, there are so few of them that it is unlikely that any should be buried and thus preserved. Thirdly, as one goes back in time toward the origin of iron nails there are fewer countries that made them so their geographic distribution becomes more restricted. If you want to find the earliest iron nails you must know where to look. Applied to fossils, these problems mean that one must be looking in the place in which the species originated. The species must have been around long enough for it to be probable that a member of the species would be buried and then not destroyed.

Preservation of Different Materials

Inorganic materials such as stone choppers can be preserved for over two million years in close to the same condition they were in when abandoned. Ceramics can be preserved for thousands of years. The earliest ceramic figurine is from 27,000 years ago and was found at Dolni Vestonice, Czechoslovakia.[18] Japanese pots have survived as long as 10,000 years.[19]

Organic remains are fragile and their preservation depends upon the environment in which they are buried. Bone carvings by Upper Paleolithic man are only found in soils that are not acidic. Acidic soils destroy bone quickly. Robert Bednarik notes,

"In considering just those allegedly female images which are allegedly of the Upper Palaeolithic and which are fully sculpted, it is evident that nearly all are of materials that would survive poorly in acidic soils, and that practically all were found in alkaline soils, i.e. loesses or cave sediments. With the exception of a few specimens, these objects are of organically derived, mineralized calcareous materials such as Tertiary limestones or ivory, which is essentially dentine, other calcium minerals and cartilage. Therefore the correlation of characteristics such as those of the raw material of surviving specimens and the conditions conducive to their survival seems more relevant than the apparent geographical distribution of the evidence. The latter can, in fact, only be relevant under certain qualifications which archaeologists have generally ignored. Some archaeologists consider that the present distribution of the figurines indicates their former geographical distribution. What they are in fact implying is that the people who deposited these images selected soil conditions that would be conducive to preservation. Perhaps Palaeolithic people were so thoughtful (and considerate to archaeologists); but before relying on such an implausible model it would need to be shown that none of the figurines was lost or discarded, rather than 'deposited'. This example illustrates the kind of logic some archaeologists have resorted to in interpreting palaeoart, and it typifies the scientific obsolescence of most interpretations in the discipline."[20] [a loess is a type of soil-GRM]

Environment of Preservation

Various environments preserve things differently. It is very difficult for an object left on the surface of the earth to become fossilized. On the earth's surface, erosion plays a major role in destroying evidence of past activities. Finding artifacts on top of a hill after 10,000 years is less likely than finding artifacts in a valley. On top of a hill, an object is very unlikely to be buried, but in the valley below, the sediment that rushes down the hillside, and the river that rises during a flood is quite capable of burying it. Burial is extremely important in the preservation of an artefact.

A waterlogged environment is very good at preserving organic remains, as long as the remains are buried under the sediment. At Somerset, England, Neolithic man built a wooden walkway across a bog. Much of the wood has been preserved.[21] Danish bogs have preserved large quantities of organic material for the past two-thousand years. Some of these include corpses, at least one of which has a noose about its neck. The oldest fibers in the world are found at the Israeli site of Ohalo II. The site was discovered when the water level in the Sea of Galilee fell dramatically in 1989. The waterlogged conditions preserved thousands of seeds, wood and fibers for over 19,000 years.[22] A wooden board was preserved by waterlogging for the past 50,000 years at Nishiyagi, Japan.[23] Very recently three seven-foot-long spears which are 400,000 years old made by *Homo erectus* were found in a waterlogged environment in Germany old.[24] The spears had been preserved over that long time because they had been continuously waterlogged. There are only three wooden artifacts from earlier than this time. As nearly as I have been able to determine, they were not preserved by waterlogging. These artifacts include two spears, a club, and a polished wooden plank. Kathy Schick and Nicolas Toth relate,

> "Of special interest was the preservation of prehistoric wood at Clacton-on-Sea on the southern coast of England, in particular the broken shaft and tip of a spear made out of yew. This is the earliest definitive evidence of wood technology in the prehistoric record. Microscopic examination by archaeologists revealed clear striations from the wood having been shaped with a stone tool about 300,000 years ago. At the Acheulean site of Kalambo Falls in Zambia, probably between 200,000 and 400,000 years old, a possible wooden club was discovered among the fossil wood specimens. And microwear analysis by Lawrence Keeley of flint tools from the English sites of Clacton and Hoxne shows clear use-wear patterns from woodworking on some implements, from hide scraping on others. Artifacts made out of wood and hide are inferred from these polishes: as previously discussed, wood could have served as spears, digging sticks, pegs, or containers, while scraped hides could have served as containers, clothing, or elements of architecture.
>
> "This meager but tantalizing evidence suggests that there was probably a range of perishable materials employed as tools, and again suggests a rich invisible technology that rarely survives in the earlier prehistoric record. Among recent Stone Age hunter-gatherers, tools made from organic materials, such as wood and hide, are very common. The stones give us the tip of the iceberg, perhaps, but an invaluable tip it is."[25]

An extremely dry environment is also a very good place for ancient organic material to be preserved. The artifacts from Ancient Egypt are exquisitely preserved by the arid environment. Deep dry deposits in Hogup Cave, Utah, preserved baskets and nets from 9000 B.P.[B.P.-Before Present][26] Colin Renfrew relates the discovery of Tocharian texts which were found early this century in the Tarim Basin in the Talikamatan Desert of Western China. He writes,

> "In and near these oases lie forgotten cities, half obscured by sand, where wood and other organic materials are remarkably well preserved in the arid conditions of the desert. This is one of the very few areas of the world where paper and wood and textiles, centuries old, have been preserved in great quantities--only on the borders of other large deserts, especially in Egypt, have comparable archives on perishable materials been preserved."[27]

The western desert of China also harbors mummies better preserved than those of the pharaohs. Burial in the hot dry sand quickly dehydrated the bodies and inhibited decay.

The important point in this discussion of how perishable materials are preserved is that it is quite rare. Perishable material is not likely to be preserved. This being the case, the conclusions we reach about fossil man are based upon a very incomplete sample of the technological material. If future archaeologists were to study our culture without any of the perishable materials available to him, his conclusions might be very doubtful. Consider that no newspaper, magazine, or book would be available. Monuments built of stone would be about the only source of information on our language. The names on the Vietnam Memorial, the sayings of our great men carved in Washington memorials, gravestones and the like are poor sources of information. Most of the metal would also rust, so any writing inscribed on metallic plaques would also erode away. The fact is that archaeologists and some Christian apologists are far too quick to condemn the cultures of ancient hominids to a subhuman status, when it might be that most of their culture involved the use of perishable materials making, them appear less than they were.

Geographic Limitations of the Record

As fossil man spread around the earth, he left evidence. In a newly inhabited region, when there were only a few humans initially, the artifacts left behind would be few and difficult to find. In such a situation, the archaeologist might never find enough evidence to know that the region had been inhabited. Archaeologists need a certain density level of artefacts before it is likely that the evidence will be found. This density level can be achieved by one of two means. If the population in a region is sufficiently high, enough people will lose a sufficient number of personal possessions, many of which will then be buried. Alternatively, a small population living in the same place for a long time will also leave behind enough objects for the archaeologist to find. This is what happens in a cave. Obviously, thousands of people cannot live in a small cave, so the population at any one time is small. But, over thousands of years, enough rubbish sometimes accumulated that one might have a very rich archaeological deposit.

Open air sites, as opposed to caves, are another limitation to finding early man. An open air site is an outdoor campsite along a river or merely in a field. Such sites are exposed to the ravages of erosion and the trampling of man and beast. These activities are not conducive to the preservation of archaeological material. It is often extremely difficult to date these sites because organic material is rapidly destroyed.[28] This destruction rapidly removes many sites from the landscape. Thus, an archaeologist will have nothing to find in a region inhabited only in the open if the sites are not in an area that can be covered. Humans may have inhabited some regions at a given time but may have left so little evidence of their presence that we may never know they were there.

Conclusion

The limitations and biases in the fossil record limit our ability to determine exactly what activities our ancestors engaged in, where they lived, when they appeared, and the level of their culture. All the archaeological record can show is the absolutely minimal level for each of these issues. In reality the culture must have been better, more widespread and older than the archaeological record is capable of revealing. This fact should be kept firmly in mind as we look, in the next chapter, at what the evidence says about fossil man.

References

1. "Egyptian Art and Architecture," The Software Toolworks Multimedia Encyclopedia, Grolier Electronic Publishing Co., 1992.
2. Jim Hamm, Bows & Arrows of the Native Americans, (Azle, Texas: Bois d'Arc Press, 1989), p. 128-130.
3. James R. Shreeve, The Neandertal Enigma, (New York: William Morrow and Co., 1995), p. 249.
4. Derek Roe, "The Handaxe Makers," in Andrew Sherratt, editor, The Cambridge Encyclopedia of Archaeology, (New York: Cambridge University Press, 1980), p. 72.
5. "Timeline," The Software Toolworks Multimedia Encyclopedia, Grolier, 1992.
6. Robert J. Horodyski and L. Paul Knauth, "Life on Land in the Precambrian," Science, Jan. 28, 1994, p.

494-498. see also "When Life First Sprouted on Land," <u>Science News</u>, March 12, 1994, p. 173

7. "Rare Fossils of Enigmatic Amphibian," <u>Science News</u>, 138, Oct. 27, 1990, p. 270.

8. Per E. Ahlberg, Andrew R. Milner, "The Origin and Early Diversification of Tetrapods," <u>Nature</u>, 368, April 7, 1994, p. 507.

9. R. D. Martin, "Bonanza at Shanghuang," <u>Nature</u>, 368, April 14, 1994, p. 586.

10. Eugene S. Gaffney and James W. Kitching, "The Most Ancient African Turtle," <u>Nature</u>, 369, May 5, 1994, p. 55.

11. Paul C. Sereno and Rao Chenggang, "Early Evolution of Avian Flight and Perching: New Evidence from the Lower Cretaceous of China," <u>Science</u>, Feb. 14, 1992, p. 845

12. Diane Edwards, et al, "Coprolites as evidence for Plant-Animal Interaction in Siluro-Devonian Terrestrial Ecosystems," <u>Nature</u>, Sept. 28, 1995, p. 329.

13. D. S. Hibbett, D. Grimaldi, and M. J. Donoghue, "Cretaceous Mushrooms in Amber," <u>Nature</u>, 377, Oct. 12, 1995, p. 487.

14. Ivan J. Sansom, M.M. Smith and M. P. Smith, "Scales of Thelodont and Shark-like Fishes from the Ordovician of Colorado," <u>Nature</u>, 379:628-630, Feb. 15, 1996, p. 628; R. Monastersky, "The First Shark: To Bite or Not to Bite?" <u>Science News</u>, 149, Feb. 17, 1996, p. 101.

15. Chongyang Cai, Shu Ouyang, Yi Wang, Zongjie Fang, Jiayu Rong Liangyu Geng and Xingxue Li "An Early Silurian Vascular Plant," <u>Nature</u>, 379, Feb. 15, 1996, p. 592

16. Paul A. Selden, "Fossil Mesothele Spiders," <u>Nature</u>, 379, Feb. 8, 1996, p. 498

17. Robert G. Bednarik, "Origins of *Homo Sapiens*," <u>The Artefact</u>, 18(1995): 84-85, p. 85

18. Chris Stringer and Clive Gamble, <u>In Search of the Neanderthals</u>, (New York: Thames and Hudson, 1993), p. 206

19. Brian M. Fagan, <u>In The Beginning</u>, 7th Ed., (New York: Harper Collins, 1991), p. 149.

20. Robert G. Bednarik,"Who're We Gonna Call? The Bias Busters!" in Michel Lorblanchet and Paul G. Bahn, <u>Rock Art Studies: The Post-Stylistic Era or Where do we go from here?</u> (Oxbow Monograph 35, 1993), pp.207-211, p. 208-209

21. Brian M. Fagan, <u>In The Beginning</u>, 7th Ed., (New York: Harper Collins, 1991), p. 150.

22. D. Nadel, et al. "19,000-Year-Old Twisted Fibers from Ohalo II," <u>Current Anthropology</u>, 35:4(1994), pp. 451-457.

23. Paul G. Bahn, Excavation of Palaeolithic Plank from Japan" <u>Nature</u>, 329(Sept. 10, 1987), p. 110.

24. Malcom Ritter, "Ancient Spears Challenge History" Associated Press, 02/26/1997 14:01 EST

25. Kathy D. Schick and Nicholas Toth, <u>Making Silent Stones Speak</u>, (New York: Simon and Schuster, 1993), p. 271

26. Brian M. Fagan, <u>In The Beginning</u>, 7th Ed., (New York: Harper Collins, 1991), p. 155.

27. Colin Renfrew, <u>Archaeology and Language</u>, (New York: Cambridge University Press, 1987), p. 64

28. Paul C. Mellars, <u>The Neanderthal Legacy</u>, (Princeton: University Press, 1996), p. 253

WHY KANGAROOS CAN'T TALK

"Now the LORD God had formed out of the ground all the beasts of the field and all the birds of the air. He brought them to the man to see what he would name them; and whatever the man called each living creature, that was its name. So the man gave names to all the livestock, the birds of the air and all the beasts of the field. But for Adam no suitable helper was found." Genesis 2:19-20 (NIV)

This enigmatic passage from the Scripture is the first indication of the acquisition of speech. If no one spoke, we would not be human. Human personalities are expressed through speech. Man's technology is taught to the next generation via speech. Man's religions could not exist without the theological concepts conveyed by language. It is the near infinite variety of messages that can be formed by human language that has allowed us to be what we are. In short, language is mainly what differentiates us from all other beings on earth.

In spite of this, we have written plays and created legends about human/animal communication. Dr. Doolittle was able to speak to kangaroos, apes, birds and other creatures. We have legends of Francis of Assisi communicating with animals. Mankind would love to be able to speak with the animals. But we can't. This chapter is about why kangaroos and other animals can't talk and why man can.

The fascinating picture shown in Genesis has God teaching man this most wondrous of skills. Naming is absolutely essential to this ability. Speech requires names for objects, otherwise there can be no subject or object in a sentence. This scene is reminiscent of some cases of people learning languages late in life. This is a very rare phenomenon because normally language must be learned quite early or the opportunity is lost. Stephen Pinker related the following interesting account of a languageless man finally grasping the concept of names and naming. He says,

> "In her recent book *A Man Without Words,* Susan Schaller tells the story of Ildefonso, a twenty-seven-year-old illegal immigrant from a small Mexican village whom she met while working as a sign language interpreter in Los Angeles. Ildefonso's animated eyes conveyed an unmistakable intelligence and curiosity, and Schaller became his volunteer teacher and companion. He soon showed her that he had a full grasp of number: he learned to do addition on paper in three minutes and had little trouble understanding the base-ten logic behind two-digit numbers. In an epiphany reminiscent of the story of Helen Keller, Ildefonso grasped the principle of naming when Schaller tried to teach him the sign for 'cat.' A dam burst, and he demanded to be shown the signs for all the objects he was familiar with. Soon he was able to convey to Schaller parts of his life story: how as a child he had begged his desperately poor parents to send him to school, the kinds of crops he had picked in different states, his evasions of immigration authorities. He led Schaller to other languageless adults in forgotten corners of society. Despite their isolation from the verbal world, they displayed many abstract forms of thinking, like rebuilding broken locks, handling money, playing card games, and entertaining each other with long pantomimed narratives."[1]

As in the biblical account, Schaller taught this man the names of objects, just as God taught Adam the names of the animals. However, one must concede the point that even language is not a prerequisite for inclusion in humanity. Ildefonso was fully human even without language. But, while not all humans speak, all who speak are human.

This is quite a different picture of what humans do versus what animals do. What is the nature of animal communication? Do animals talk in the same sense that humans talk? Contrary to recent claims about animal speech, their communication repertoire is quite limited. Communicative displays by various animal species were studied by E.O. Wilson and he found that they were limited to between 10 and 37 different displays. Even when an ape is taught an artificial language, its vocabulary remains quite limited. Wilson reports,

> "Sarah, a chimpanzee trained with plastic symbols by David Premack at the University of California at Santa Barbara, acquired a vocabulary of 128 'words,' including a different 'name' for each of eight individuals, both human and chimpanzee, and other signs representative of 12 verbs, six colors, 21 foods and a rich variety of miscellaneous objects, concepts, adjectives and adverbs. Although Sarah's achievement is truly remarkable, an enormous gulf separates this

most intelligent of the anthropoid apes from man. Sarah's words are given to her, and she must use them in a rigid and artificial context. No chimpanzee has demonstrated anything close to the capacity and drive to experiment with language that is possessed by a normal human child."[2]

Things have not changed over the past 30 years since Wilson wrote that. Despite claims by some investigators and the press, chimps taught American Sign Language are not really learning to speak. Pinker relates,

"To begin with, the apes did *not* 'learn American Sign Language.' This preposterous claim is based on the myth that ASL is a crude system of pantomimes and gesture rather than a full language with complex phonology, morphology, and syntax. In fact the apes had not learned *any* true ASL signs. The one deaf native signer on the Washoe team later made these candid remarks:

'Every time the chimp made a sign, we were supposed to write it down in the log....They were always complaining because my log didn't show enough signs. All the hearing people turned in logs with long lists of signs. They always saw more signs than I did....I watched really carefully. The chimp's hands were moving constantly. Maybe I missed something, but I don't think so. I just wasn't seeing any signs. The hearing people were logging every movement the chimp made as a sign. Every time the chimp put his finger to his mouth, they'd say 'Oh, he's making the sign for drink,' and they'd give him some milk...When the chimp scratched itself, they'd record it as the sign for *scratch*...When [the chimps] want something, they reach. Sometimes [the trainers would] say, 'Oh, amazing, look at that, it's exactly like the ASL sign for *give!*" It wasn't.'

"To arrive at their vocabulary counts in the hundreds, the investigators would also 'translate' the chimps' pointing as a sign for *you*, their hugging as a sign for *hug*, their picking, tickling, and kissing as signs for *pick, tickle,* and *kiss*. Often the same movement would be credited to the chimps as different 'words,' depending on what the observers thought the appropriate word would be in the context. In the experiments in which the chimps interacted with a computer console, the key that a chimp had to press to initialize the computer was translated as the word *please*. Petitto estimates that with more standard criteria the true vocabulary count would be closer to 25 than 125.

"Actually what the chimps were really doing was more interesting than what they were claimed to be doing. Jane Goodall, visiting the project, remarked to Terrace and Petitto that every one of Nim's so-called signs was familiar to her from her observations of chimps in the wild. The chimps were relying heavily on the gestures in their natural repertoire, rather than learning true arbitrary ASL signs with their combinatorial phonological structure of hand shapes, motions, locations, and orientations."[3]

Petitto goes so far to say that each observer was entirely free to create his or her own individual standard for what was and what wasn't a sign. This means that two observers watching the same event would see different things.[4] Petitto states,

"It is clear that Patterson tolerated variations in the form of an individual sign which in ASL could change its meaning entirely."[5]

But a vocabulary count even in the hundreds does not begin to mimic what humans do with their language. A veritable explosion of vocabulary takes place in children leading to absolutely huge numbers of words which could be used. Three years ago, I began trying to master the Mandarin language. At one point early in my studies, I wrote down every Chinese word I knew. It amounted to around 700 words. Even with 700 words, I was unable to communicate except on matters like where the bathroom was, the sun is shining, how much money is that and how to tell a taxi driver the name of my hotel. It was almost worse than knowing nothing, because the frustration level was higher. Yet vocabularies of chimpanzees never exceed 125, according to Pinker.

Steven Mithen relates that one chimp, named Kanzi had been able to master 150 words after six years of training.[6] With humans it is quite different. A six-year-old will have mastered on average around 13,000 words and a high school graduate will know around 60,000.[7] Chimpanzees are not even in the same language league with a six-year-old human.

Mithen agrees with Pinker that language is a hardwired feature of our brains. Mithen describes the brain as made up of various mental modules, each of which is architecturally suited to a particular task. Each of these mental modules is innate and is content-rich, in that they provide structure to the outside world. Mithen presents evidence for humans having at least seven types of intelligence which work together to provide man with his marvelous flexibility and rapidly learned skills. These intelligences are: "linguistic, musical, logical-mathematical, spatial, bodily-kinesthetic and two forms of personal intelligence, one for looking at one's own mind, and one for looking outward toward others."[8] One could incorporate these "intelligences" into four mental modules, which are hardwired by the brain. Mithen calls these mental modules general intelligence, technical intelligence, social intelligence and natural history intelligence. Mithen would also have one overarching module which integrates the various pieces of consciousness.

This view of the mind explains why the child has little difficulty in rapidly acquiring a language. He is using his preexisting linguistic mental module. The chimp, on the other hand, must work very hard to achieve minor results. Mithen cites this disparity in result as evidence that chimpanzee minds are not architecturally suited for language and that it is not simply the lack of vocal cords that prevents them from communicating.[9] They do not have a linguistic mental module and no amount of badgering will cause them to develop one.

Laura Petitto was involved in the training of Nim Chimpsky to learn American Sign Language. After what appeared to be initial success, Petitto rejected the notion that the chimp learned the language. She writes:

"The largest corpus of utterances from any signing ape, that of Terrace, Petitto, and Bever shows that their subject, a chimpanzee named Nim Chimpsky, did, in fact ,combine each vocabulary sign with a large number of other signs. Although each of the resulting combinations could be interpreted metaphorically, a simpler interpretation is that he merely combined signs randomly."[10]

Pinker points out that it is entirely inappropriate to try to teach a chimpanzee to speak our language. This effort is an attempt to bring humanity down a notch by involving our pestering a nonverbal species to mimic our instinctive abilities of speech. If chimps had the ability to pester us into mimicking their hooting and shrieking, would we thereby understand their "language"? Of course not. I can meow to my cat, and occasionally she will meow to me in return, but no true conceptual or even emotional communication has taken place. Just because an occasional sign looks somewhat like American Sign language does not mean that the chimp understands what it is saying. To posit language understanding in the mind of a chimp goes far beyond what we can really know. In fact some researchers believe that the animals have no ability to use symbols as referents.[11]

Another factor Pinker notes is that, unlike children, whose average sentence length grows rapidly, the average length of a chimp's sentence remains constant, even after years of intensive training. He notes,

"Recall that typical sentences from a two-year-old child are *Look at that train Ursula brought* and *We going turn light on so you can't see*. Typical sentences from a language-trained chimp are:

Nim eat Nim eat.
Drink eat me Nim
Me gum me gum.
Tickle me Nim play.
Me eat me eat.
Me banana you banana me you give.
You me banana me banana you.
Banana me me me eat.
Give orange me give eat orange me eat orange give me eat orange give me you."[12]

The qualitative difference is immediately obvious. The apes lack grammar. They also lack articles, auxiliaries and prepositions.[13] Because of all this Stephen Pinker of the Massachusetts Institute of Technology has come to the conclusion that human language is an instinct. Only the form of the language is learned, not language itself. Pinker cites several lines of evidence to support this conclusion. First, he cites the fact that not a single human culture has ever been found which lacked language. Secondly, a full-fledged language with grammar and syntax can be created within a single generation. No gradual transition is required. Thirdly, there are limited ranges of syntax which are used in these languages.

One of the most amazing things in the past few hundred years was the fact that everywhere the western Europeans went during the great voyages of discovery, they found the world inhabited. Mankind was the only being on the planet with such a universal distribution. Clive Gamble in his book, Timewalkers, remarks on how unremarkable the Europeans found this nearly universal distribution of humanity. He writes:

> "The nearly universal distribution of humanity, which so exceeded the geographical range of any other mammal, never drew their attention. One of the great Victorian naturalists, Alfred Wallace, who arrived at the principle of natural selection independently of Darwin, expressed this indifference perfectly. In his preface to *Island life* in 1880, a work that founded the study of biogeography, he commented that, while there was a great deal to understand and describe about the geographical distribution of all other species, with *Homo* all that could be said was 'the bare statement- 'universally distributed'... and this would inevitably have provoked the criticism that it conveyed no information.' So he left us out."[14]

Not only was mankind in all these diverse places he was also fully articulate. Steven Pinker tells of the fascinating discovery of the interior of New Guinea.[15] Surprisingly, this did not happen until the 1930s. For centuries, even with European habitation of the New Guinean coast, no one had traveled inland. The reason was that there was a rugged mountain range all along the southern coast of New Guinea. No one lived on that mountain range so no one saw a reason to travel north. Similarly, along the northern coast of that large island was a similar rugged mountain range and no one on the northern coast saw a reason to travel south. To all appearances, it looked like the interior of the island was one treacherous mountain range. Even the local New Guineans knew nothing about what lay beyond the mountains. This lack of local knowledge was considered proof that no one lived there.

During the late 1920s gold, that universal motivator of human migration, was discovered along a tributary to one of the main rivers. This caused an Australian prospector, Michael Leahy, to set off on May 26, 1930, in search of the gold's source. Leahy, a fellow prospector, and a few lowland natives made camp on a hill overlooking a huge plateau between the *two* mountain ranges—one following the northern coast and one following the southern coast. It was obvious that the land below them was habitable. Leahy and his friends thought they had discovered a major uninhabited region, which would make them rich. As the afternoon turned to night, their glee turned to fear and horror. The night revealed thousands of campfires on the land below. This land was already inhabited, and Leahy and his few friends did not know what to expect. They spent a sleepless night, made a crude bomb, and loaded their weapons before making first contact with these unexpected inhabitants of the land below. When Leahy's lowland New Guinean guides saw that the highlanders brought their women with them, they informed Leahy there would be no fighting. This relieved a lot of their fear. Leahy waved at the natives, trying to indicate that they should come closer. The natives were as frightened of Leahy as he had been of them. They were fearful of Leahy's white appearance. When Leahy took off his hat, they fell back in fear until an old man came up and touched Leahy to see if he was real. Quickly, the highlanders lost their fear and, as Leahy wrote in his diary, began "jabbering" and pointing to all the strange things the lowlanders had brought with them.

The "jabbering" was one of the 800 languages that would eventually be found in the highland region. Nowhere, even in the last great isolated spot on earth, was a human culture found which had no language. This remarkable fact argues strongly that language is innate, only the form of the language is learned.

But if the specific language is learned, a language with its own syntax and grammar, how can it be instinctive? This is where Pinker's second and third points come into play. Pinker's second point was that a full-blown language can arise within one generation. This phenomenon has been observed during the past century and a half. On Hawaiian pineapple farms, workers from a variety of nationalities and languages were thrown together to perform the work of growing and harvesting the crop. Each worker figured out a unique way to talk to his fellow workers. Since the workers had no common language, a form of pidgin communication was necessary. Often people of different nationalities would marry and raise children. The children would have no consistent linguistic model to follow.[16] There was no consistent syntax no grammar modeled for them. Even in these communities, children born to parents who speak the same language still have a huge pressure to abandon the parents' language and adopt the language of their peers. Given this difficulty, the children of such a community create their own grammar and syntax. They grow up with a new, completely developed language.

Derek Bickerton of the University of Hawaii has studied such creole languages, which are found all around the world. These languages were created from the union of a whole variety of parent languages. There is Haitian Creole, Negerhollands, Lesser Antillean Creole, Guyanese Creole, Djuka, Sranan, Guanais, Palenquero Gullah, Maritian Creole, Tok Pisin, etc. The one fascinating thing that came out of Bickerton's work was the discovery that grammars in creole languages from around the world are amazingly similar![17] Bickerton believes that an innate grammar is imprinted on the brain, which, if not suppressed by having to learn an existing language, comes through in a creole grammar.

This leads to Pinker's third point. The brain does have a grammar that is instinctive. Pinker cites some interesting evidence for this. Languages do not have an infinite variation in grammar. Most languages have a structure with an order of subject-verb-object or subject-object-verb. A few have verb-subject-object. Less than 1% are verb-object-subject or object-subject-verb. None seem to have object-subject-verb. These characteristics vary even within a language family and can change quickly, in less than a millennium.[18] A subject-object-verb language has postpositions instead of prepositions. But if the language flips to a subject-object-verb language, the postpositions also flip to prepositions. This is indicative of a universal switch in the brain circuitry. The switch that controls the order of subject, object and verb also controls the placement of prepositions either before or after the object it modifies. If the language has the order *Eat your salad* it is called a head first language. If the language says, *Your salad eat* it is a head-last language. This choice is important to all sorts of grammar decisions. He writes:

> "Chomsky suggests that the unordered super-rules (principles) are universal and innate, and that when children learn a particular language, they do not have to learn a long list of rules, because they are born knowing the super-rules. All they have to learn is whether a particular language has the parameter value head-first, as in English or head-last, as in Japanese. They can do that merely by noticing whether a verb comes before or after its object in any sentence in their parent's speech. If the verb comes before the object, as in *Eat your spinach!*, the child concludes that the language is head-first; if it comes after, as in *Your spinach eat!*, the child concludes that the language is head-last. Huge chunks of grammar are then available to the child, all at once, as if the child were merely flipping a switch to one of two possible positions. If this theory of language learning is true, it would help solve the mystery of how children's grammar explodes into adultlike complexity in so short a time. They are not acquiring dozens or hundreds of rules; they are just setting a few mental switches."[19]

Now, if language is truly an instinct, it raises the question: what is it in our brains which gives rise to this language instinct. The human brain has two unique structures called Broca's area and Wernicke's area. Broca's area was first discovered in 1861, by Paul Broca, who noticed, when doing autopsies of people with speech disorders, that they almost always had damage to one part of the left hemisphere. Those who had suffered what became known as Broca's aphasia could understand language just fine, but they could not speak or write it well. They could produce a given word, but the endings were often omitted, and the sentences were not very understandable. Sometimes the damage extended to the motor cortex and these aphasiacs then also suffer paralysis on the right side.[20] Broca became convinced that this part of the brain was involved in the motor control of articulate speech. Today we know that this is a simplistic view and that Broca's area is also involved in planning and making hand gestures,[21] and we know that it is involved in the sequencing of the fine motor movements which are required for speech.[22] If damage occurs to the outside portion of Broca's area, the only damage is to motor control. If the damage goes deeper, then language production itself is destroyed.[23]

A second area, just to the rear of Broca's area, is the region known as Wernicke's area. This region also is involved in speech but seems to be involved in looking up words and sending that information elsewhere in the brain.[24] Damage to this area leaves the patient unable to understand language, but they able to produce grammatically correct streams of nonsense. Pinker gives an interesting example of the speech of a Wernicke's aphasia patient. With Wernicke's area cut off from the rest of the brain, there is no way that definitions can be accessed. Thus, Wernicke's aphasia, Pinker suggests, is caused by Broca's area being cut off from its constraint on meaning. Broca's area, without the dictionary provided by Wernicke's area, produces nothing but endless streams of grammatically correct nonsense.[25]

Both Broca's area and Wernicke's area are necessary for speech. Do other animals have these regions in their brains? The answer is yes and no. There are homologous regions of the brain in all animals, but the closest analogues of these areas are found in monkeys. The monkey's Broca's area is involved in producing hand gestures but is *not* involved in the production of their vocalizations. Monkeys use their Wernicke's area for distinguishing the calls of other monkeys from those they themselves make[26]. While there are similarities in function of the monkey and human brains in the production of hand gestures, there is also the significant difference that animal vocalization is not controlled by Broca's area. In monkeys, as in all other animals, the neural circuitry which controls vocalization is deep within the brain, in the brain stem and limbic system, not on the outer cortex as in man.[27] These structures are heavily involved in emotions. We still use this part of the brain for vocalization at times when we stub our toes or slam our thumbs with a hammer. At these times our unfortunate ability to engage in deeply felt profanity is tied to this region of our brain and its primal vocalization mechanism.

Thus we find that there are four factors which prevent animals like apes or kangaroos from developing speech. First they have a very limited language memory. Their natural calls number in the tens and if taught, the most words they seem to be able to learn number less than 200. Secondly, they do not use symbols as humans use symbols.[11] Thirdly, they do not have the innate grammar humans have. And finally, they use a different part of their brain for vocalization. For these reason, kangaroos can not talk.

This difference in the location of human speech and animal vocalization raises an interesting question concerning the evolution of language. Deacon notes,

> "It is generally accepted that there is a functional and neurological dichotomy between the brain systems involved in the production of human language and those involved in the production of primate calls. This presumed dichotomy poses a serious problem for explanations of the origins of language. If indeed these systems are anatomically and functionally separate then there can be no evolutionary continuity from the vocal communication of our ape ancestors to human language. If this is the case, then language must have arisen independent of other communicative vocalizations. It is also difficult to conceive of transitional or intermediate stages in which control of vocal communication was shifting between the neural substrate for primate calls and that for linguistic communication."[28]

This difficulty of providing a pathway for language to have evolved has generated a controversy. Some, like Elizabeth Bates, are concerned with the implications this gap has for a Creator. She argues against such a distinction between animals and men as far as language is concerned. However, the lack of a known mechanism for this type of transition from animal vocalization to human language does not mean that there is none. The gap is supportive of the view many Christians have for the origin of language, but placing a large amount of faith in the current inability of scientists to explain this gap is a God of the Gaps view. This is the use of gaps in scientific knowledge to prove the need for a Creator. When this approach is taken, and science does finally figure out some explanation, the need for a Creator is diminished in the minds of large segments of the scientific population. As science has explained more and more phenomena, the gaps into which God can be placed shrank.

However, with this having been said, there is indeed physical evidence for the advent of language in the hominids. In man, Broca's area is much larger than it is in monkeys. In fact, in man alone it is so large that it leaves an impression on the inside of the cranium. The markings left by Broca's area can then be found long after death by simply examining the inside of the skull. Secondly, language seems to be associated with the asymmetry of the left and right hemispheres of the brain and the asymmetry is also associated with handedness.[29] People are predominantly right-handed because of our language ability. Because of this relationship we can examine stone tools and determine whether they were made by a right- or left-handed individual. From the habilines on, tools are made by predominantly right-handed individuals.[30] The significance of finding Broca's area, handedness in tool manufacture and brain-lateralization in an endocast is that this would indicate the time at which the brain structures for language were in place.

Amazingly, the earliest skull with these characteristics clearly outlined in an endocast is KNM-1470, a skull found by Richard Leakey that is 2 million years[31] old. This has some difficult implications for Christian apologetics. Language is one of the things which we have shown separate us from the animals. As shown at the start of this chapter, the Bible indicates that language was taught to man by God. Theologically, I do not believe we can ignore this important fact. Biblically, any being that speaks must be human. If these beings known as *Homo habilis* have language, then biblically, they *must* be mankind! Yet these beings are morphologically quite different from the people alive today. The *habilis* skeleton is different. We have chins jutting from the lower jaw; habilines, like apes, have no chin and a simian shelf in the inside of the lower jaw which performs the same function as our chin. The habiline brain size averages only half our brain size. In short, they are different enough that if they are men, then evolution is required to unite us with them! Yet, these small brained habilines show every sign of having the physical mechanism of language. If language is an instinct for modern humans, then one must wonder if it was also an instinct for the habilines. It is not just the habilines which show evidence of Broca's area. *Homo erectus*, and Neanderthal also have an enlarged Broca's area, brain asymmetry and even in the case of Neanderthal evidence of a modern vocal tract.[32]

Walker and Shipman report that the spinal cord in the earlier *Homo erectus* was significantly smaller than that in *Homo sapiens*.[33] They then use this fact to infer the speechlessness of *Homo erectus*. They claim that there were not enough nerves going to the diaphragm for *Homo erectus* to have sufficient control over the breathing during speech. At the end of their book, they write of an imaginary meeting with Nariokotome boy, an *erectus*,

> "But then, as I approached him closely, preparing mentally to hail him and at last make his acquaintance in person, it was as if he turned and looked at me. In his eyes was not the expectant reserve of a stranger but that deadly unknowing I have seen in a lion's blank yellow eyes. He may have been our ancestor, but there was no human consciousness within that human body. He was not one of us."[34]

This dramatic passage goes far beyond the evidence. Terrence Deacon, of Boston University, rebuts this view. He says,

> "In their recent book on human evolution, the Johns Hopkins paleontologists Pat Shipman and Alan Walker suggest that the evolution of speech may be contingent on this enlargement [of the spine--GRM]. They looked for this feature in the relatively complete vertebral column from a fossil Homo erectus boy, but could not find a corresponding thoracic enlargement. They conclude from this that speech had not yet evolved at this stage in human evolution. Given the neurodevelopmental information suggesting that cortical takeover of breathing control is a quantitative allometric effect--and not likely the result of a pair of all-or-none mutations that added new neurons to this part of the cord and retargeted cortical axons to match--I suspect that an intermediate interpretation is more likely. With a relative brain size intermediate between modern apes and modern humans, this erectus boy likely also had an intermediate level of cortical control over respiration, supported in part by an increase in cortical projections to thoracic motor neurons as well as to other higher respiratory centers."[35]

and

> "This means that if this erectus boy and his contemporaries did communicate using something like a language, it may well have been one that was more reliant on orally produced sound variations than on laryngeally produced ones. In other words, it likely employed fewer vowels and rapid tonal variations, relied more heavily on consonants and oral clicks, may have been limited to short phrasing, and so probably required more nonverbal support as well."[36]

The important item to notice is that, contrary to Walker and Shipman, *Homo erectus* would have been able to speak. Whether he did or not is obviously open to question, although I do believe he spoke. Theologically, speech is a hallmark of humanity. There is no biblical requirement that Adam be able to speak the Queen's English or the King's Hebrew, or an existing language or even to speak as well as you or me. Adding such a requirement clearly adds to what the Bible actually says.

Over the past 30 years, some authorities have attempted to prove that Neanderthal did not have a vocal tract which was capable of making all of the sounds which we are capable of today. Philip Lieberman has been at the forefront of this hypothesis. In his 1984 book Lieberman proposed that the Neanderthal larynx was located higher in the throat than it is in modern adult humans. With a larynx higher in the throat the vocal tract would be like that of an infant *Homo sapiens*. Lieberman notes,

> "Despite these acrobatic maneuvers, the reconstructed Neanderthal supralaryngeal vocal tract could not generate the formant frequency patterns of vowels like [a], [u], and [i] or the formant transitions that define the stop consonants [k] and [g]."[37]

Lieberman further stated that this defect would cause the Neanderthal to speak very slowly. This would cause communication to be difficult. Lieberman used this data to argue that language first appeared around 40,000 years ago at the same time as the Upper Paleolithic artistic explosion.

Christian apologists have taken this view. Wilcox claims that Neanderthal is not human based upon his supposed lack of speech. Wilcox claims that they showed no sign of language.[38] John Wiester, another old-earth creationist, also doubts the language ability of Neanderthal.[39] And given Hugh Ross' view that Neanderthal was merely a bipedal mammal, not spiritual, one could infer that he, too, believes that he didn't talk.[40]

Is this view correct? Lieberman's hypothesis made two predictions. First if the larynx was indeed higher in the Neanderthal throat one could be sure that the hyoid bone (the Adam's apple bone) would be shaped differently than modern humans. Secondly, Lieberman's hypothesis required that those with this defect would talk slowly. Both of these predictions are contradicted by the evidence. In 1989, Yoel Rak discovered a fossil Neanderthal hyoid bone which was identical in shape to that of modern humans.[41] this meant that the Neanderthal vocal *was* like those of modern men. Neanderthal could speak as most authorities now agree.[42] Secondly, but now less important, even if Neanderthal lacked these properties it probably would not have hurt communication abilities. Hayden writes,

> ". . . as Fremlen has so effectively point out in response to the minimalist view of Neandertal laryngeal areas:
>
> '...et seems emprebeble theth ther speech wes enedeqwete bekes ef the leck ef the three vewels seggested. The kemplexete of speech depends en the kensenents, net en the vewels, es ken be seen frem the generel kemprehensebelete ef thes letter.'"[43]

Their language would have been different, but it would have been a language and they would have been men in the theological sense..

Of course this is not what all Christian apologetical leaders teach. The young-earth creationists do include Neanderthal and *Homo erectus* in the family of man. Marvin Lubenow, the author of <u>Bones of Contention</u>, says that the morphological differences between modern man and *Homo erectus* are not sufficient to classify them in a separate species.[44] Lubenow[45] also includes KNM-ER 1470, the earliest skull found with a Broca's area, as fully human as does John W. Cuozzo,[46] and initially Gish.[47] By doing this, Lubenow ascribes humanity to a being with a brain size only two times larger than that of a chimpanzee.[48] He also minimizes the extent of morphological change required and eludes the evolutionary implications of what they are doing by turning that evolution into microevolution. Other young-earth creationists, apparently realizing the contradiction in that position, reject all but Neanderthal and *Homo sapiens* from humanity. Other authors taking this approach include Gish[49] and Peterson.[50] An interesting twist of fact occurs with one of Henry Morris treatment of the issue of erectus' humanity. He makes the statement that *Homo erectus* is fully human then cites the discovery of a human-like brain in *Homo habilis* as support for his statement about *Homo erectus*. He writes,

> "There is no doubt that *Homo erectus* ('erect man') had an upright posture. That he was truly human, rather than an erect ape, has recently been confirmed by studies of the brain endocast from the skull known as '1470,' discovered a number of years ago by Richard Leakey.
>
> 'An endocast from the Kenya National Museum, a *Homo habilis* specimen known as ER 1470, reproduces a humanlike frontal lobe, including what appears to be Broca's area.'"[51]

This inclusion of fossil man in the human family by the young-earth creationists, is generally more in accord with the osteological and cultural data than is the view of the old-earth creationists, both theistic evolutionists and progressive creationists. These two groups usually draw the humanity/nonhumanity line at either the appearance of anatomically modern humans or the supposed cultural explosion seen in the artifacts of southwest Europe around 35,000 years ago. Beings like Neanderthal are occasionally included as humans, but are often rejected. Almost all theistic evolutionists and progressive creationists reject *Homo erectus*, *Homo habilis*, and the australopithecines.

Wiester rejects them because they do not show the traits that we consider uniquely human.[52] However, his argument is circular. He lists the traits he considers characteristic of humanity in the following,

> "Humans alone possess a 'soul,' a 'spirit,' and have the quality of self-conscious reason. They are distinctly human gifts. From them come the desire and ability to know about ourselves, artistic awareness of beauty, the power of conceptual thinking, the capability to grasp the difference between right and wrong, and the knowledge of and the possibility of a relationship with the Creator."[53]

However, these traits are very difficult to see in the fossil record. In his discussion of Neanderthal, Wiester cites the compassionate treatment of an injured Neanderthal as evidence of this human spirit and claims that this was the first evidence of "compassion and tenderness" in the fossil record. It is ironic that a year before Wiester's book was published, a case of *Homo erectus* compassion and tenderness was described in the pages of Nature.[54] While there is little way that a person could glean the details of compassionate care from the pages of this report, it is interesting that ultimately the case of vitamin A poisoning in a *Homo erectus* is now the earliest evidence of compassion and tenderness in the fossil record. We will discuss this case later, but, briefly, this woman was extremely sick and was cared for by her fellow tribesmen for several weeks before her death from vitamin A poisoning. The point is that in spite of the evidence for language in hominids back to the habilines Wiester does not consider them as human (Since Wiester does not consider language as a human characteristic).

Hugh Ross, a progressive creationist, often has stated that if mankind were more than 60,000 years old the Bible would be wrong.[55] Thus, Broca's area in KNM-ER 1470 cannot be considered a human characteristic, although only humans have this feature. Ross, then, also excludes Neanderthal from humanity even though this fossil also has a well-developed Broca's area indicative of language.[56] One further difficulty with Ross' position is that it also excludes anatomically modern men, men who look like you and me, from humanity, because anatomically modern men are found in strata dated to 120,000 years ago[57]--prior to when Ross believes it is biblically acceptable for mankind to be around. Thus, an obvious logical deduction from Ross' position leads one to conclude that some men who look like us, are morphologically identical with us, who have Broca's area, speak and make music and musical instruments, may not really be spiritual humans worthy of human respect. Such a view of men, looking like us, but believed to be sub-human has resulted in many atrocities this century. To me, this is the worst implication of Ross' position. It could lead to a conclusion that I find repugnant. One, I might add, that I am sure Ross himself would be loathe to accept.

Phillip Johnson does not include in the human family any fossil man older than 200,000 years.[58] He bases this on the now discredited[59] molecular "Eve" theory, in which one woman was believed to be the mother of all living humans. It was thought that a lineage could be derived from the molecular data, but because there are so many lineages, no one can know whether the 200,000 year old lineage is true or not. Language is not on his list of human traits either. Davis and Kenyon also reject all fossil men prior to Neanderthal.[60] This means that they do not believe that *Homo erectus* and *Homo habilis*, with their obvious Broca's area, were really human. David Wilcox denies the humanity of Neanderthal and erectus because he also denies them language.[61] Wilcox does not say why they have Broca's area nor does he discuss the implications of that anatomical feature.

The difficulties with the theistic evolutionist and progressive creationist treatment of fossil man are twofold. First, their facts are not correct and secondly, new discoveries continue to destroy their assertions. For instance, Wilcox, in the summer of 1996, published this:

> "The only 'cultural' evidence we have of Homo erectus is the Acheulian bifacial 'handaxe' industry which appeared suddenly and remained more or less unchanged by time and location for a million years."[62]

The first problem is that the technology of *Homo erectus* is not static. The oldest hand axe from West Natron,

Tanzania is shown in Figure 8.[63] As can be seen it has only eight relatively large flakes removed from around its edges, and it is not very symmetrical. One of the later hand axes is shown in Figure 9.[64] It is a work of art as is shown in the fine workmanship, the numerous small flakes removed from it. It is also much more symmetrical than the West Natron hand axe. Deacon points out that some of the later *erectus* tools were made from volcanic glass with a technique that required multiple steps and multiple chipping tools. Deacon says, "Thus, reports of *erectus*' mental stagnation are likely exaggerated.[65] It is simply not true to say that technology "remained more or less unchanged by time and location for a million years."

The second problem is that of recent discoveries. Less than a year later, beautifully carved wooden spears, made to the same specifications as modern javelins, including the same balance point, were found in an Acheulean context, a context I might add includes the skeleton of *Homo erectus*.[66] Years prior to Wilcox's publication, much was known about *Homo erectus* culture. The site of Bilzingsleben, Germany yielded remains of *Homo erectus* as long ago as 1978.[67] Work there was revealing that *erectus* of this time had villages similar to those of primitive man. Mania, Mania and Vlcek write,

> "The home base of early man from Bilzingsleben was situated on a shore terrace close to the outflow of a karst spring into a small lake. Previous excavations revealed a division of the camp site into different activity areas and outlines of three simple shelters with hearths and workshops set up in front of them. Five to 8 m from the dwelling structures, an artificially paved area with a diameter of 9 m was found. According to the archaeological evidence, special cultural activities may have been carried out there.
>
> "Along with large pebble tools (choppers, chopping tools, and hammerstones), small specialized tools of flint appear. Basic standard forms are knives, scrapers, denticulates and notches, simple points which are pointed-oval, Tayac and Quinson points, borers, and core-like tools. Edge retouches predominate, but also unifacial and bifacial retouches occur. Large scrapers, knives, chisel-shaped tools, wedges, bodkins, and work supports were manufactured from the compact bone, preferably of the straight-tusked elephant. Mattock- and cudgel-shaped tools were made from cervid antlers. Specific, deliberate manufacturing activities are recognizable in the workshops. Apart from the dissection of the animal prey, these tools served for the working of predominantly organic materials which in turn were used for the manufacture of other tools and objects of daily use. Wood was also a frequently used raw material. Numerous calcified remains of wood artifacts were found at the site. Some bone tools display deliberately engraved sets of lines which we regard as expressions of abstract thinking, perhaps as graphic symbols."[68]

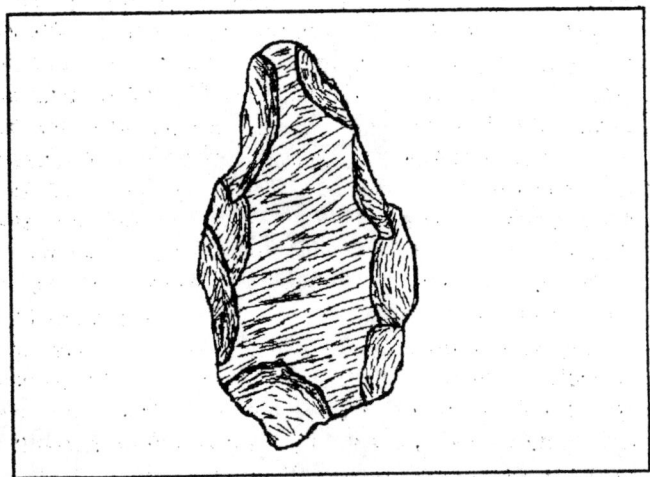

Figure 8 West Natron Hand axe 1.4 Million years old.

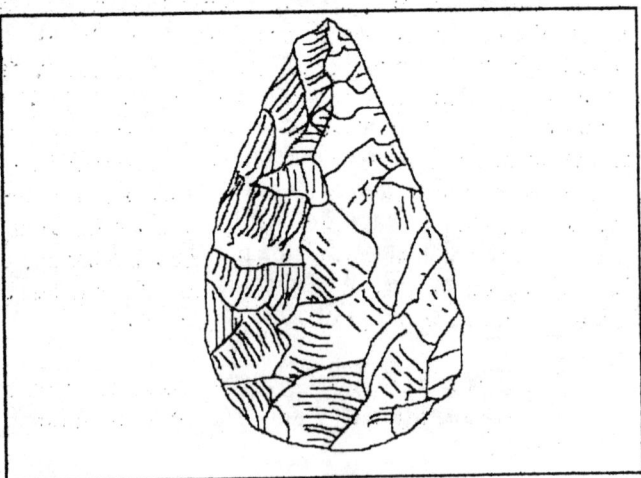

Figure 9 Late Hand axe Middle Paleolithic

This type of complexity in lifestyle requires a language, culture and other accoutrements of humanity.

Having looked at language, we will now turn our attention to religion. This is an activity humanity does not share with any other species on earth.

References

1. Steven Pinker, The Language Instinct, (New York: Harper/Perennial, 1994), p. 68
2. E. O. Wilson, "Animal Communication", Scientific American, 227:3(Sept. 1972):52-60, p. 60.
3. Steven Pinker, The Language Instinct, (New York: Harper/Perennial, 1994), p. 337-338.
4. Laura A. Petitto, "On the Evidence for Linguistic Abilities in Signing Apes," Brain and Language 8, 12-183(1979), p. 165.
5. Laura A. Petitto, "On the Evidence for Linguistic Abilities in Signing Apes," Brain and Language 8, 12-183(1979), p. 165.
6. Steven Mithen, The Prehistory of the Mind, (New York: Thames and Hudson, 1996), p. 86
7. Steven Pinker, The Language Instinct, (New York: Harper/Perennial, 1994), p. 150-151
8. Steven Mithen, The Prehistory of the Mind, (New York: Thames and Hudson, 1996), p. 40
9. Steven Mithen, The Prehistory of the Mind, (New York: Thames and Hudson, 1996), p. 86
10. Laura A. Petitto, "On the Evidence for Linguistic Abilities in Signing Apes," Brain and Language 8, 12-183(1979), p. 163.
11. Terrence W. Deacon, The Symbolic Species, (New York: W.W. Norton, 1997), p. 50
12. Steven Pinker, The Language Instinct, (New York: Harper/Perennial, 1994), p. 339-340.
13. Derek Bickerton, Language & Species, (Chicago: University of Chicago Press, 1990), p. 107-108
14. Clive Gamble, Timewalkers, (Cambridge, Mass.: Harvard University Press, 1994), p. 2
15. Steven Pinker, The Language Instinct, (New York: Harper/Perennial, 1994), p. 25-26.
16. Derek Bickerton, "Creole Languages," Scientific American, 249:1(July, 1983):116-122, p. 119.
17. Derek Bickerton, "Creole Languages," Scientific American, 249:1(July 1983):116-122, p. 121.
18. Steven Pinker, The Language Instinct, (New York: Harper/Perennial, 1994), p. 235.
19. Steven Pinker, The Language Instinct, (New York: Harper/Perennial, 1994), p. 112
20. Dean Falk, Braindance,(New York: Henry Holt and Co., 1992), p. 68-69
21. Erik Trinkaus and Pat Shipman, The Neandertals, (New York: Alfred Knopf, 1993), p. 353-355.
22. Clive Gamble, Timewalkers, (Cambridge: Harvard University Press, 1994), p. 172; Alan Walker and Pat Shipman, The Wisdom of the Bones, (New York: Alfred Knopf, 1996), p. 268-269.
23. Steven Pinker, The Language Instinct, (New York: Harper/Perennial, 1994), p. 309.
24. Steven Pinker, The Language Instinct, (New York: Harper/Perennial, 1994), p. 311
25. Steven Pinker, The Language Instinct, (New York: Harper/Perennial, 1994), p. 311
26. Steven Pinker, The Language Instinct, (New York: Harper/Perennial, 1994), p. 350; Terrence W. Deacon, "Human Brain Evolution: 1. Evolution of Language Circuits." Intelligence and Evolutionary Biology, ed. H. J. Jerison and I. Jerison, (New York: Springer-Verlag, 1988),p. 363-381, p. 364.
27. Terrence W. Deacon, "Human Brain Evolution: 1. Evolution of Language Circuits." Intelligence and Evolutionary Biology, ed. H. J. Jerison and I. Jerison, (New York: Springer-Verlag, 1988),p. 363-381, p. 364; Steven Pinker, The Language Instinct, (New York: Harper/Perennial, 1994), p. 334
28. T.W. Deacon, "The Neural Circuitry Underlying Primate Calls and Human Language," Human Evolution, 4:5(367-401), p. 367.
29. Kathy D. Schick and Nicholas Toth, Making Silent Stones Speak, (New York: Simon and Schuster, 1993), p.103; James R. Shreeve, The Neandertal Enigma, (New York: William Morrow and Co., 1995), p. 274-275
30. James R. Shreeve, The Neandertal Enigma, (New York: William Morrow and Co., 1995), p. 274-275; Clive Gamble, Timewalkers, (Cambridge: Harvard University Press, 1994), p. 172; Dean Falk, Comments, Current Anthropology, 30:2, April, 1989, p. 141-142.
31. Iain Davidson and William Noble, "The Archaeology of Perception," Current Anthropology 30:2, April 1989, p. 135-136; and Dean Falk, Comments, Current Anthropology, 30:2, April, 1989, p. 141-142.
32. Alan Walker and Pat Shipman cite the work of Marcus Raichle's group as showing that Broca's area does not indicate speech. But Raichle's group only very slightly redefined the location of Broca's area. Instead of it being on the very surface of the cortex, Raichle's images show it to lie just below the surface. To use this redefinition misses the point. It is the bump preserved in the inner mold of the skull which is indicative of speech because the bump is made by an enlargement of Broca's area which lies just below the cortex. Walker and Shipman describe it as a cat lying beneath a sheet. The cat makes the bump, not the sheet. The bump on the bed is as indicative of the cat lying below as the bump on the brain is for the possession of an enlarged Broca's area which indicates speech. See Alan Walker and Pat Shipman, Wisdom of the Bones, (New York: Alfred Knopf, 1996), p.

267-269 and S.E. Petersen et al, "Positron Emission Tomographic Studies of the Processing of Single Words," Journal of Cognitive Neuroscience 1:2(1989):153-170.

33. Alan Walker and Pat Shipman, The Wisdom of the Bones, (New York: Alfred Knopf, 1996), p. 263-265
34. Alan Walker and Pat Shipman, The Wisdom of the Bones, (New York: Alfred Knopf, 1996), p. 294
35. Terrence W. Deacon, The Symbolic Species, (New York: W.W. Norton, 1997), p. 252
36. Terrence W. Deacon, The Symbolic Species, (New York: W.W. Norton, 1997), p. 252-253
37. Philip Lieberman, The Biology and Evolution of Language, (Cambridge: Harvard University Press, 1984), p. 318
38. David L. Wilcox, "Adam, Where Are You? Changing Paradigms in Paleoanthropology," Perspectives on Science and Christian Faith , 48:2(June 1996), p. 93
39. John Wiester, The Genesis Connection, (Nashville: Thomas Nelson Publishers, 1983), p. 180
40.   "From a biblical standpoint, I see Neanderthals as one of the *nephesh,* soulish (not spiritual) creatures God made before he made humans. In other words, the Neanderthals must have been a bipedal mammalian species created a few tens of thousands of years before Adam and Eve. Neanderthals became extinct, possibly as the result of some climactic upheaval, at least several thousand years before the creation of Adam and Eve."~Hugh Ross, "Link with Neanderthals Cut by Computer," Facts & Faith, 9:3, 3rd Qtr. 1995, p. 22
41. B. Arensburg, "A Middle Palaeolithic Human Hyoid Bone," Nature, 338(April 27, 1989):758-760
42. Donald Johanson and Blake Edgar, From Lucy to Language, (New York: Simon and Schuster, 1997), p. 106; Terrence W. Deacon, The Symbolic Species, (New York: W. W. Norton, 1997), p. 372
43. Brian Hayden "The Cultural Capacities of Neandertals ", Journal of Human Evolution 1993, 24:113-146, p. 131
44. Marvin L. Lubenow, Bones of Contention, (Grand Rapids: Baker Books, 1992), p. 120
45. Marvin L. Lubenow, Bones of Contention, (Grand Rapids: Baker Books, 1992), p. 162;
46. John W. Cuozzo, "Skull 1470--A New Look," Creation Research Society Quarterly, 14:3(December, 1977), pp 173-176.
47. Gish writes: "...Leakey's latest report lends considerable support to creationists, who maintain that man and the apes have always been contemporary." This of course implies that KNM-ER 1470 is human. See Duane Gish, "Richard Leakey's Skull", Impact #11, February, 1974, in Henry M. Morris and Duane T. Gish, editors, The Battle for Creation (San Diego: Creation-Life Publishers, 1976), p. 198.
48. Compare the 800 cc for KNM-ER 1470 with the 400 cc brainsize for chimpanzees. see J. W. K. Harris, "Early Man," in Andrew Sherratt, editor, The Cambridge Encyclopedia of Archaeology, (New York: Cambridge University Press, 1980), p. 66-67; Bernard Campbell, Human Evolution, (Chicago: Aldine Publishing Co., 1974), p. 272.
49. Duane T. Gish, Evolution: The Challenge of the Fossil Record,(El Cajon: Creation-Life Publishers, 1985), P. 179
50. Dennis R. Petersen, Unlocking the Mysteries of Creation, (South Lake Tahoe: Creation Resource Foundation, 1990), p. 116-119
51. Dean Falk "The Petrified Brain," Natural History, vol 93, September 1984, p. 38 cited by Henry M. Morris, Creation and the Modern Christian, (El Cajon, California: Master Book Publishers, 1985), p.183-184
52. John Wiester, The Genesis Connection, (Nashville: Thomas Nelson Publishers, 1983), p. 188.
53. John Wiester, The Genesis Connection, (Nashville: Thomas Nelson Publishers, 1983), p. 178-179
54. A. Walker, M.R. Zimmerman, and R.E. F. Leakey, "A Possible Case of Hypervitaminosis A in *Homo Erectus,*" Nature, 296, March 18, 1982, p. 248-250.
55. Hugh Ross, "Art and Fabric Shed New Light on Human History," Facts & Faith, 9:3 (1995)p. 2; Hugh Ross, "Chromosome Study Stuns Evolutionists," Facts & Faith, 9:3,(1995) .p. 3
56. Hugh Ross, "Link with Neanderthals Cut by Computer," Facts & Faith, 9:3, 3rd Qtr. 1995, p. 2
57. B.Bower, "Early humans Make their Marks as Hunters," Science News, April 12, 1997, p.222
58. Phillip E. Johnson, Darwin on Trial, 2nd ed. (Downer's Grove: NavPress, 1993), p. 85.
59. For an excellent historical discussion, see Erik Trinkaus and Pat Shipman, The Neandertals, (New York: Alfred Knopf, 1993), p. 394-396; Alan R. Templeton, "The 'Eve' Hypothesis: A Genetic Critique and Reanalysis," American Anthropologist 95(1): 51-72. p. 52
60. Percival Davis and Dean H. Kenyon, Of Pandas and People, 2nd edition (Dallas: Haughton Publishing Co., 1993), p. 110
61. David L. Wilcox, "Adam, Where Are You? Changing Paradigms in Paleoanthropology," Perspectives on Science and Christian Faith , 48:2(June 1996), p. 93
62. David L. Wilcox, "Adam, Where Are You? Changing Paradigms in Paleoanthropology," Perspectives on

Science and Christian Faith, 48:2(June 1996), p. 91
63. Thomas Wynn, "Handaxe Enigmas," World Archaeology, 27(1995):1:10-24, p. 13
64. Another example is seen in Ashley Montagu, Man: His First Two Million Years, (New York: A Delta Book, 1969), p. 78
65. Terrence W. Deacon, The Symbolic Species, (New York: W. W. Norton, 1997), p. 369
66. Hartmut Thieme, "Lower Palaeolithic Hunting Spears form Germany," Nature, 385(Feb. 27,1997), p. 810; for the *Homo erectus* remains see Derek Roe, "The Handaxe Makers" in Andrew Sherratt, Editor, The Cambridge Encyclopedia of Archaeology; (Crown Publishers, 1980), p. 77; Robert G. Bednarik, "Art Origins", Anthropos, 89(1994):169-180, p. 170; Emanuel Vlcek, "A New Discovery of *Homo erectus* in Central Europe,"Journal of Human Evolution, (1978) 7:239-251, p. 247; D. Mania and U. Mania and E. Vlcek, "Latest Finds of Skull Remains of Homo erectus from Bilzingsleben (Thruingia)", Naturwissenschaften, 81(1994), p. 123-127, p. 124;
67. Emanuel Vlcek, "A New Discovery of *Homo erectus* in Central Europe,"Journal of Human Evolution, (1978) 7:239-251, p. 250
68. D. Mania and U. Mania and E. Vlcek, "Latest Finds of Skull Remains of Homo erectus from Bilzingsleben (Thruingia)", Naturwissenschaften, 81(1994), p. 123-127, p. 124

A WORLD FULL OF RELIGION

" In the course of time Cain brought some of the fruits of the soil as an offering to the LORD. But Abel brought fat portions from some of the firstborn of his flock. The LORD looked with favor on Abel and his offering, but on Cain and his offering he did not look with favor. So Cain was very angry, and his face was downcast. "[1]

In the above passage, Scripture tells of the first religious ritual. We saw in the previous chapter how John Wiester defined man as having the knowledge of God and knowledge of the possibility of having a relationship with Him. Anthropologists have a similar definition of religion to Wiester's. Steven Mithen defines religion as the belief in the existence of nonphysical beings.[2] He to cites several near universals among religions. These are: the belief that a nonphysical part of the individual can survive death and remain with beliefs and desires, certain people are able to have special communicational abilities with the god, and finally, the performance of certain rituals will change the physical world or its circumstances.[3]

The religious activities of Cain and Abel fits Mithen's criteria for religion perfectly. Cain and Abel were offering sacrifices to a nonphysical being, God. They were performing a ritual which when done wrong by Cain, was not accepted. Since only one of the brothers was able to commune with God, it illustrates the fact that only certain people are able to enter a state of communication with a Divinity. And when God said, " Your brother's blood cries out to me from the ground."(Genesis 4:10 NIV), this is an indication of Abel surviving death in another realm. This early event in human history illustrates clearly the fact that religion is the *sine qua non* of the image of God. Assuming we can find evidence of religion in the archeological record, it would be strong evidence of the spiritual nature of those men and strong evidence that these are men covered by the plan of salvation. This issue is not one which will sit well with many Christians, because it raises the possibility of humanity and salvation to roughly dressed, or maybe naked, savages who were poor in the material possessions, and who do not look like us at all. The only difference between these ancient primitives and those in the modern world is that we are able to send missionaries to those who live today and we can't send any to those in the past.

How do we recognize religious life in the past? This, indeed, is quite a problem. First, belief in a nonphysical being does not fossilize. Beliefs held by a man are lost when he dies. Remember from a previous chapter that anything made from perishable material will not survive far into the future. This means that any wooden idols, straw dolls, body painting etc. are very unlikely to be preserved. Thus, if a group of people, like the western Australian aborigines of today, leave no nonperishable record of their religious life, we will be hard pressed to find it. James Shreeve writes,

> "In western Australia today there are aborigines who make very crude-looking stone tools. But their wooden implements are very elaborate, with fancy painting on bark, and beautiful spearthrowers and shafts. They also have extremely complex social systems, cosmology, and narrative traditions. If you were to dig up one of their sites a thousand years from now, however, all you would see would be the clunky stone tools. Does this mean those aborigines were technologically inferior? Not at all. They were simply relying on perishable materials."[4]

Similarly, we cannot say that lack of evidence for a religious life is evidence for the absence of religious life. If an ancient people were to have used only wood and animal skin for their religious rites, no evidence of their religious life would remain. During the 1970s, 1980s and early 1990s, anthropologists discounted the concept that men 300,000 years ago could possibly be big game hunters. They viewed these people as scavengers. Then, in 1997 three wooden javelins were found in strata dated to 300,000 years ago which bore the evidence of superior workmanship, planning and knowledge of aerodynamics. These seven-foot-long spears could only be used for throwing at big game. Suddenly the issue of man, the big-game hunter, was back on the table. Anthropologists had made the mistake of believing that the lack of evidence for spears was evidence that early man lacked spears. We must not do the same for religion.

Of course, this does not mean that we can make up evidence which is lacking. So how do we approach the issue? The only possible means is to look for those rare instances of religious life put into preservable forms. It is obvious that someone, somewhere, was the first person to carve an idol. This may or may not have occurred a long time before everyone else began carving idols in stone. Several things prevent us finding evidence of this first idol. First, the odds are that this first effort was lost to posterity. Erosion, burial or just plain bad luck in searching for it are going to prevent us from finding the very first idol. Thus whenever we find the "first" idol, we can be fairly certain that it is not really the very first one ever made. Secondly, carving shapes in hard stone is hard work and requires much time. If you don't have a certain level of wealth, allowing you to spend the required time, you are not going to engage in this behavior. Middle and Lower Paleolithic peoples were very, very poor technologically, at least as far as we can tell. If they had to spend larger amounts of time finding food than their Upper Paleolithic descendants, there would, by necessity be fewer stone idols. This would especially be true if a wooden idol was sufficient for worship.

What kinds of objects should we search for? Objects similar to what modern people use in their rituals should be included in this list. These include crystals, art objects of veneration, iconographic symbols, evidence of body painting or tattooing, shamanistic costumes, evidence for sanctuaries, ritualistic treatment of animal remains, cannibalism and treatment of the dead. One can find these types of objects in archeological sites, but proving their religious usage is more difficult. What we will do is examine each of these items and see if the entire behavioral pattern would look like religion if these artefacts were produced by modern people.

Crystals
Art Objects of Veneration
Iconographic Symbols
Body Painting
Tattooing
Shaman Costumes
Sanctuaries
Ritual Treatment of Animals
Cannibalism
Burial of Dead

Table 2 Tangible Evidences for Religion

Crystal Power

One can go into bookstore today and find books on the power of crystals to influence one's life. Crystals are believed to exert power and influence over certain situations. At the company where I work we once had an Exploration Manager whom we called "Crystal Man." This otherwise intelligent scientist once distributed quartz crystals on a string to a group of business men and told them that stirring their tea with these crystals would be beneficial. By his actions, he was displaying his religious belief in nonmaterial forces and influences. Such use of crystals is not unique. During the Middle Ages scrying, or crystal gazing, became widespread. This was based on the belief that the future could be foretold by studying the reflections on the stone. However, crystal gazing was not invented by Medieval Europeans. The behavior has been observed among the Apaches, Maoris, Zulus, Incas, Iroquois, Canadian Indians, Polynesians and the Malagasy of Madagascar.[5] Primitive man may have interpreted such reflections, which would be in constant motion near a flickering fire, as images or visions of the spirit world[6].

The earliest evidence that man collected crystals is from Zhoukoudian Cave near Beijing, China, the site of the famous Peking Man. Twenty pieces of quartz crystal were found. One of these (Cat. No. Q2:25) was said to be a "perfect crystal" 6 centimeters long. The crystal specimen was smoky in color with all the crystalline faces complete.[7] The collectors of these crystals were *Homo erectus*. We know several things about these collectors of crystals. First, they controlled fire. Hearths in Zhoudoudian Cave are filled with 7 feet of ash.[8] As cold as it gets in northern China during the winter they *needed* some means of keeping warm. Without that they could not have survived, after all, Beijing is the same latitude as southern Nebraska. In many cultures, the control of fire is surrounded by mysticism. Secondly, there is evidence of cannibalism among the *Homo erectus* at Zhoukoudian, and while extremely grisly, we shall see below that cannibalism is often indicative of deep spiritual beliefs.

Other ancient sites also contain quartz crystals which appear to have been carried in from elsewhere. One of them is Singi Talav, India, where six quartz crystals were recovered from Acheulean strata. Remember that the Acheulean technology is that which was generally associated with *Homo erectus*. Acheulean technology extends from 200,000 to 1.5 million years ago.[9] The Singi Talav crystals were too small to be used as tools,[10] and mineralogically they are quite different from each other, leading to the conclusion that they came from different places and were not part of the same quartz crystal flowers. Quartz crystals were also found in Acheulean strata of Gudenus Cave in Austria, and the Acheulean strata at Gesher Benot Ya'aqov, Israel, which dates between 240-750 thousand years ago. It was at the latter site that a man-made polished wooden plank was found.[11]

It would appear that the collecting of crystals was widespread in the ancient world and that the crystals collected had no known utilitarian purpose. Were they jewelry? Probably not. They were not drilled with holes for strings and there is no mention of any residual glue left on the crystals for attachment to strings.

Religious objects

Objects of veneration, such as idols, have gone hand in hand with the various religions of man since time immemorial. The entire Roman pantheon of gods was portrayed in stone in the form of human images. The durability of Greek and Roman stone statues has left a huge legacy of religious icons. Christians have also left a large statuary depicting the saints, Madonna and child, and other religious figures. Many pre-Christian religions also depicted a type of Madonna, a goddess representing Mother Earth, which was venerated by various religions far back into the past. The oldest still-standing buildings on earth appear to be the temples on Malta, built 5500 years ago. On Malta, as in many Mediterranean civilizations, statues of naked or semi-naked women with large buttocks are found associated with Temples and worship. On Malta, a large Mother Goddess is depicted sitting, standing, and even sleeping. One at Hal Tarxien was over 7 feet high and is said to resemble naked headless figurines found elsewhere. A clay model of a woman sleeping on a couch, naked from the waist up, having large breasts and large buttocks has been found at Hal Saflieni, Malta.[12] Since this temple had many cubicles where people could sleep, this goddess has been interpreted as the goddess who could interpret dreams. This Mother Goddess, like later and earlier goddesses, is a recurring theme in the religion of the past.

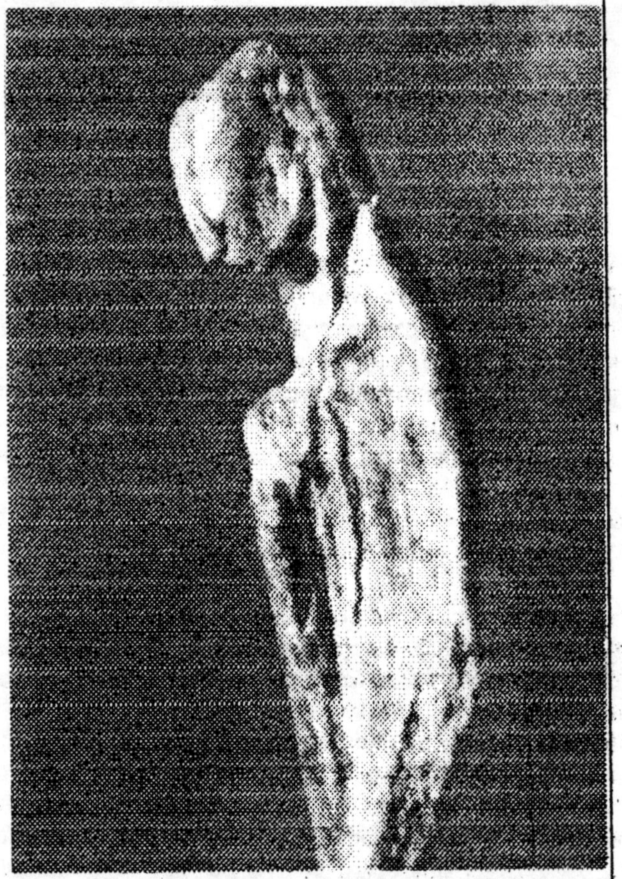

From Vinca, Yugoslavia, 6500 years ago, female statues, naked from the waist up with large buttocks, are found. These statues have loincloths rather than the skirt of the Maltese goddess from a thousand years later.[13] This artistic trend can be followed even farther back in time. At Megali Vrisi, Tirnavos, Thessaly, Mother Goddess figurines are found which are totally naked, with extremely large buttocks and large breasts. This archaeological site dates to 8000 years ago.[14] From 15,000 years ago, a different style of Venus figurine is found at Ma'alta, Siberia.[15] This tradition of Mother Goddess continues back in time to the Magdalenian and Aurignacian of the Upper Paleolithic. Twenty-three thousand years ago someone in northern Italy carved a naked Venus figurine (Mother Goddess) with huge buttocks and huge dangling breasts. The woman was obese, as was the Mother Goddess at Malta.[16] One of the oldest Venus figurines is the recently discovered Galgenberg figurine that dates to 29,000 years ago.[17] Since this figure has a hand resting on the hip, it required the boring of a hole to form the space between the arm and the torso. The technical skill required is amazing.

Figure 10 Pseudo-Venus from Wildenmannlisloch Cave.

But these are not the oldest figurines that have been found. We have traced the evidence of worship of a Mother Goddess back nearly 30,000 years, but there are two more ancient cases of Venus figurines—one made by Neanderthal and one made by *Homo erectus*. On October 21, 1926 at the Wildenmannlisloch cave, workmen found a bone figure that resembled a human figure with eyes, nose and mouth. This figure had been placed into another fragment of bone, which became the stand, allowing the human-like figure to stand erect. This object was then placed in a niche in the cave wall (see Figure 10). The maker of this object was Neanderthal man. Ivar Lissner writes:

"How for instance can we to explain the discovery, in a carefully protected niche in one of the chambers of the Wildenmannlisloch, of a small figure resembling a female sculpture? Made out of the lower jaw of a cave bear, it may be either an artifact or a freak of nature. One thing is certain: the flattened planes of its 'head' were rubbed smooth by some human agency; perhaps, as Emil Bachler suggests, because the bone was originally used as an instrument for smoothing animal skins. This may be the reason why certain portions of the so-called 'pseudo-Venus' appear to have been polished. Bachler is of the opinion that the figure came into being accidentally, as a result of continual friction due to use, not as a deliberate attempt to reproduce the shape of the human head. I have examined the figure closely. The closed eyes, delicate mouth, small forehead, slim neck and back all convey an impression of careful workmanship. A second 'Venus' discovered in the same hiding place has smooth patches but no recognizable head.

"Even if the pseudo-Venus was not actually made by Stone-Age man, the cave dweller must have noticed its resemblance to the figure of a girl. Why else would he have put it to one side and preserved it so carefully? The prehistorian Friedrich Behn in his book *Vorgeschichte Europas*, asserts that the people of the Neanderthalian race were lacking in any form of artistic impulse. The celebrated Venus statuettes of the Stone Age belong to the Aurignacian, a far later period. The pseudo-Venus may, therefore, be unique in its period, the earliest portrayal of the human figure known to have been made, or at least recognized as such, by man. It is probably the most remarkable evidence of prehistoric activity or comprehension in the world. Between four and five inches tall the Venus was found on October 21, 1926, and reposes today in the Heimatmuseum at Saint Gallen, a Paleolithic Sleeping Beauty waiting to rejoice the eye of the occasional visitor."[18]

While we cannot be sure of this figure's role as an object of worship, it was found in a cave just a few miles from Drachenloch Cave, an extremely controversial and problematic cave which gave evidence for religious ritual among the Neanderthals.

The oldest carved object and possible religious figurine is the Berekhat Ram figurine, also known as the Golan Venus. The figurine is a crude carving of a female. This object was first reported in 1986 and was found at Berekhat Ram, Israel, by Goren-Inbar.[19] The lake of Berekhat Ram lies in the remains of an ancient volcanic caldera. In the northwest part of the lake two basalt flows are separated by a two-meter thick, brownish-red ancient soil horizon(paleosol). This paleosol contains stone tools. The basalt above the ancient soil has been potassium-argon dated to 233 ± 3 thousand years.[20] The underlying basalt dates to 470 thousand years and thus the age of the intervening soil profile is bracketed by two dates. The level at which the figurine was found can be estimated to be about 300 thousand years old.

The fact that this figurine is that of a female appears to continue the extremely long tradition of creating Mother-Goddess images. Needless to say, some have challenged the validity of this object. Pelcin suggested that the object was nothing more than a geological curiosity.[21] The figurine is made of scoria, which is a volcanic rock made of welded ash. Such objects can take on strange shapes. Unfortunately, Pelcin had not actually examined the object. Pelcin also did not address why this particular piece of scoria was the *only* piece of scoria found at the site.[22] This means that the object was most likely carried to the site from elsewhere. Only objects of value are carried around. When Alexander Marshack, a leading authority on Upper Paleolithic art and the originator of microscopic examination of Paleolithic art, examined this object, he concluded that it had been intentionally modified. He wrote:

"In his note on the Berekhat Ram figurine, excavated from a late Acheulean level and dated at ca. 230,000 B. P. from a late Acheulean level and dated at Ca. 230,000 B. P., Pelcin argues that the figurine is scoria, as it was generically described in the initial publication. He documents the fact that scoria can acquire odd shapes and natural grooving and therefore recommends that the Berekhat Ram figurine be subjected to microscopic analysis. I performed such microscopic analysis in the summer of 1994 and am preparing the results for publication. When I presented the results to Sergio Peltz of the Geological Survey of Israel, Jerusalem, a specialist in scoria and the pyroclastic materials of Israel, he examined the figurine and reported (personal communication, October 23, 1994) that 'the material of the figurine was part of the matrix of a welded scoria deposit, but specifically the figurine is not a scoria.'

"Peltz reported that it was clear that 'human hands had worked a fragment of pyroclastic rock, namely an indurated tuff.' The illustrations and arguments presented by Pelcin therefore do not apply. To complement my microscopic analysis, Peltz and N. Goren-Inbar are preparing an analytical paper on the geology of the site and the pyroclastic nature of the figurine. Until publication of these analyses, the debate on possible pre-Upper Paleolithic symboling may perhaps best be addressed not by suppositions at a distance but through the microscopic analysis of a late Middle Paleolithic incised composition from the site of Quneitra, Israel. I pointed to the Quneitra analysis in my recent criticism of the Eurocentric presumption that there was a punctuated, apparently genetic 'species' shift in symboling capacity at the Middle/Upper Paleolithic transition."[23]

What this means is that the object was a figurine manufactured by human hands more than 230,000 years ago! And Marshack confirmed his earlier conclusion in a more thorough examination of the object reported this year.[24] Because of its resemblance to later religious objects the Berekhat Ram figurine is of great importance.

With the Berekhat Ram figurine, our look at the mother Goddess religion comes to an end. It is unlikely that there was no manufacture of Venus figurines prior to 300,000 years ago because it is unlikely that the Golan Venus was the very first attempt at carving Mother Goddesses. It may be that the earlier ones remain undiscovered or there was a period of carving wooden figures in these earlier times.

In either case, however, this evidence for religious activity extends much further back into the past than some Christians are willing to admit. The Golan Venus is more than twice as old as the earliest anatomically modern human. It was not manufactured by modern men. The creator of the Golan Venus was either *Homo erectus* or archaic *Homo sapiens*. In any event, it was *not* created by anatomically modern man. Christian apologetics needs to do more with such data than ignore it.

Sanctuaries

A sanctuary is a hallmark of man as we saw above. The earliest still-standing religious sanctuary, on Malta, was built around 4000 B.C. The earliest ruins in Mesopotamia, which used to be religious sanctuaries, were built in 6000 B.C.[25] Oates claims that this is the beginning of religious worship but she is wrong. Shrines and sanctuaries go much farther back in time, possibly as long as 80,000 years ago.

The earliest evidence for a religious sanctuary built by anatomically modern men is the 14,000-year-old site of El Juyo in northern Spain. A half-human, half-feline face was carved into a rock. Freeman and Echegaray state,

"From the description given above and an examination of the photograph and drawing, the reader will realize that the stone face represents a being whose nature is dual, although the two sides of its character have been harmoniously integrated into one single face. The proper right side of the face is that of an adult male human with moustache and beard. The proper left side is a large carnivore, with oblique eye, large lachrymal, and a moderately long nose, ending in a good depiction of a naked rhinarium. The chin is triangular, and a sharply pointed tooth projects above the mouth. On the muzzle there are three subparallel lines of black spots suggesting the bases of whiskers or vibrissae, a characteristic feature of felids. Taken as a whole, these features represent a large cat, probably a lion or leopard (both existed near El Juyo in Magdalenian times)."[26]

Other features at the cave suggest a religious purpose. Two piles of dirt were constructed with alternating designs and rosettes of vividly colored clay. During the construction of the piles, red ochre was sprinkled over the whole pile at various intervals. Since this construction was invisible to any visitors and could only have been seen by the builders, it must have had some religious significance. The two mounds are connected by a low trench into which some greasy material was poured and then burned.[27] But this is not the only example of religious activities of fossil man. There is are two possible Neanderthal sanctuaries

In Bruniquel Cave, France, a 47,600 year old square structure was found several hundred meters from the cave entrance. Since anatomically modern men had not arrived in western Europe at this time, the builder must have been Neanderthal. The site shows several interesting features of Neanderthal life. Measuring 13 by 16 feet, the structure, made from stalagmites, was located in a total darkness. This means that Neanderthal possessed the technology of making torches or lamps. But even more curious was the burnt bear bone found inside the structure.[28] What the Neanderthals were doing may never be known but engaging in ritual would fit the data. It is unlikely that the structure was a living site. Torches would burn out overnight and make it difficult for them to return to the surface. Using the site for a picnic seems unlikely. If you wanted merely to barbecue a bear the surface--in the light--would be much more convenient. In order for someone to enter a dark and dangerous cave for the purpose of burning a bear, one need a powerful incentive. Religion would not be out of the question. Men have long entered caves to engage in religious ritual. It very well might be that Bruniquel was a Neanderthal sanctuary--part of the bear cult which apparently was part of the Paleolithic cultural landscape.

There is one possible religious sanctuary built by *Homo erectus*. At Bilzingsleben, Germany, an *erectus* village with the remains of several huts has been excavated by Dietrich and Ursula Mania. They found a paved area among the ruins of the huts. Gore relates,

> "But Mania's most intriguing find lies under a protective shed. As he opens the door sunlight illuminates a cluster of smooth stones and pieces of bone that he believes were arranged by humans to pave a 27-foot-wide circle.
>
> "'They intentionally paved this area for cultural activities,' says Mania. 'We found here a large anvil of quartzite set between the horns of a huge bison, near it were fractured human skulls.'"[29]

Were these men engaging in a well-known form of worship--human sacrifice?

The Bear Cult

The Mother Goddess is not the only religion for which there is evidence from these ancient times. There is also evidence of a bear cult. Well into this century, many circumpolar peoples engaged in a religion which had the bear as the central part of the ritual. Such cults are found scattered among the Lapps and Finns, across northern Eurasia to the Bering Sea. There are only a few sites in Alaska and Western Canada which show evidence of the bear cult, but from the region surrounding Hudson Bay east to the Atlantic, more activity is found.[30] The groups include Finns, Lapps, Voguls, Ostyaks, Yurak-Samoyedes, Yeniseians, Karagasses, Tungus, Orochi, Gilyaks, Yukaghir, Chuckchee, Koryaks, Kamchadals, Tlingit, Kwakiutl Hutka, and the Algonkins.

Ivar Lissner was a German anthropologist, born in Riga, Latvia. He spent several years studying the circumpolar cultures of the N.E. Siberians, including the Tungus, Ostyaks, Samoyeds, Gilyaks, Negidals Orochi and Olcha tribes. He extended his study to some of the North American Indians and Eskimos. Through these studies, he became familiar with the bear cults of the circumpolar region. Much of what we know about the bear cults come from the people of Northeast Asia whom Lissner studied.

The bear cults have several features in common. They believe that the bear in some way is human. The ancient Finns believed that the bear had a human soul.[31] The Orochi are equally convinced of this fact.[32] Eskimos believe that humans and bears can be transformed into one another.[33] These cultures also have a bear festival at which the bear is sacrificed. The bear is viewed as a mediator between God and man.

Such a festival has been observed in the activities of the Gilyaks, a tribe of Northeastern Asians. They believe in a supreme being named Kur, who is the essence of goodness.[34] This god is invisible and so cannot be depicted by image. Kur decides the fate of men. Gilyaks use the bear to send a message to God. They capture a young bear and raise him in their village for two or three years. The bear, destined for the great bear festival, is treated as an honored guest. They feed him, take him for walks on a chain, bathe him and treat him as a friend. When the festival day draws near, everyone in the village helps in the preparations. A location for the sacrifice is prepared and effigies of the minor deities are hung from poles in pairs representing a man and his wife. The animal is taken from his cage and led to each yurt (or house) in the village. At each yurt the bear is greeted with laughter and rejoicing as each family shows its respect to the bear. Some men will grab the bear by the sides of its head and kiss it. If the bear strikes and claws the man, the scars are viewed as a badge of honor. The bear is then led to the favorite fishing holes so that he can bless these sites and ensure good fishing for the rest of the year. Finally the bear is led three times around the house of the family which raised him.

The bear is then led to the place of execution and tied to strong posts. The families go back to their yurts to celebrate while the *narch-en*, the executioners, prepare the bows and arrows. Invariably, the *narch-en* are guests who have been invited to the festival for the purpose of killing the bear. After a while the men, but none of the women come back to the bear and sacrifice it. A few wounding shots are fired by young *narch-en* while their young boys throw stones at the bear. Finally the oldest *narch-en* draws his bow and dispatches the animal. The Gilyaks watching this wish the bear a good journey to God and tell the bear to inform God how well he was treated by the Gilyaks. The reason that the *narch-en* are strangers or guests is so that the bear will think well of the Gilyaks, having been killed by members of another tribe. This ensures a good report to God.

The dead bear is then laid out, head to the west, so that it will not see the rising sun before it sees God. The bear is skinned, his meat is cooked, and the *narch-en* are required to eat all the bear. The Gilyaks can only eat some soup made from the bear but are forbidden to eat the bear themselves, out of respect for having raised him. The feasting may last for days and when the guests finally go home they take all the remaining food from the bear with them.

Another tribe, the Ainu of northern Japan, also engages in a bear ritual that is quite similar to that of the Gilyaks. The excellent treatment of the young bear however goes even a step further. One missionary to the Ainu, John Batchelor, reported that he once preached at one end of a hut while five women passed a bear cub from lap to lap, each allowing the cub to suckle from her breast.[35] The very young cub is even taken to bed with the family. Prior to the sacrifice, the Ainu would ask the bear's forgiveness and say something like, "You were brought into the world for us to hunt. We have reared you with great love. Now that you are grown, we send you to your father and mother. When you reach them, tell them how good we were to you. And come again so that we may sacrifice you again."[36] The Finns are also reported to have told the bear to relay to God how well treated the bear had been and the Chippewas of North America, like the Ainu, would apologize to the bear.[37]

One of the most interesting items concerning the bear cult is the hint of the origin of some of our constellations. After the bear is sacrificed, he is known as the "*chinukara-gur*, which means 'prophet' or 'guardian.'"[38] They also use this same term to refer to the pole star. This is where the bear's spirit goes after it is sacrificed. Another Northeastern Asiatic tribe, the Udehe, also looked to the pole star for good hunting after their yearly sacrifice of a bear. The pole star is in the constellation *Ursus minor*, which is Latin for Little Bear. We also call this constellation the Little Dipper. The primeval Mediterranean and European cultures as well as the Ainu and Udehe, see that constellation as being associated with a bear. Apparently, in the case of the Mediterranean cultures, the goddess Artemis was believed to have been a bear at one time. This is one more manifestation of the bear cultists' tendency to see bears as human. The constellation of the bear was named in her honor.[39] The Big Dipper or Ursa Major was associated with the Nymph Callisto, whom Zeus placed into the heavens in the form of a bear.[40]

After these festivals, all these cultures treat the bear remains with respect. Some like the Ostayaks place the skeleton on a raised platform. The Chippewa would hang the skull from the trees, as do the Orochi and Manegi of Siberia. In 1913, B. Zitkov reported seeing a huge mound of bear skulls which the Samoyeds had been ceremonially collecting at that spot for over 100 years.[41] It appears that the ceremonial treatment of the bear's skull was always attended to.

Given this modern day example of the bear cult, what evidence is there that this cult extends into the distant past? There is, admittedly, less evidence here than there is with the Mother Goddess, but occasionally artifacts are found which look amazingly similar to those we know bear cult peoples produce today. At Mas de Azil in the French Pyrenees, between 9,500 and 12,500 years ago,[42] a broken bone disc was found with two pictures engraved on it.[43] On one side was a masked figure dancing in front of a bear's paw, which is striking at the dancer. The other side of the disk portrays a naked man wearing a bear mask and holding a pole which appears to be taunting the bear on the lost piece of the disc. A bear's paw again reaches out to the man. This scene is amazingly similar to a drawing published by Lissner of an Ainu ceremony where two men are threatening a bear with stick or spears.[44] One figure found at the cave of Les Trois Freres, which dates at

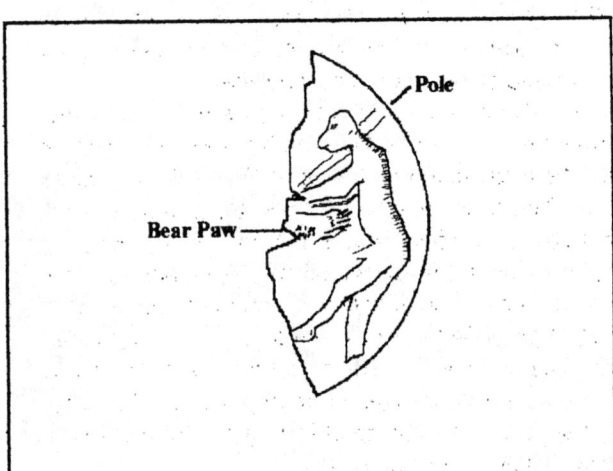

Figure 11 Side 1 Mas d'Azil plaque

18,000 years ago, has been interpreted to illustrate the death of a bear similar to the fashion in which the Gilyaks killed their bear. The bear figure has small circles drawn on it that are believed to be stones hitting it. Furthermore, there are darts which appear to stick into the bear and radiating lines come out from it, indicating bleeding.[45] At the site of Massat, a short walking distance from Les Trois Freres, another bone was found engraved with a bear's head with lines radiating out of his mouth similar to the bear painted in Les Trois Freres. These lines have been interpreted as bleeding from the mouth. There is a similar picture found at Santimamine Cave in northern Spain.[46]

Probably the most impressive evidence of a bear cult from the Upper Paleolithic is that of Montespan Cave in the Pyrenees. In 1923, Norbert Casteret and his assistant Henri Godin discovered a headless bear statue with a bear skull lying between its forepaws.[47] In the middle of the shortened neck was a triangular hole which is believed to have held a short wooden stick to which the bear skull was formerly attached. The sculpture was two feet high and four feet long. The body of the bear was riddled more than 30 times with spears or sharp sticks and was apparently used for ritual. Count Begouen, Abbe Breuil and Miss Garrod, archaeological experts of the day, were called to examine the discovery. Count Begouen established that the bear sculpture was worn all over by the friction of some garment or skin draped across it. It was believed therefore, from this data, that Paleolithic man would drape the skin of a bear, with the head and skull attached, over the sculpture and then attack it with spears or sharpened sticks. This ritual object dates to the Magdalenian period, which is around 15,000 years ago. The oldest Upper Paleolithic evidence of the bear cult comes from the 30-34,000 year old Chauvet Cave site. A cave bear altar was discovered there in which a bear skull was placed on a rock where a small fire had been kindled.[48]

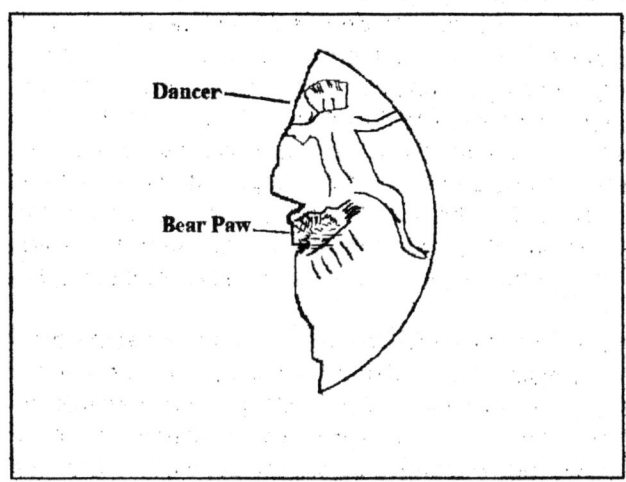

Figure 12 Side 2 Mas d'Azil plaque

Earlier than this, there are a few evidences of fascination with bears. In 1996 a stone structure was found in the French cave of Bruniquel. The structure was far from the entrance of the cave which meant that the builders required torches to see while they were constructing it. The structure, measuring 13 by 16 feet, also contained a burnt bear bone. While there is no certainty of the purpose of the structure deep in the cave, this situation does carry some resemblance to an altar upon which bears were sacrificed. So who made this structure? It was Neanderthal and dates to c. 46-48,000 years ago, long before modern man entered Europe.[49] It seems difficult to conceive of why Neanderthal would carry a bear deep into the cavern to cook him when he could be more easily cooked on the surface, unless there were a religious reason for doing so. But this is not the only evidence of a Neanderthal bear cult.

From the 80,000-year-old Regourdou site there is a discovery relevant to our discussion. A young adult Neanderthal man was found buried in a rectangular stone-lined pit covered by a large stone slab. The man was buried at one end of this cyst and the carefully arranged skeleton of a brown bear at the other end of the pit.[50] The pit also contained 20 brown bear skulls. This site, like all sites which suggest any human or human-like activity among Neanderthals, has been questioned. Many authorities absolutely refuse to consider any evidence which would suggest a religion among these most denigrated of ancient men. Recently Brian Hayden, reviewing the cultural activities of Neanderthal, wrote of Regourdou,

> "After discussing the Regourdou Neandertal and bear remains with Eugene Bonifay and Bernard Vandermeersch, and after examining photographs of features as well as artefacts from the site, I am also convinced that both the human and at least some of the bear remains were intentionally placed in the cave and buried."[51]

Other authorities, like Kurten,[52] Barbaza,[53] Shreeve,[54] Smirnov,[55] and many others accept the basic validity of Regourdou, that 20 bear skulls were placed into an intentionally constructed rectangular pit. This type of behavior indicates some type of ritualistic behavior. Kurten states,

"All this goes to show that, at least since interglacial times—80,000 years ago—man has been interested in the brown bear, the species that is still with us; interested enough to hunt it, to paint and engrave its likeness on the walls of his caves, and, at least in more recent times, to make it an object of cult and ritual."[56]

When describing this site, Bruce Dickson uses a term, "stone cyst", that was used by an earlier discoverer of an interaction of Neanderthal and bear. Dickson wrote,

"At the Regourdou site, a single individual was buried along with grave goods under a low mound or cairn of rocks. In the vicinity of the mound—and perhaps associated with it—a number of trenches were found as well as stone cysts containing brown bear (*Ursus arctos*) remains. The discovery of these Neanderthal graves provides rare insight into the expansion of the mind taking place during the Paleolithic."[57]

If this is true, that Neanderthal man constructed rectangular stone pits and placed bear skulls in them, then a reappraisal of a discovery earlier this century is in order. If Neanderthal engaged in this type of activity at 80,000 years ago, then there is absolutely no reason to believe that he was unable to do this 70,000 years ago. The discovery at Drachenloch, in the Swiss Alps, has been discounted by many authorities for over half a century. The reader needs to be aware of this rejection by the majority, but by no means all, of archaeologists before we discuss this find. F. Clark Howell,[58] J. M. Coles and E. S. Higgs[59] accepted Drachenloch as did Lissner. As is the case with most of the archeological discoveries early in the twentieth century, there are problems of documentation that are validly problematical. Besides, the standards of scientific control were less stringent during this early period and this applied to every other excavation. The problems at Drachenloch have been used as a basis for the rejection of the conclusion, when at other equally early and equally problematical sites the conclusions are not questioned. Drachenloch, if true, sheds a fascinating light on the religious aspects of Neanderthal and the problems of ideas born before their time.

In 1911, Marcellin Boule published a four part monograph on the La Chapelle-aux-Saints skeleton which was destined to securely fix him as one of the leading anthropologists and to distort the view of Neanderthal for many years to come. Boule had painted a picture of Neanderthal in which he was apish, stooped, slow-witted and shuffling. Pictures drawn in the popular press at the time show an extremely hairy, vicious, animal-like creature. This was not the sort of creature to which one would attribute any spirituality. This picture, drawn by Boule, ignored the fact that the skeleton he had studied was preserved through the ages largely because it had been *intentionally* buried. Among modern primitive peoples, burial is always associated with a belief in an afterlife.[60] This is a religious belief, not something that would be expected from the sort of creature that Boule described. Boule's view also ignored the fact that the cranial capacity of the Neanderthal was equal to or greater than ours. Yet the people in the early twentieth century preferred an apish view of Neanderthal.

Emil Bachler reported on his excavations at the cave at Drachenloch in 1921. What he reported certainly did not fit into the picture of Neanderthal which had just been published ten years earlier by Marcellin Boule.[61]

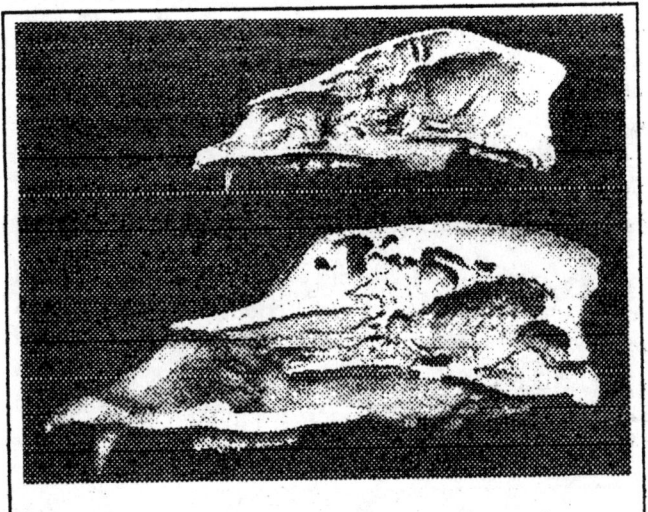

Figure 13 Brown Bear (above) Cave Bear (below)

Bachler believed that Drachenloch was a sacrificial or ritual site for Neanderthal man. Bachler had found a cave situated 7335 feet above sea-level. Bachler claimed that his workmen found a rectangular stone cyst or pit with seven cave bear skulls piled up. Cave bears are unusually large as Figure 13 shows. A second stone chest held the leg bones of the animals. He gave inconsistent information on which direction the skulls were facing. He claimed to have found Mousterian flint tools and fireplaces. He claimed that ancient man had carried the cave bear bones up to that height and left them there.

Figure 14 Drachenloch Cave

He claimed to have found bones aligned in peculiar patterns. His critics would have nothing to do with this. Today this cave is high on a rocky mountain with no trees at that level.[62] In Figure 14, the white dots lead up to

Drachenloch.

Over the last 20 years, as archaeology has downplayed the abilities of both *Homo erectus* and Neanderthal, Drachenloch has been forgotten. Bednarik tells of this denigration,

> "After the Neanderthals were recognized as a hominid group almost 140 years ago, they were initially considered to be brutish, primitive and rather ape-like creatures, an opinion prompted by deformed skeletal remains, evolutionary paradigms and inadequate reconstructions. During the twentieth century, their position on the ladder of evolutionary recognition improved gradually, culminating in the pronouncements relating to finds in Teshik Tash (Tadjikisthan) and Shanidar Cave (Iraq). But in the 1970s and 1980s came a skeptical reaction and the human status of the Neanderthals waned in response to a change in archaeological fashion. During the reign of the 'New Archaeology' it became popular to question all of their hard-earned cultural and cognitive capacities, and the most daring critics of previous work suggest that the Neanderthals had no art, hence could not have had any language, and really belonged to the apes rather than the humans."[63]

Drachenloch is rarely mentioned in the anthropological literature today, and in those in where it is mentioned it is always quickly dismissed. If Neanderthal was merely a scavenger, someone totally unlike us, then any evidence of his religious activity was ignored. The reasons given by anthropologists for the rejection of Drachenloch by the anthropological community are a study in sloppy scholarship. Bruce Dickson wrote:

> "However, recent comparative analysis of similar bone material from cave bear sites that were never occupied by hominids has led Kurten (1976:83-107) to the conclusion that Drachenloch was never anything but a cave-bear lair and Bachler's piles of cave-bear 'trophy heads' probably accumulated naturally in it *without* human agency. Binford (1983), studying bone attrition and patterns of bone accumulation by both modern people and animals, rejects Bachler's Neanderthal cave-bear cult hypothesis for similar reasons."[64]

Dickson's reference to Binford's 1983 <u>In Pursuit of the Past</u>, is quite interesting. Dickson makes it sound like Binford reports on a study of the Drachenloch bones in this book, but this is not true. A glance at Binford's index reveals *no references to Drachenloch, Bachler or a study of the taphonomy of bones in a cave!*. There is only one reference to bear on page 55. On that page Binford talks about how grizzly bears destroyed some of his experimental wolf kill sites. There is nothing on Drachenloch or cave bears. Dickson has another book by Binford in his bibliography, <u>Bones, Ancient Men and Modern Myths</u>.[65] Thinking that Dickson might have mixed up his references and that this might be the book which had Binford's careful study of Drachenloch, I checked it out. Binford does mention Drachenloch in this book but he provides no study of the material. His only mention of Drachenloch is part of a review of the lines of reasoning defining human behavior. He actually never says that he rejects the evidence of religion among the Neanderthals at Drachenloch, although his tone would suggest this. On page 28 of Binford's <u>Bones:</u> book, he says that all interpretations of bear cults, cannibalism and hunting of elephants by *Homo erectus* at Torralba, Spain, are unproven because the arguments are circular. But this is not to say that they are wrong. When we find a seven-foot-long wooden spear stuck in the ribs of an elephant, as occurred at Lehringen, Germany in 1950,[66] what are we to think, that the elephant grew a spear? While it is true that religious intent can never be proven beyond a shadow of a doubt, such a hypothesis is consistent with the evidence.

Dickson's reference to Kurten was even more enlightening, because while Kurten does indeed reject Drachenloch as evidence of religion among the Neanderthals, he did not actually examine the Drachenloch material. Kurten criticizes what he says are terrible inconsistencies in the Bachler's two published reports on Drachenloch. Bachler published two reports, one in 1921[67] and one in 1940.[68] Kurten points out that the drawings of the cross section of the cave showed crucial differences that he felt invalidated Bachler's conclusions.[69] Kurten's criticisms are based upon the cross-sections and are only valid if one assumes that a cross section is supposed to show *only* those items absolutely cut by the cross section. Most cross-sections are not so constructed. A north-south cross-section attempts to show the characteristics of the layers in a north-south direction. They can include items which are slightly east or west of the plane of the cross section. Cross sections are used to convey a stratigraphic framework for the deposit and in Bachler's day were not usually used to state precisely the location of each artifact. While this is modern practice, it was *not* earlier practice. Kurten seems to want to hold Bachler, who excavated in the 1910s and 1920s, to the standards required in the 1980s. This is unfair since other early sites are not so criticized. At Le Moustier, another early archaeological site, a body was excavated, re-buried and re-excavated several times for financial supporters.[70] Yet most accept the data from that cave. Kurten cites the types of problems that plagued all early anthropological sites, even those whose validity have not been questioned. Obviously this makes it easy to attack the validity of what was found under these circumstances. This is a pervasive problem in the early excavations that were little more than high-minded grave-robbing. Yet archeology will accept this early data *if* it supports the downplayed view of these ancient men.

Figure 15 1921 Drachenloch cross section

Kurten says that the "1923" picture shows two stone chests, one holding two skulls facing south, the smaller one holding leg bones.[71] Kurten points out that the 1940 drawing shows six or seven skulls facing east. This is true and Bachler hurts his case here. Kurten also points out that the size of the empty stone chest in the 1940 drawing is smaller than in the 1921 drawing, although there is no scale on the 1921 drawing. This is somewhat of a nitpick. I measured the height of the two stone chests in both photos. The ratio of the smaller to the larger is .78 in the 1921 drawing and .72 in the 1940. The relative widths of the two chests change from .6 to .5. This is not much of a difference, especially considering the lack of scale in the 1921 drawing. Another nitpick is that Kurten criticizes the shape of the stones in the two drawings. He further claims that the longbones and skull shown in level IV in the 1921 drawing disappear in the 1940 drawing. Kurten claims "..the skull in layer IV has vanished."[72] This is true, it vanished in the drawing. Since there is a *photograph* (plate 16) of the level IV skull in the 1921 report, I would presume that this should be documentation enough for the skull's existence, and for the fact that Bachler didn't presume to make the cross sections show only what cut the plane of the cross section itself. It should also serve to demolish Kurten's criticism of the vanishing skull. If Kurten had examined more than the two drawings, he would have seen that photo. Kurten also criticizes a wall that was drawn differently between the two reports. The area behind the wall contained more bones in the 1940 drawing. However, there are bones in the same location in the 1921 drawing.

Even with all these problems, Bachler's character was such that there was no question of fraud as critic Kurten attests.[73] Kurten says that the larger pieces of cave bear skull will be pushed to the side of the cave by subsequent cave occupants making it look like the skulls were placed in niches. Thus Kurten says that Drachenloch is nothing more than a chance arrangement of stones that fell from the ceiling intermixed with bears that died during hibernation. Two facts argue against this interpretation. First, Bachler claimed the skulls were found in a square stone chest *in the center of the cave,* not at the cave wall.[74] Second, rocks falling from a cave roof are extremely unlikely to form a rectangular box when they land. There was a collection of bear skulls at the back of the cave which probably collected in the manner suggested by Kurten, but Bachler never claimed that those were in a stone chest. The skulls Bachler claimed were indicative of intentional burial were in a rectangular chest in the middle of the cave.

I outlined the issues involved with Drachenloch very rapidly so that the reader can understand what they are as we now go back to look at them in more detail. Ivar Lissner, in the 1950s, studied the material at the St. Gallen Museum and came away convinced that there was sacrificial activity carried out at Drachenloch. He gave several lines of evidence supporting this.

First, as mentioned above, the cave is above the treeline today. The chances are that 70,000 years ago, it was also above the treeline because it was colder during that interglacial than it is today. This means that any humans who went to Drachenloch would have had to carry their firewood with them. This would be a strenuous activity requiring lots of determination. To go and stay at Drachenloch would require a very good reason. Yet we know that Neanderthal visited there because there are hearths in the cave. Some of the hearths had charred pinewood still in the fireplace.[75] Between chamber 2 and chamber 3, a flat square stone slab a foot and a half wide was found covering a hearth. The underside of this slab was stained with smoke, and underneath it was charred wood and burnt bones. While Neanderthal probably didn't need as much wood to keep warm as modern humans would, logs would still have burned at the same rate as today and to keep a fire going requires a constant supply of wood.

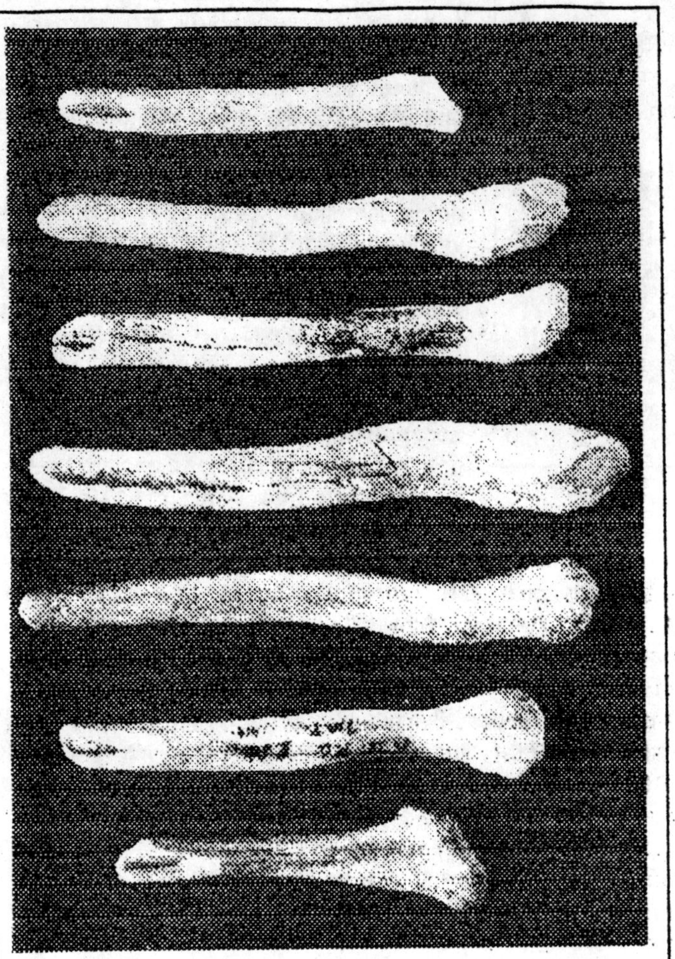

Lissner would also cite the rectangular chest. Today many authorities will dismiss this chest as being the chance alignment of rocks that have fallen from the ceiling.[76] It is a well established fact that regular shapes, like rectangles, are extremely unlikely to occur by chance. Rocks of any length can fall at any angle. It is easy to tell the difference between random arrangement and human construction. Lissner stated, "Only a human intelligence could have been responsible for this. Nature does not build rectangular chests out of flat slabs of stone, nor does she spirit seven bears' skulls into them."[77] Now that Regourdou has been found as a second example of this behavior, it is much more likely that what Bachler found *was* a stone chest filled with bear skulls.

Then there is the arrangement of thirty fibulas placed on a stone slab, aligned with the jointed ends pointing one direction and the broken end pointing the opposite.[78] This too would be difficult to explain based upon chance alignment. An examination of plate 21 and plate 22 of Bachler's 1921 report shows 14 of these bones, many of which seem to have been cut and then rubbed along a rock creating a flat surface on the broken end. The angles which the ground-down surfaces make with the bone are often quite shallow and are polished. Normally when a tibia breaks, the end is jagged. Few of the Drachenloch tibia fit that description. These can be seen in Figure 16 . Are we to believe that the bears broke their own limbs, ground them smooth and placed them in this fashion? Only man polishes bones.

Figure 16 Fibulas from Drachenloch have polished ends.

The critics of Drachenloch, like Kurten, suggest that too many bears would have died in a cave like Drachenloch over the eons. Kurten cites the estimate that 30-50,000 cave-bears died in Dragon Cave near Mixnitz Austria, which was also claimed to be a bear cult site, yet only 76 bears were found. The rest, Kurten says, would have been ground to fragments. Applying this to Drachenloch, Kurten concludes that the skulls were not placed there by human agency. The problem with this line of reasoning is that the 30-50,000 bears which Kurten says are there, *are not there*. They are only an estimate based on the idea that one bear per year died in the cave. Since the cave was open for approximately 30,000 to 50,000 years, the estimate follows directly. Kurten argues that the skulls survived due to accidents of preservation whereby skulls that happened to be in niches in the cave would not be smashed by falling rocks or by subsequent bears. Kurten's analogy with Mixnitz is flawed for another reason. Mixnitz shows more evidence of human activity in the form of a bird bone flute.[79] Surely the bears didn't drill the holes and manufacture this flute.

Is there a way to tell whether the bears in the cave were the product of man or of nature? I believe there may be. Much of the evolution of mammals has been due to the preservation of mammalian teeth and jaws.[80] The teeth and jaws are hard, and are usually the very last part of the animal to decay. This being the case, there is an extremely powerful argument that supports Bachler's contention that Neanderthal man carried the bear skulls to the cave. If the deposit were due to the natural collection of cave bear skeletal parts year after year, one would expect there to be a normal representation of bear mandibles and teeth. Under normal circumstances, one would expect there to be more mandibles than skulls, yet this is not the case. Lissner relates that almost all the unbroken cave-bear skulls did not have a lower jaw! This lack of mandibles is amazing because of the durability of lower jaws. I went through the photos of cave bear parts in both the 1921 and 1940 reports by Bachler. Pictured are 15 complete bear craniums but only three complete lower jaws.

This data is consistent with what Neanderthal man did with bear mandibles in other locations. At Taubach, near Weimar, Germany, the remains of at least forty-three brown bears were found. Not a single complete skull was found. Natural processes seem to have damaged them all. But what is interesting is that on 80 percent of the mandibles, Neanderthal man had broken the canines off. All other teeth on the mandibles showed normal wear, but the canines were intentionally broken off.[81] Why? One can only speculate that maybe they were used for pressure flaking of their stone tools. At Dragon Cave near Mixnitz, Austria bear canines were used as tools but the cave was also lacking in mandibles.[82] In any event, Neanderthal did have an interest in bear teeth. If a similar interest prevailed at Drachenloch, it would, or could, explain what happened to the mandibles of the 76 skulls. The theory of natural accumulation, which says that the 76 skulls are the lucky survivors of all the rock falls and footfalls of subsequent cave-bears, cannot explain why the mandibles were so unlucky. Surely a mandible would have an excellent chance of surviving if it were in a niche. One cannot expect only cave bears lacking lower jaws to have visited the cave. If Neanderthal had removed the mandible prior to the time the bones were placed, this effect would be expected.

Other examples of ritual treatment of bears have been found elsewhere and have been equally denounced by archaeologists. Of Petershohle in Germany, Kurten states,

> "'In the Petershohle in Germany a rock niche, situated like a cupboard in the rock, contained five Cave Bear skulls, two thigh bones, and one brachial bone,' state Professor Abel in 1935. He went on to say, 'All these pieces must have been put into this niche by Ice Age Man, as a deposit formed by water is quite out of the question.'
> "Of course it does not have to be a niche in the wall. Any kind of protecting rock will do."[83]

With that Kurten dismisses Petershohle. What he does not address is the fact that the bear skulls were in a niche "four feet above the floor of the cave," along with two femurs and one humerus![84] Are we to believe that the cave bears placed their own skulls into such a high recess after their death? The critics need a more thorough discussion of the facts.

While I don't often like to disagree with the consensus of a given scientific discipline, I feel that with the discovery of Regourdou, Drachenloch should receive another look.

Burial and the Afterlife

As is often noted, burial is associated with religious belief. Christians have taken a variety of approaches when discussing the implications of human burials. Davis and Kenyon believe that burial is indicative of humanity. They write of *Homo erectus*:

> "It had significant anatomical differences from modern man that have prevented its classification as Homo sapiens. It also left no evidence that it buried its dead, no signs of art, or other recognizably human culture."[85]

By this they imply that burial is an evidence of humanity. Wilcox seems to agree with this position.[86] Hugh Ross would disagree. He advocates,

> "Evidence of man's spiritual dimension would include divine worship, shown by religious relics, altars, and temples. From the Bible's perspective, decorating, burial of dead, or use of tools would not qualify as conclusive evidence of the spirit."[87]

So what are the implications and the purpose of burial? Does it indicate religion? If there is no burial does this mean that religion is lacking? Birdsell notes of burial,

> "Among living primitives, burial is always associated with a belief in life after death. Like Neanderthalians, modern peoples often provide food offerings to assist the dead in reaching the next world, and tool kits appropriate to the individual's role in life are frequently placed in the grave."[88]

Of course one can believe in a life after death and still not bury granny. What this means is that burial is evidence of belief in an afterlife; lack of burial is not evidence of the lack of religious belief. In point of fact, burial seems to be a facet of sedentary life.[89] Most nomadic, hunter-gatherers dispose of bodies in a variety of ways. Some simply abandon the body and the living site and move on to a new locality until the evil has dispersed from the old campsite. Sometimes they simply insert the corpse into natural caves, crevasses, or hollow trees. Cremation is sometimes carried out, as well as burial. But often the burial consists of nothing more than piling stones on top of the body. This type of burial will not generally preserve the body long enough for the archaeologist to find it tens of thousands of years later.

It is a little self-centered of Christians to believe that only our form of disposal of the body implies spirituality. But it is our form of body disposal that is most likely to preserve the remains for future archaeologists to find. Other cultures dispose of the dead in deeply spiritual ways without burial at all. The Siberian tribes believe that man's soul remains with the body and to bury the body is to bury the soul. Burning the body in their eyes will also destroy the soul. Thus they place the body in a tree or on a platform raised above the ground.[90] Parsees, or Zoroastrians, expose their dead on high platforms above the earth in Towers of Silence. The vultures are encouraged to eat the mortal remains of grandma so that she does not pollute the ground.[91] When a chieftain of the Yukaghirs, a northeast Asian tribe, dies, his body may be defleshed. Members of the tribe cut all the flesh off the bones and hang the flesh from a tree. They then distribute the bones of the dead man among the family who often wear the bones as amulets. When they have problems, they will consult with the bones of the dead chief. In this way the wisdom of the chief is passed on to the tribe.[92] Obviously, this means of disposal of a dead body is religious in the sense that the Yukaghirs believe in the chief's afterlife, but would not count as evidence of spirituality among some of the Christian apologists, because, as Davis and Kenyon believe, there is 'no evidence that [they] buried [their] dead' It is equally obvious that this method of body disposal will leave nothing for the archeologist to find.

Given the various methods of disposing of the dead, other than burial, We need to reassess how we determine that a being found in the fossil record is spiritual. In view of the above, nearly all the widely believed positions are wrong. Hugh Ross is correct that spirituality is separate from burial, but is wrong in believing that burial does not imply spirituality, since it is always associated with a belief in an afterlife. Ross states that such things as burial are not indicative of being human at all. In this he is mistaken. He wrote:

> "Moreover, nonspirit creatures such as bower birds decorate their nests, elephants bury their dead, and chimpanzees use tools.[93]

In point of fact *NO* other animal buries its dead as humans do. The claim that animals treat the mortal remains of their fellows with the respect shown by humans is unsupported. The supposed burial behavior by elephants is nothing more than covering a dead elephant with branches and broken brush, which they do occasionally, but not very often. This is *not* the digging of a trench and placing the body into it. Iain and Oria Douglas-Hamilton cite several instances of elephant 'burying behavior'. Two of the cases involve burial, not of other elephants, but of humans. In both of those instances, the 'burying' behavior was a response to anger, not to mourning. In one of these cases a bull elephant killed a tourist and then covered his body with branches and bushes. In the other, a bull elephant in a zoo was mad at his keeper and threw grass and straw at the keeper until he was nearly covered. This is not a case of sadness. In two other cases, the elephants 'buried' a rhinoceros, in one, and a buffalo in the other. The rhino had been dragged quite some distance before being covered. What do these activities mean? I don't know, but this is not correlative behavior to human burial with grave goods and the digging of a burial pit. Ross' use of this to reject the humanity of Neanderthal seems inappropriate since the two behaviors are not the same.

Wilcox, Davis and Kenyon are wrong to believe that burial is a necessary condition of being human, because there are many human cultures today which do not bury their dead. Yet people of these cultures are fully human and subject to the atonement of Christ. Are we to withdraw missionaries from societies that do not practice burial? Of course not. Neither should apologists use lack of burial as a reason to exclude a fossil human from humanity. *Homo erectus*, who apparently did not bury his dead, may have disposed of dead bodies in a fashion similar to that of many modern tribes. There would be little or no evidence left to the archeologist of their burial rituals.

Intentional disposal of dead bodies via burial and defleshing may go quite far back into the past. The evidence for the intentionality of the burial diminishes with time, as does all data. There are several features of an intentional burial according to Yuri Smirnov. He defines the mortuary complex as consisting of the human remains, a burial structure which are found in every burial and the grave goods and associated features which are variable.[94] Grave goods consist of food offerings and items associated with the individual during his lifetime. The associated features might be things like an altar, a tombstone or a fireplace. These items can be either within the grave or adjacent to it. The most important indicator of intentionality is the structure containing the body.

These features can be seen in the Egyptian Pyramids where the pyramid is the structure, the pharaoh provided the body, and food, furniture, gold etc. were left as grave offerings and associated items. Between 25,000 and 22,000 years ago, six spectacular burials took place at Sungir, near Vladimir, Russia.[95] The grave goods consisted of 3000 ivory beads bedecking a single man. An 8-foot long ivory spear weighing up to 44 pounds was also laid alongside the man and a statue of a mammoth was beneath him. Red ochre was sprinkled on many objects in these graves, apparently as a symbol of life and blood. This was an impressive grave. The people were extremely wealthy for such an early period.

Going back to the time of the Neanderthal burials, we find similarities with those of later times. We find pits dug into the substratum, bodies intentionally placed and positioned. Some of the bodies are flexed prior to burial. This means they are tied into a fetal position. The body had to be intentionally handled to achieve this effect. Just as with the modern humans at Sungir, the Neanderthals at Le Moustier Cave sprinkled red ochre over the body.[96] We also find grave goods in Neanderthal graves. A Neanderthal infant was found with the jaw of a red deer placed over its pelvis.[97] Such grave goods are not as pretty as are those at Sungir but they are all that the economically poorer Neanderthals had to contribute. Grave goods include flint tools, horns, tusks and food.

While rare, the graves of Neanderthal are intentional. This can be deduced from several facts. First, red ochre is not normally a part of caves and to find it sprinkled on the grave site but not on other parts of the cave implies a purposeful burial.[98] Second, the flexing of many Neanderthal bodies required that someone tied them up. While bodies will bend as the ligaments dry up, the extent of the flexing can only be done by intentional preparation. The flexing of bodies is very important as an indicator of spiritual belief. In many modern societies, bodies are flexed (tied up) to prevent the person's spirit from wandering.[99] Finally, Neanderthal bodies are preferentially oriented in an east-west direction.[100]

There are some anthropologists who are not convinced that Neanderthal engaged in any burial ritual. But often this criticism seems to be of the nature that absolutely *no* evidence will be sufficient. Schepartz says,

> "Citing examples where the evidence for burial is equivalent for both samples but where 'burial' is questioned for the Middle Paleolithic, they argue that Middle Paleolithic burials are subjected to much more rigorous scrutiny because of an attitude reflecting 'a refusal to accept the possibility that hominids other than *H. sapiens* reached the level of symbolic sophistication expressed, among other things, in intentional burials . . . it [recognition of burial] appears to rely on the physical anthropological and cultural context . . . intentional burials are recognized only when Middle Paleolithic human remains are anatomically modern individuals, since on the basis of their biological resemblance to modern humans they are granted the capacity for complex behaviors such as mortuary practices' (Belfer-Cohen and Hovers, 1992:468-469)."[101]

I would agree with this. Consider the case of Mithen, who dismisses ritual activity in Neanderthals by saying,

> "This is a puzzle because while there is clear evidence that Neanderthals were burying some individuals in pits, there is no evidence of graveside ritual accompanying such acts, nor of the placing of artifacts within the pits/graves along with the dead, as is characteristic of Modern Humans. Isolated burials of Neanderthals have been found in several caves, such as Teshik Tash, La Ferrassie and Kebara. It was once believed that a 'flower burial' had occurred in

Shanidar Cave, high pollen frequencies in the soil seeming to indicate that the body of a deceased Neanderthal had been covered with a wreath of flowers. But this pollen is now believed to have been blown into the cave, or even brought in on workmen's boots.[102]

Mithen doesn't even mention Le Moustier cave, where a man was found sprinkled with red ochre, his body flexed, his head resting on a pillow of flints with burnt wild cattle bones which may have been a food offering. This is evidence of a ritual Mithen does not mention Regourdou cave discussed above where a bear was laid out in the same grave as the human. Mithen does not mention the burial of the Neanderthal infant mentioned above where a red deer jaw was placed on top of it. This is evidence of ritual. The excavator, Yoel Rak, interpreted this as the offering of grieving parents.[103] Concerning the 'flower burial', Mithen dismisses it by citing an article by Gargett in which 9 of 11 commentators on the article felt the science was atrocious. One went so far as to say that there was no science at all in the work. But what is amazing is the dismissal by Mithen of the flower burial totally ignores the details of the burial. Arlette Leroi-Gourhan, one of the excavators of Shanidar, wrote of this concept that the pollen was brought in by workmen.

> "The 3 samples from the dark-brown loamy soil directly beneath Shanidar IV were unique in the cave in containing numerous (145) clusters of anthers. None of the other 50 samples from occupational deposits from Mousterian to Mesolithic contained any anthers, even though more than 6,000 pollens were identified.
> "Pollens are transported in two ways: by wind and by animals (mostly insects, sometimes birds). Wind-transported pollens enter caves only if there is a draft. Animal-transported pollens come from brightly colored or perfumed flowers and are carried into caves on animals' fur or feet. Anthers may be found in rock shelter sediments where the plants were near or within the cave entrance, carried there by rodents along with fruits. Gargett imagines the wind's having blown the flowers just into the Neandertal burial soil and having chosen bright-colored flowers belonging to five different genera. It is a pity that he has constructed his argument without considering the dispersion of pollens and without reading my paper on the subject instead of Solecki's."[104]

To which Gargett replied,

> "With due respect to Mme. Leroi-Gourhan, I do not see how the criticisms I raised regarding the context of Shanidar 4 or of the plant macrofossils said to have been associated with it are diminished by the color or number of flowers that found their way into the cave."[105]

This totally misses the point. The anthers, which are part of a flower are quite numerous (145) in the grave but absent in the rest of the cave. Why is this not evidence of ritual? Of course it is. Mithen has an unfortunate habit of rejecting or ignoring any data which contradicts his thesis of how the mind evolved. The type of flowers which were found in the grave falsifies his view. Marshack notes that most of the seven species of flowers which were found in the grave have been used during historical times in folk medicine as curative herbs.[106] Mithen does not mention any of the many observations which falsify his thesis that Neanderthals are not like modern humans.

Speaking of anatomically modern humans, Mithen says that they differed from the Neanderthal in not placing parts of animal carcasses in graves.[107] But once again he ignores the Regourdou and Le Moustier burials. He also ignores the 1992 discovery of the 50,000-year-old burial a Neanderthal infant. Yoel Rak reported that the body was laid out with the arms at its side with the jawbone of a red deer placed over the child's pelvis. Rak said, "It was an intentional offering, though it is not clear whether it was meant as food for the afterlife, or as some more symbolic gesture. However, this was a primary grave, and the baby was deliberately put in it."[108] One can, of course, close his eyes and reject the evidence, but one cannot say there is none.

One more example of an unwarranted rejection of ritual among the Neanderthals bears mentioning. Akazawa et al,[109] described a Neanderthal child that had been buried with a triangular piece of flint on its chest. Akazawa et al relate,

> "The infant was found *in situ* in the Mousterian deposit, lying on its back with arms extended and legs flexed, indicating an intentional burial. A subrectangular limestone slab at the top of the head and a small piece of triangular flint just on the infant's heart was found in the most sterile layer

of the burial fill. A limestone slab of this type is rare in the Dederiyeh Cave deposits."[110]

They showed a drawing of the find, described the find, and discussed the probability of this being a natural deposit (the limestone type is rare). Mithen questioned Akazawa's interpretation by saying,

> "They claim that a piece of flint was intentionally placed over the heart of the child, but they provide insufficient data to support such an assertion."[111]

Consider the problem here. How does one leave any physical evidence of the non-physical property of "intentionality"? If I open a coffin and find a Bible on the chest of the skeleton, I would conclude that it was placed there intentionally by the survivors. Why? Because Bibles are not naturally placed in that position when everyone dies. Similarly, when a piece of flint, which is foreign to the cave is found on a child's chest, why must we conclude differently? Does the fact that the child is a Neanderthal influence the evaluation of the intentionality of this act? Probably it does.

Burials are rare. As one goes further back in time, the quantity of burials decreases. Graveyards are unknown in the Middle East prior to around 12,000 years ago.[112] This is most likely because as we have noted earlier, burial suggests a sedentary lifestyle, which was a requirement at the time when agriculture was just beginning. As the population began to increase, there were far too many people dying in too short a time frame for the people to move away from the dead body. Hanging numerous bodies in the open air would also begin to become odiferous. This leaves the burial of the loved one's remains as the only solution to the problem. Prior to 35,000 years ago burials are known only from caves. No burials are known from any open-air site. This may be due to the greater preservation of corpses inside of caves versus their preservation in the open. In Europe and Asia there are only 60 known burials between 70,000 and 40,000 years, an average of only two per millennium.[113] And this only occurred in the region of Europe and the immediately surrounding parts of Asia. Burial was not a widespread phenomenon. Only four possible intentional burials are known from Africa during this same time.[114]

The scarcity of burials may not imply that mankind did not revere his dead prior to that time. Nor does it mean that there were no spiritual dimensions to life at that time. As noted, other methods of body disposal may have been predominant. A case in point may be the burial at Krapina. There is evidence that someone between 115,000 and 130,000 years ago was treated like the Yukaghir chief. He was defleshed. There are cut marks on the bones which indicate that the flesh was removed. Schepartz writes:

> "The earliest evidence of burial treatment in the paleoanthropological record comes from the archaic *sapiens* populations of the Middle Paleolithic. Russell proposes that the hominid bones from Krapina rockshelter in Croatia display cutmarks produced by the defleshing of decomposing bodies. If this is accepted as evidence of burial treatment [as opposed to cannibalism . . .] then Krapina is the earliest evidence of burial treatment, as the site may date to the Last Interglacial, preceding other European or Middle Eastern burial sites. "[115]

Somewhere prior to 300,000 years ago, men living in Spain began to dispose of their dead by carrying the corpses into a cave and them tossing them into a pit some 50 feet deep. The site is Sima de los Huesos in the Sierra Atapuerca. The excavator, Juan-Luis Arsuaga, "believes that over several generations bodies were carried into the cave from an entrance, now lost, near the pit and tossed into the shaft in a form of mortuary ritual that may point to some embryonic religious belief."[116] The site is so rich in human bones that since 1990, this single mortuary site has produced more than 90 percent of all the pre-Neanderthal human bones ever found in Europe. Over 1600 bones from more than 32 individuals have been found after the removal of only two percent of the sediment. The fact that there are no herbivore bones proves that this was not a human living site and the lack of tooth gnaw marks on the bones proves that the human bones were not left by carnivores. These people had to master their fear of dark places, the fear of the very large cave bears and the fear of falling into the pit themselves in order to dispose of their fallen comrades. Today, to get to the pit one must go several hundred yards into the cave, pass through an enormous cave, then climb over some rocks and down a clay slope, and crawl through a small space between two stalagmites. Finally, 1600 feet deep into the cave is a 65 foot tall chamber. One must then climb down a rope 50 feet into the pit. A side chamber is where the bodies are found.[117] Many modern people don't go to this much trouble to dispose of their friend's bodies. The Atapuercans must have had an important reason for engaging in such behavior, perhaps a religious reason.

One friend, who has strong anti-evolutionary feelings, discounted the actions of the Sima people as nothing more than throwing garbage into a pit. His claim was that this was not evidence of any sort of mortuary ritual. What he did not know was that many cultures of modern humans performed exactly the same sort of burial. Brady et al, discovered Talgua Cave in Honduras which contained over 200 Maya bodies. Their relatives had carried them

> "Caves have played an important sacred role in Mesoamerican religion and culture. Since they penetrated the earth, they were believed to be entrances to the underworld where the souls of the dead resided. Many indigenous groups today tell stories about the final journey that the soul must make; some of them describe its entry into a cave. Burial within a cave would speed the soul's journey to the afterlife and guard against the frightening possibility of it wandering the earth causing mischief." [118]

Now, the cave drawing in the article shows that the Maya carried their dead as deep as 1200 feet back into the cave, and as much as 40 up into an upper chamber. The upper chamber would be very dark. Both the Maya and the Sima people took their dead into the cave. Both needed torches. The physical similarities are remarkable.

But even today, our burial rituals in which the individual is placed in a coffin and dropped into a manmade pit is a simulation of these ancient forms of burial. By using the coffin we create a small cavern in the earth in which the body resides. Thus, the symbolic evidence left by our burial rituals is quite similar to that left by the Sima people. When we find this type of evidence one has to go to great lengths to exclude the Sima people from the ranks of humanity.

There are two earlier possible example of religious treatment of the dead, both are the intentional defleshing of the bones of humans. First, there is the possible scalping of the man at Bodo, Ethiopia, 600,000 years ago.[119] Tattersall remarks,

> "Interestingly, we also begin to pick up evidence for the intentional defleshing of human remains, as evidenced, for example, by stone tool cut-marks on the forehead and within the orbit of the *Homo heidelbergensis* [archaic *Homo sapiens*--GRM] cranium from Bodo. Whether this scalping indicates cannibalism (which seems a little unlikely) or some other ritual behavior is anybody's guess. From this point on, too, stone tools began to decrease in size, although some of the older forms were preserved; for example, small handaxes were made on flakes, rather than on large cores broken off much larger boulders as was the practice in Acheulean times."[120]

The importance of scalping is that, at least to some tribes who engaged in this practice, it was part of an effort to placate the spirit of the dead. This is a religious reason.

The second case comes from Gran Dolina, Spain, 780,000 years ago. The skulls of *H. heidelbergensis*, found there show clear cut marks at the attachment sites for cranial muscles. Similar cut marks were found on toe bones also.[121] Under the microscope, these cut marks have the characteristically v-shaped form of those made by stone tools. Carnivore teeth marks are more U-shapes.

What does cannibalism or defleshing of humans tell us about religious beliefs? All modern cultures which have been reported to engage in such behavior use it for ritual purposes such as gaining the strength of their victim.[122] These evidences indicate that there was religious beliefs in men three-quarters of a million years ago. Christians who claim that there is no evidence of ritual mortuary behavior among pre-humans need to consider these examples.

Shamanism

Shamanism is a widespread religious belief among primitive peoples. A shaman experiences trances and possession and communicates with the supernatural world. He is identified metaphysically with the animals and plants which the primitive people use to feed and clothe themselves. The shaman tries to appease the souls of the animals he kills as well as to control them so that the hunt will be good. He may use his trance to attempt to locate a missing member of the tribe. He is also the medicine man in many situations. He knows the physical world and its powers.

Shamans will often wear costumes of animals whom they believe are either spiritual guides or animals they are trying to control. Lissner notes of the Altaics of Siberia:

> "The whole costume, which is worn only on official occasions, is modeled on the external appearance of a bird, reindeer or roebuck. It is made of deer or reindeer hide and hung with a large number of leather thongs, not to mention the diverse magical symbols characteristic of shamanic robes in general. Uno Harva lists many of these appurtenances: for instance twelve ermine pelts attached to a strap which hangs down the shaman's back, bells beneath his armpits, eagle owl's talons, snakes and iron hands on his sleeves, bear's foot bones on his footwear and iron bear's claws on his gloves and boots. Shamanic robes sometimes varied between different Altaic tribes."[123]

Dickson further states:

> "According to [Lommel], the shaman's costume is essential to his performance and this garment is generally an animal disguise. This leads Lommel to the conclusion that depictions of men disguised as animals are very probably meant to represent shamans."[124]

Such costumes can be quite elaborate. Often the costumes include antlers which symbolize wisdom. A Mongol holy man drawn during Czar Nicholas II's trip through Asia, wore a beautiful multicolored robe with beads and other items worn on top. On the man's head was a full reindeer head with a ten-point rack of antlers. The effect is quite startling because he looks like a bipedal reindeer.[125] Similar items have been found in other cultures. In 1705 a Siberian Tungus shaman was drawn wearing a head dress with antlers and bear costume including bear hands and feet.[126] At Spiro Mound, Oklahoma, a wooden carved mask with antlers is dated to around 1200 AD. Wooden masks like this one would be unlikely to survive several thousand years since, as we have pointed out, wood decays rapidly on an archeological time scale. A wood and antler sculpture was recovered from a 4th century BC grave showing a man with antlers and a long tongue. The antlers represent the acquisition of wisdom and the tongue--the teaching of wisdom.

Drawings of men wearing costumes and having antlers are common in Upper Paleolithic art. As of 1960, fifty-five man/animal chimerical paintings had been discovered.[127] An engraved baton was found in the Upper Magdalenian site of Abri Mege dating to around 13,000 years ago.[128] Sixteen thousand years ago at Lascaux Cave, someone drew a shaman, wearing a bird-head mask. This shaman was being killed by a bison. The shaman has an erect phallus which is often interpreted as proof that this was a real event, because men being killed by animals are known to have erections just before death. Eighteen thousand years ago at Le Trois Freres Cave, someone drew three drawings of shamans. One is wearing deer antlers, two are wearing bison costumes. All shamans are partly naked.[129]

But this is not the earliest example of a shaman costume. The tradition extends back to 50,000 years to the Neanderthals, 50,000 years ago.[130] James Shreeve relates:

> "But the Neandertals' true humanity revealed itself in the actions of their souls. At the 50,000-year-old site of Hortus in southern France, two French archaeologists in 1972 reported the discovery of the articulated bones of the left paw and tail of a leopard. Their arrangement suggested that the fragments were once the remnants of a complete leopard hide worn as a costume."[131]

The details of the find are highly indicative of this being a shaman's cape. The fact that only the paws and the tail bones of the leopard remain indicates that the rest of the animal had been removed. As noted above, the jaws and teeth of a mammal are usually the last skeletal items to decay, yet there is no evidence that the teeth were buried with the Neanderthal. This is not a case of a human accidentally buried on top of a leopard skeleton; it is a human buried wearing an animal cape. Contra Mithen, this is another piece of evidence of ritual among Neanderthals.

Once again, there is good evidence of a religious life among the Neanderthals. People who chose to ignore this data will never see the Neanderthal for what he was, a spiritual being with primitive technology.

Personal ornamentation

Another form of ritual among modern humans is body painting, scarification, tattooing and evulsion. Scarification is the creation of scars on the body in a pattern which conveys personal identity or tribal identity. Unfortunately little evidence of such ritual can survive since the skin, upon which these marks are made, decays at the death of the individual. Evulsion is the removal of front teeth sometimes done during initiation rites. Modern peoples use these techniques for decoration and to provide magical protection.[132]

Evulsion was apparently practiced by the Mungo man in Australia around 32,000 years ago.[133] Fagan has no doubt about the use of red ochre for body paint among the Upper Paleolithic peoples because of their widespread use of pigments.[134] Ochre powder was made at Becov, Czechoslovakia 250,000 years ago.[135] Ochre powder is almost certainly used in rituals such as body painting. At Terra Amata, 300,000 years ago, various colors of ochre were created by heating the ochre. De Lumley believes these were created for body painting. But the earliest evidence of possible body painting was the discovery in the 1.8 million year old Olduvai Gorge Bed II of a red pigment which created a red powder when rubbed.[136]

Animal Ritual

Nahr Ibrahim, Lebanon was a Neanderthal site[137] A ritually placed antelope was found. Marshack writes:

> In the Mousterian cave shelter of Nahr Ibrahim in Lebanon the bones of a fallow deer (*Dama mesopotamia*) were gathered in a pile and topped by the skull cap. Many of the bones were unbroken and still articulated. Around the animal were bits of red ochre. While red ochre was common in the area and so may have been introduced inadvertently, the arrangement of the largely unbroken bones suggests a ritual use of parts of the animal."[138]

There is one example of ritual behavior with *Homo erectus*. At Toralba, Spain, dating 400,000 years ago, an elephant carcass was laid out--well, not exactly a whole elephant, but just one half of an elephant. It had been laid out in an anatomically correct position in Area 1 of the excavation. Freeman and Butzer write,

> "The most striking feature in this level is the half skeleton of a straight-tusked elephant lying in the extreme northern sector of the exposure. The bones of this individual, spread over 50 sq. m, were found in semi-articulated position, lying skin-side up, head to the west. Only the bones of the left side of the animal and some of the vertebrae seem to be represented. The cranium is missing, except for the tusks. The pelvis is also missing. The mandible, broken across both ascending rami, is present but it was found on the east side of the main bone accumulation, which comprises the hind part of the skeletal distribution. A single *Equus* molar is the only bone in Area 1 from an animal other than *Elephas*. In the fact that most of the contents of the cluster are bones from a single individual, Area 1 differs from adjacent areas."[139]

This type of display is obviously intentional and required much forethought. It is ritualistic, but what it meant no one knows. At Ambrona, nearby, elephant leg bones were found end to end in two parallel lines.[140]

Conclusion

Contrary to what many Christian apologists believe and teach about the fossil record, ancient man left much evidence for ritual practices. Some of this goes back to *Homo erectus*. If modern man, Neanderthal, archaic *Homo sapiens* and late *Homo erectus* engaged in ritual and religion, then the Christian can do nothing except to conclude that they carry the image of God. This then raises the question of what to do with the early *Homo erectus*, prior to 400,000 years ago, who at present have shown no evidence of ritual? If late *Homo erectus* are spiritual, then either all *Homo erectus*' back to 1.8 million years are spiritual, or Adam was a *Homo erectus* living around a half-million years ago. The data would force one of these conclusions. The data does not support the generally advocated view of the progressive creationists or theistic evolutionists who believe that no one other than modern man show evidence of religious activity. We Christians can live in the world with the facts as they are, and deal with those facts, both the problematical as well as the supportive, or we can believe only the facts that we prefer to believe. Integrity requires the former course of action.

One thing is certain. we need to stop talking about a cultural explosion occurring at 30-40,000 years ago. Pinker puts it best,

> "Modern *Homo sapiens*, which is thought to have appeared about 200,000 years ago and to have spread out of Africa 100,000 years ago, had skulls like ours and much more elegant and complex tools showing considerable regional variation. It is hard to believe that they lacked language, given that biologically they *were* us, and all biologically modern humans have language. This elementary fact, by the way, demolishes the date most commonly given in magazine articles and textbooks for the origin of language: 30,000 years ago, the age of the gorgeous cave art and decorated artifacts of Cro-Magnon humans in the Upper Paleolithic. The major branches of humanity diverged well before then and all their descendants have identical language abilities; therefore the language instinct was probably in place well before the cultural fads of the Upper Paleolithic emerged in Europe. Indeed, the logic used by archaeologists (who are largely unaware of Psycholinguistics) to pin language to that date is faulty. It depends on there being a single 'symbolic' capacity underlying art, religion, decorated tools, and language, which we now know is false (just think of linguistic idiot savants like Denyse and Crystal from Chapter 2, or, for that matter, any normal three-year-old)."[141]

By giving the 30-40,000 year date, Christians create a huge theological problem. The widespread dispersal of humanity long prior to that time means that human beings, who lived in all parts of the world, were specially infused with the ability for language, religion, art etc. This means thousands of Adams, not just one! If the ability for all these human traits suddenly appeared all over the world, there is no other conclusion. There is no evidence that there was sufficient time for reproduction to put these traits everywhere.

Next we will look at the curse and its effect on humanity. There is clear evidence that the effects of the curse were already affecting *Homo erectus* 1.8 million years ago.

References

1. Genesis 4:3-5, New International Version.
2. Stephen Mithen, The Prehistory of the Mind, (New York: Thames and Hudson, 1996), p. 174
3. Stephen Mithen, The Prehistory of the Mind, (New York: Thames and Hudson, 1996), p. 176
4. James R. Shreeve, The Neandertal Enigma, (New York: William Morrow and Co., 1995), p. 249
5. Victor Barnouw, An Introduction to Anthropology: Ethnology, Vol. 2, (Homewood, Illinois: The Dorsey Press, 1971), p. 228
6. Crystal Gazing", Encyclopedia Britannica, Vol. III, (Chicago: Encyclopedia Britannica, 1982), p. 273
7. W.C. Pei, "Notice of the Discovery of Quartz and Other Stone artifacts in the Lower Pleistocene Hominid-Bearing Sediments of the Choukoutien Cave Deposit," Bulletin of the Geological Society of China, 11:2:1931:109-146, p.120
8. John H. Relethford, Fundamentals of Biological Anthropology, (Toronto: Mayfield Publishing Co., 1994), p. 271
9. Ian Tattersall, The Fossil Trail, (New York: Oxford University Press, 1995), p. 26.
10. Robert G. Bednarik, "On Lower Paleolithic Cognitive Development," 23rd Chacmool Conference Calgary 1990, pp 427-435, p. 432.
11. S. Belitszky et al, "A Middle Pleistocene Wooden Plank with Man-made Polish," Journal of Human Evolution, 1991, 20:349-353.
12. E. O. James, The Cult of the Goddess, (New York: Barnes and Noble, 1994), p. 43
13. E. O. James, The Cult of the Goddess, (New York: Barnes and Noble, 1994), p. 40; Marija Gimbutas, The Language of the Goddess, (San Francisco: HarperSanFrancisco, 1989), p. 81
14. Marija Gimbutas, The Language of the Goddess, (San Francisco: HarperSanFrancisco, 1989), p. 81
15. Marija Gimbutas, The Language of the Goddess, (San Francisco: HarperSanFrancisco, 1989), p. 198
16. Marija Gimbutas, The Language of the Goddess, (San Francisco: HarperSanFrancisco, 1989), p. 230
17. Paul G. Bahn, "New Advances in the Field of Ice Age Art," in M.H. Nitecki and D.V. Nitecki, eds. Origins of Anatomically Modern Humans, (New York: Plenum Press, 1994), p. 124
18. Ivar Lissner, Man, God and Magic, (New York: G. P. Putnam's Sons, 1961), p. 189-191

19. N. Goren-Inbar, "A Figurine from the Acheulian Site of Berekhat Ram," Mi' Tekufat Ha'Even 19(1986):7-12
20. G. Feraud et al, "$^{40}Ar/^{39}Ar$ Age Limit for an Acheulian Site in Israel," Nature, July 21, 1983, p. 263
21. Andrew Pelcin, "A Geological Explanation for the Berekhat Ram Figurine," Current Anthropology, Dec. 1994, 35:5, p. 674-675.
22. L. A. Schepartz, "Language and Modern Human Origins," Yearbook of Physical Anthropology, 36:91-126(1993), p. 117
23. Alexander Marshack, "On the "Geological' Explanation of the Berekhat Ram Figurine," Current Anthropology, 36:3, June, 1995, p. 495.
24. Alexander Marshack, "The Berekhat Ram Figurine: A Late Acheulian Carving from the Middle East," Antiquity, 71(1997):327-337.
25. Joan Oates, "The Emergence of Cities in the Near East," in Andrew Sherratt, editor, The Cambridge Encyclopedia of Archaeology, (New York: Crown Publishers, 1980), p. 114.
26. L. G. Freeman and J. G. Echegaray, "El Juyo: A 14,000-year-old Sanctuary From Northern Spain," History of Religion, Aug. 1981, p. 15-16.
27. L. G. Freeman and J. G. Echegaray, "El Juyo: A 14,000-year-old Sanctuary From Northern Spain," History of Religion, Aug. 1981, p. 15.
28. Robert G. Bednarick, "Neanderthal News," The Artefact 1996, 19:104
29. Rick Gore, "The First Europeans," National Geographic, July, 1997, p. 110
30. Ivar Lissner, Man, God and Magic, (New York: G.P. Putnam's Sons, 1961), p. 162-163
31. Ivar Lissner, Man, God and Magic, (New York: G.P. Putnam's Sons, 1961), p. 160
32. Ivar Lissner, Man, God and Magic, (New York: G.P. Putnam's Sons, 1961), p. 154
33. Steven Mithen, The Prehistory of the Mind, (New York: Thames and Hudson, 1996), p. 48
34. Ivar Lissner, Man, God and Magic, (New York: G.P. Putnam's Sons, 1961), p. 160
35. Ivar Lissner, Man, God and Magic, (New York: G.P. Putnam's Sons, 1961), p. 238
36. This is a modernized form of what is reported by Ivar Lissner, Man, God and Magic, translated by J. Maxwell Brownjohn, (New York: G.P. Putnam's Sons, 1961), p. 240
37. Victor Barnouw, An Introduction to Anthropology: Physical Anthropology and Archaeology, Vol. 1, (Homewood, Illinois: The Dorsey Press, 1982) p. 156-157
38. Ivar Lissner, The Living Past, translated by J. Maxwell Brownjohn, (New York: G. P. Putnam's Sons, 1957), p. 207
39. Bjorn Kurten, The Cave Bear Story, (New York: Columbia University Press, 1976), p.106
40. "Ursa Major", Encyclopedia Britannica, 1982, Vol. X, p. 304
41. Ivar Lissner, Man, God and Magic, translated by J. Maxwell Brownjohn, (New York: G. P. Putnam's Sons, 1961), p. 165
42. D. Bruce Dickson, The Dawn of Belief, (Tuscon: The University of Arizona Press, 1990), p. 82
43. Alexander Marshack, The Roots of Civilization, (New York: McGraw-Hill Book Co., 1972), p. 274
44. Ivar Lissner, Man, God and Magic, translated by J. Maxwell Brownjohn, (New York: G. P. Putnam's Sons, 1961), figure 87 between pp 224-225.
45. Alexander Marshack, The Roots of Civilization, (New York: McGraw-Hill Book Co., 1972), p. 236-237
46. Bjorn Kurten, The Cave Bear Story, (New York: Columbia University Press, 1976), p.93
47. Bjorn Kurten, The Cave Bear Story, (New York: Columbia University Press, 1976), p. 95-96
48. Jean-Marie Chauvet et al, Dawn of Art, (New York: Harry N. Abrams, 1996), p. 50
49. Alexander Marshack, "The Berekhat Ram Figurine: a Late Acheulian Carving from the Middle East," Antiquity, 71(1997):327-337, p. 327; Mark Berkowitz, "Neandertal News," Archaeology, Sept./Oct. 1996, p. 22; Robert G. Bednarick, "Neanderthal News," The Artefact 1996, 19:104
50. Michael Barbaza, "From the Middle Paleolithic to the Epipaleolithic in the Old World," in Jean Guilaine, Prehistory,, (New York: Facts on File, 1991), p. 59-60
51. Brian Hayden "The Cultural Capacities of Neandertals ", Journal of Human Evolution, 24(1993):113-146, p. 121
52. Bjorn Kurten, The Cave Bear Story, (New York: Columbia University Press, 1976), p. 104-105
53. Michael Barbaza, "From the Middle Paleolithic to the Epipaleolithic in the Old World," in Jean Guilaine, Prehistory,, (New York: Facts on File, 1991), p. 59-60
54. James Shreeve, The Neandertal Enigma (New York:William Morrow and Company, 1995), p. 52
55. Yuri Smirnov "Intentional Human Burial: Middle Paleolithic (Last Glaciation) Beginnings," Journal of World

Prehistory, 3:2(1989), pp 199-233, p. 220
56. Bjorn Kurten, The Cave Bear Story, (New York: Columbia University Press, 1976), p. 105-106
57. D. Bruce Dickson, The Dawn of Belief, (Tuscon: University of Arizona Press, 1990), p. 49
58. F. Clark Howell, Early Man, (New York: Time-Life Books, 1965), p. 126 cited in Lewis R. Binford, Bones: Ancient Men and Modern Myths, (Orlando: Academic Press, 1981), p. 10.
59. J. M. Coles and E. S. Higgs, The Archaeology of Early Man, (New York: Praeger, 1969):286-287 cited in Lewis R. Binford, Bones: Ancient Men and Modern Myths, (Orlando: Academic Press, 1981), 10-11
60. J. B. Birdsell, Human Evolution, (Chicago: Rand McNally & Co., 1972), p. 285
61. Erik Trinkhaus and Pat Shipman, The Neandertals, (New York::Vintage Books, 1992), p. 180-194
62. Ivar Lissner, Man, God and Magic, translated by J. Maxwell Brownjohn, (New York: G. P. Putnam's Sons, 1961), p. 187
63. Robert G. Bednarik, "Neanderthal News," The Artefact 1996, 18:104
64. D. Bruce Dickson, The Dawn of Belief, (Tuscon: The University of Arizona Press, 1990), p. 51
65. Lewis R. Binford, Bones: Ancient Men and Modern Myths, (Orlando: Academic Press, 1981).
66. Hallam L. Movius, Jr., "A Wooden Spear of Third Interglacial Age from Lower Saxony," Southwestern Journal of Anthropology, 6(1950):139-142.
67. Dr. Emil Bachler, Das Drachenloch, (St. Gallen:Druck der Buchdruckerei Zollikofer & Cie., 1921).
68. Emil Bachler, Das Alpine Palaolithikum der Schweiz, Monographien Zur Ur-Und Fruhgeschichte der Schweiz, (Basel: Verlag Birkhauser & Cie, 1940).
69. For an excellent discussion (in English) of the problems with Bachlers work see, Bjorn Kurten, The Cave Bear Story, (New York: Columbia University Press, 1976), p. 84-85
70. Chris Stringer and Clive Gamble, In Search of the Neanderthals, (New York: Thames and Hudson, 1993), p.159-160
71. Bjorn Kurten, The Cave Bear Story, (New York: Columbia University Press, 1976), p. 84-85
72. Bjorn Kurten, The Cave Bear Story, (New York: Columbia University Press, 1976), p. 86
73. Bjorn Kurten, The Cave Bear Story, (New York: Columbia University Press, 1976), p. 86
74. See plate 22 in Emil Bachler, Das Alpine Palaolithikum der Schweiz, BD II, Monographien Zur Ur-Und Fruhgeschichte der Schweiz, (Basel: Verlag Birkhauser & Cie, 1940).
75. Ivar Lissner, Man, God and Magic, translated by J. Maxwell Brownjohn, (New York: G. P. Putnam's Sons, 1961), p. 187
76. Peter Rowley-Conwy, "Was There a Neanderthal Religion?" in Goren Burenhult, The First Humans, (San Francisco: Harper SanFrancisco, 1993), p. 70
77. Ivar Lissner, Man, God and Magic, translated by J. Maxwell Brownjohn, (New York: G. P. Putnam's Sons, 1961), p. 187
78. Ivar Lissner, Man, God and Magic, translated by J. Maxwell Brownjohn, (New York: G. P. Putnam's Sons, 1961), p. 183
79. J.V.S. Megaw, "Penny Whistles and Prehistory," Antiquity XXXIV, 1960, pp 6-13, p. 7-8
80. E. C. Olson, The Evolution of Life, (New York: Mentor Books, 1965), p. 36.
81. Bjorn Kurten, The Cave Bear Story, (New York: Columbia University Press, 1976), p. 105
82. Ivar Lissner, Man, God and Magic, (New York: G. P. Putnam's Sons, 1961), p. 192
83. Bjorn Kurten, The Cave Bear Story, (New York: Columbia University Press, 1976), p. 88
84. Ivar Lissner, Man, God and Magic, translated by J. Maxwell Brownjohn, (New York: G. P. Putnam's Sons, 1961), p. 191
85. Percival Davis and Dean H. Kenyon, Of Pandas and People, 2nd edition (Dallas: Haughton Publishing Co., 1993), p. 110
86. David L. Wilcox, "Adam, Where Are You? Changing Paradigms in Paleoanthropology," Perspectives on Science and Christian Faith, 48:2(June 1996), p. 93
87. Hugh Ross, The Fingerprint of God, (Orange: Promise Publishing, 1991), p. 159-160.
88. J. B. Birdsell, Human Evolution, (Chicago: Rand McNally & Co., 1972), p. 285
89. Carleton S. Coon, The Hunting Peoples, (Boston: Little, Brown and Co., 1971), p. 331-332; and D. Bruce Dickson, The Dawn of Belief, (Tuscon: The University of Arizona Press, 1990), p. 195.
90. Ivar Lissner, Man, God and Magic, translated by J. Maxwell Brownjohn, (New York: G. P. Putnam's Sons, 1961), p. 271
91. Ivar Lissner, Man, God and Magic, translated by J. Maxwell Brownjohn, (New York: G. P. Putnam's Sons,

1961), p. 271
92. Ivar Lissner, Man, God and Magic, translated by J. Maxwell Brownjohn, (New York: G. P. Putnam's Sons, 1961), p. 169
93. Hugh Ross, The Fingerprint of God, (Orange: Promise Publishing, 1991), p. 159-160.
94. Yuri Smirnov,"Intentional Human Burial: Middle Paleolithic (Last Glaciation) beginnings," Journal of World Prehistory, 3:2(1989):199-233, p. 210-211.
95. Olga Soffer, "Sungir: A Stone Age Burial Site," in Goren Burenhult, editor, The First Humans, (San Francisco: HarperSanFrancisco, 1993), p. 138-139
96. James Shreeve, The Neandertal Enigma, (New York: William Morrow and Co., 1995), p. 53
97. Chris Stringer and Robin McKie, African Exodus, (New York: Henry Holt and Company, 1997), p. 87-88
98. James Shreeve, The Neandertal Enigma, (New York: William Morrow and Co., 1995), p. 53
99. Alexander Marshack, "Early Hominid Symbol and Evolution of the Human Capacity," in Paul Mellars, ed. The Emergence of Modern Humans, (Ithica: Cornell University Press, 1990), p. 489
100. Brian Hayden, "The Cultural Capacities of Neandertals: a Review and Re-evaluation," Journal of Human Evolution, 24(1993):113-146, p. 121
101. L. A. Schepartz, "Language and Modern Human Origins," Yearbook of Physical Anthropology, 36:91-126(1993), p. 113-115
102. Steven Mithen, The Prehistory of the Mind, (New York: Thames and Hudson, 1996), p. 135-136.
103. Chris Stringer and Robin McKie, African Exodus, (New York: Henry Holt and Company, 1997), p. 87-88
104. Arlette Leroi-Gourhan, "Comments" Current Anthropology, 30:2(April 1989), pp 157-190, p. 182
105. Robert H. Gargett, "Reply", Current Anthropology, 30:2(April 1989), pp 157-190, p. 185
106. Alexander Marshack, "Some Implications of the Paleolithic Symbolic Evidence for the Origin of Language," Current Anthropology, 17:2, June, 1976, p. 276 Note 2.
107. Steven Mithen, The Prehistory of the Mind, (New York: Thames and Hudson, 1996), p. 180
108. Yoel Rak quoted in Christopher Stringer and Robin McKie, African Exodus, (New York: Henry Holt, 1997), p. 87-88
109. T. Akazawa, et al, "Neanderthal Infant Burial,", Nature 377:585-586.
110. Takeru Akazawa et al, "Neanderthal Infant Burial," Nature, 378, Oct. 19, 1995, p. 586
111. Steven Mithen, The Prehistory of the Mind, (New York: Thames and Hudson, 1996), p. 249.
112. Clive Gamble, Timewalkers, (Cambridge: Harvard University Press, 1994), p. 192
113. Yuri Smirnov,"Intentional Human Burial: Middle Paleolithic (Last Glaciation) beginnings," Journal of World Prehistory, 3:2(1989):199-233, p. 203
114. Yuri Smirnov,"Intentional Human Burial: Middle Paleolithic (Last Glaciation) beginnings," Journal of World Prehistory, 3:2(1989):199-233, p. 203
115. L. A. Schepartz, "Language and Modern Human Origins," Yearbook of Physical Anthropology, 36:91-126(1993), p. 113
116. Paul G. Bahn, "Treasure of the Sierra Atapuerca," Archaeology, January/February 1996, p 47.
117. Paul G. Bahn, "Treasure of the Sierra Atapuerca", Archaeology, January/February, 1996, pp 45-48, p. 47
118. James E. Brady, George Hasemann and John H. Fogarty, "Harvest of Skulls & Bones," Archaeology, May/June 1995, p. 40
119. Recent dating has shown this site to be 600,000 years old. Donald Johanson and Blake Edgar From Lucy to Language, (New York: Simon and Schuster, 1997), p. 194.
120 Ian Tattersall, The Fossil Trail (New York: Oxford University Press, 1995), p. 244.
121. Donald Johanson and Blake Edgar, From Lucy to Language, (New York: Simon and Schuster, 1997), p. 93; E. Carbonell, et al, "Lower Pleistocene Hominids and Artifacts from Atapuerca-TD6(Spain)," Science 269(1995):826-830.
122. Ivar Lissner, Man, God, and Magic, (New York: G. P. Putnam's Sons, 1961), p. 28; Dean Falk, Braindance, (New York: Henry Holt and Co., 1992), p. 182-183
123.Ivar Lissner, Man, God and Magic, translated by J. Maxwell Brownjohn, (New York: G.P. Putnam, 1961), p. 272
124. D. Bruce Dickson, The Dawn of Belief, (Tuscon: The University of Arizona Press, 1990), p. 131.
125. see the drawing on page 83 of Joan Halifax, Shaman, (New York: Thames and Hudson, 1982),
126. Joan Halifax, Shaman, (New York: Thames and Hudson, 1982), p. 82
127. D. Bruce Dickson, The Dawn of Belief, (Tuscon: The University of Arizona Press, 1990), p. 131.

128. Alexander Marshack, The Roots of Civilization, (New York: McGraw-Hill Book Co., 1972), p. 260
129. Alexander Marshack, The Roots of Civilization, (New York: McGraw-Hill Book Co., 1972), p. 272-273
130. For more information see Alexander Marshack, "Early Hominid Symbol and Evolution of the Human Capacity," in Paul Mellars, ed. The Emergence of Modern Humans, (Ithica: Cornell University Press, 1990), pp. 457-498, p. 478 and M.-A. Lumley, and H. de Lumley, La Grotte de l'Hortus (Valflaunes Herault), Etudes Quaternaire, 1. (Marseilles: Laboratoire de Paleontologie Humaine et de Prehistoire, 1972)
131. James R. Shreeve, The Neandertal Enigma, (New York: William Morrow and Co., 1995), p. 52
132. "Tattooing," Encyclopedia Britannica, IX, (Chicago: Encyclopedia Britannica,1982), p. 841
133. Clive Gamble, Timewalkers, (Cambridge: Harvard University Press, 1994), p. 167; L. Luca Cavalli-Sforza, Paoli Menozzi and Alberto Piazzi, The History and Geography of Human Genes, (Princeton: Princeton University Press, 1994), p. 344
134. Brian Fagan, The Journey From Eden, London: Thames and Hudson, 1990, p. 171
135. D. Bruce Dickson, The Dawn of Belief, (Tuscon: The University of Arizona Press, 1990), p. 42-43
136. D. Bruce Dickson, The Dawn of Belief, (Tuscon: The University of Arizona Press, 1990), p. 44
137. This site is listed as a Neanderthal site in Past Worlds, (New York: Crescent Book, 1995), p. 65
138. Alexander Marshack, "Early Hominid Symbol and Evolution of the Human Capacity," in Paul Mellars, ed. The Emergence of Modern Humans, (Ithica: Cornell University Press, 1990), p. 481
139. L.G. Freeman and K. W. Butzer, "The Acheulean Station of Torralba (Spain): A Progress Report," Quaternaria 8(1966):9-21, p. 15-16.
140. Donald Johanson and James Shreeve, Lucy's Child, (New York: William Morrow and Co., Inc., 1989), p. 221
141. Steven Pinker, The Language Instinct, (New York: Harper/Perennial, 1994), p. 353-354

A CURSE UPON YOUR HOUSE

"To the woman he said, 'I will greatly increase your pains in childbearing; with pain you will give birth to children.'" Genesis 3:16

"To Adam he said, '...By the sweat of your brow you will eat your food until you return to the ground, since from it you were taken; for dust you are and to dust you will return.'" Genesis 3:17-19

"The LORD God made garments of skin for Adam and his wife and clothed them." Genesis 3:21

These verses tell of the curses God placed upon mankind when Adam and Eve sinned. When we look at the fossil record of mankind, any being which shows signs of being afflicted by these curses must be a son of Adam. This is the surest way to determine theologically which of the fossil men were children of Adam. The three curses, pain in childbirth, sweat, and clothing, are all possible consequences of a single cause. The cause is surprising.

What is the cause of pain in childbirth? The main cause is the size of the baby's head, which distends and dilates the soft parts of the birth canal. Other mammals do not experience such pain during birth because the size of their offspring's heads is small compared to the size of the birth canal. The size of the human birth canal is restricted because of the need for bipedal walking.[1] Walking upright requires narrow hips and a narrow birth canal.[2] Human birth requires a large birth canal. These competing tendencies are balanced by females having wider hips than men, but they can't be too wide or walking would be impossible. No matter what, the head size of infants is just barely able to pass through the birth canal and this causes great pain. This gives the first hint that all the curses are the result of a big head.

The Christian knows that sweat was part of the curse, but even in cursing us, God granted us a blessing. He gave us a set of gifts that have allowed us to inhabit even the most hostile environments. These gifts are a large brain, a mechanism to live in hot environments and the ability to be active during the heat of the day and still maintain the health of our large brain. He gave us clothing with which we can handle cold environments. As I have outlined in a previous book,[3] one can easily incorporate an African postflood habitat for early man. In such an environment, the heat load can be quite high. In a hot environment like the African savanna, walking on all fours exposes more surface area of the body to the sun's radiant heat. Standing upon two legs reduces this exposure, especially at midday when the sun is directly overhead and is at its hottest. If mankind were to inhabit the world and be active at all times of the day or night, the items in the curse were a necessity. The story begins with the demands a large brain makes upon its owner's physiology.

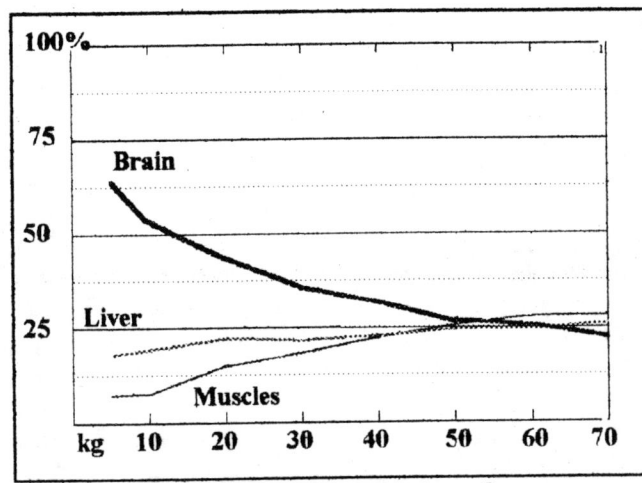

Figure 17 Total Energy Use for Various Organs.

The brain uses a huge percentage of the human metabolism given its relatively small size (Figure 17). Adults use 20% of their metabolic energy powering the brain.[4] Mithen says that the brain requires 22 times the amount of energy that muscles do when at rest.[5] A 50 kg woman will use approximately 30% of her metabolism supporting her brain, 25% supporting her liver and about 20% supporting her muscles.[6] What this means is that large quantities of heat are generated in a small volume inside the skull of a large-brained human. If this heat were not removed, the brain would overheat and we would die. I must be careful here to distinguish between heat and temperature. Heat is energy; temperature is a measure of how much energy an object contains. A high temperature means the object has a high energy content; a low temperature means a low energy content. The reason for this distinction is that an object can generate a large quantity of energy and remain at the same temperature *if* the large quantity of energy is removed as rapidly as it is generated. An everyday example of this would be the humble toilet. One can pour water into the toilet at a prodigious rate but the toilet will not overflow and the water level will remain relatively flat. In fact you can stick a garden hose into it and turn the water on full blast (as long as you hold the hose to keep the rocket-like repulsion from pushing the hose out of it). The reason for this is that at the toilet's base the water flows over into the sewer line. This means that the water you pour in will force an equal amount of water out into the sewer. Similarly, the brain can be a prodigious producer of heat (analogous to the water hose) but the temperature of the brain will not rise, since the body can remove this heat as rapidly as it is formed.

Animals living in a hot environment must develop mechanisms to keep the brain cool. Human brains are delicate. If the brain temperature rises to 106 degrees Fahrenheit, the brain will be damaged. Other animals also have a similar problem but the temperature at which damage occurs is different. Camels are able to tolerate temperature variations as much as 11-14 degrees Fahrenheit without brain damage compared to the human seven degree toleration. Elands, oryxes and some other mammals have a special organ known as the *retia mirabilia* which helps cool the brain. The blood, which has been cooled by the evaporation from the moist nasal passage lining, leaves the nasal passages and enters a region in which the blood flow splits into a fine network of veins. This fine venous network is mingled with another fine network of arteries. The arteries lead to and nourish the brain. The intermingling of the cool blood leaving the nasal passage with the warmer arterial flow has the effect of cooling the arterial blood prior to its entering the brain. This is such an effective cooling system that the brain of a gazelle can be up to 2.9^0 C cooler than the blood leaving the heart.[7]

Men are at a disadvantage; we do not possess a *retia mirabilia* to keep our brain cool. Yet we have the largest brain for our body size which means that it generates a great amount of heat that must be removed. But, lacking the *retia mirabilia* what is the mechanism? Under normal circumstances, when the daytime temperature is less than 85 degrees Fahrenheit, the arterial flow is quite sufficient to remove heat from the human brain.[8] When we must exercise, engage in long periods of muscular activity, or live in an environment where the daily temperatures rise above our body temperature, some mechanism must help remove the excess body heat. From an evolutionary perspective, if the brain grows as large as our brain has, it has to develop a means of cooling itself. Dean Falk, a leading authority on hominid brains, has advocated a widely respected "radiator theory" of brain growth. She says,

> "It was an 'aha' experience, if ever I've had one, and the weirdest combination of events led to it. First, the engine in my 1970 Mercedes needed major surgery. I took it to Walter Anwander (a whiz) in Lafayette, Indiana, who completely rebuilt the engine. One day, while enumerating the wonders beneath the hood (about which I definitely needed schooling), Walter pointed to the radiator and told me 'the engine can only be as big as that can cool.' I didn't think much about it at the time."[9]

The brain, like the engine, can be no bigger than the cooling mechanism's ability to handle the average period of high heat production. Overheating a car engine ruins it. Overheating the brain does the same. In the brain, the blood acts like a coolant. If chasing after game each day caused the body to overheat to such an extent that the brain is thermally cooked, a person would not be able to survive. Similarly, if once a month, the species engages in an activity level that the cooling system is unable to handle, the species will die. The conclusion must be that the cooling system needs to be able to handle the average high stress event during a lifetime. Every summer in Texas several people die because of heat stress. These are people who have placed themselves in a situation in which the body's cooling system is unable to remove the heat rapidly enough.

Humans have developed two unique ways of coping with the heat in high temperature or high exertion environments. First, we sweat profusely. Sweating lowers the temperature of the entire body and keeps the brain cool. While all animals sweat, humans have carried sweating to an extreme. Instead of saying "Sweating like a pig," we should say, "Sweating like a human." P. E. Wheeler, a biologist at Liverpool Polytechnic, says that humans have the "most effective cooling system of any living mammal."[10] Falk believes that this is part of the radiator system which helps keep our very large brains cool. As we will see, the human sweating system is unique among the mammals.

Sweat glands fall into two categories: the apocrine glands and the eccrine glands. The apocrine glands act as scent glands and are used as sexual attractants. The apocrine glands also respond to stress. Arthur Custance relates the effect of the apocrine glands during paratroop drops in World War II. Just prior to a parachute drop, the cabin of the airplane would develop an overpowering, unpleasant odor from the sweat of the nervous troops.[11] The apocrine glands in man are distributed mainly under the arms, around the navel, nipples, ear and in the anal and genital regions.[12] The glands in the armpits of men are more numerous per area than on any other animal. However, even with the greater density of glands under the arms, if apocrine glands were the only glands under the arms, the armpits would be rather dry since they secrete a milky, viscid whitish or reddish fluid that contains only small quantities of water.[13]

The eccrine glands serve two purposes: to moisten gripping surfaces and to remove heat. Apparently, in primates, their initial function was to moisten the volar pads (the friction surfaces) on the hands and feet. This was needed to ensure a good grip when grabbing onto branches, since there is less friction when the palms are dry. On primates the eccrine glands are predominantly found on the hands, feet, the hairless part of the prehensile tail, and the knuckles of chimpanzees and gorillas. They have eccrine glands under their fur but these glands do not secrete as much water as human eccrine glands do. The Encyclopedia Britannica states,

> "Sweat glands on the hairy skin of subhuman primates probably function subliminally or not at all, although they are structurally similar to those of man. The skin of monkeys and apes remains dry even in a hot environment. Profuse thermal sweating in man, then, seems to be a new function."[14]

Campbell believes that the sweating mechanism in man arose as a selective response to the thermal effects of intense muscular activity and the need to keep the large brain cool at those times.

The second, and more recently evolved, function of eccrine glands is thermoregulation. Humans have between two and five million sweat glands covering their bodies and these glands can put out more water than any other animal.[15] The effectiveness of the human sweating system is truly remarkable. The evaporation of a single gram of water removes 580 calories of heat from our body. This is a tremendous amount of heat, yet the human body can deliver 1 liter, 1000 grams of water, to the skin surface in a single hour![16] What this means is that the human is more efficient at removing heat via sweat than even the desert mammals. When it is hot, the average eland loses less than 10% as much water per pound as a human.[17] The oryx, another desert animal, is even more parsimonious with water, sweating about 5% of the water per pound as a man would.[18] A camel sweats about as much as an oryx on a per pound basis. Among all animals on earth, man is the champion sweator.

The second mechanism mankind has developed to maintain a constant brain temperature comes into play only during periods of high activity. These are the reversible emissary veins in the cranium. These veins really call to mind the words God spoke to Adam, "By the sweat of your brow you will eat your food..." (NIV). The brain has several emissary veins that go from the interior of the skull to the skin of the face. These veins are part of the "radiator" system. When a person is cold, blood flows from the cranium outward in these veins. When a person exercises and becomes overheated, the blood flow reverses and blood flows into the cranium. Zenker and Kubik report,

> "Furthermore, Cabanac refers to the high cooling capacity of the skin of the head (particularly rich in sweat glands), whence venous blood, cooled by sweat evaporation, flows via emissary veins towards the brain in hyperthermic situations."[19]

The skin of the face (the brow included) acts as a radiator, cooling the blood which then enters the brain to cool that organ. Adam earned his food by the sweat of his brow!

Dean Falk discovered this system and its implications for the evolution of a large brain. Some of the veins are preserved in the skulls of extinct hominids (and man) in the form of emissary foramina[20] (a foramina is a hole in the skull). Thus, a record of the size and number of emissary foramina are preserved in ancient skulls for anthropologists to examine. The amazing thing was that when Falk plotted the number of emissary veins against the size of the brain, she found that as the brain got bigger there were more emissary veins. Falk notes:

> "It was beautiful. For the past two million years, the increase in frequencies of emissary foramina kept *exact* pace with the sharp increase in brain size in *Homo*. Clearly, the brain and the veins had evolved rapidly and together. I saw that Cabanac's letter was right and that I had unwittingly charted the evolution of a radiator for the brain in my earlier work on emissary

foramina. As Anwander had said about my car, the engine can only be as big as the radiator can cool. Apparently, the same is true for heat-sensitive brains."[21]

In other words, the sweat of the brow, during hard work, helps keep the large brain cool.

There are other implications which a sweat-based cooling system has for man. Our relative hairlessness follows from this cooling system. If it were not for the loss of hair, mankind would have to spend the heat of the day in unproductive sitting.[22] This is due to several factors. First, even dry fur is an excellent insulator which has implications for an animal that wants to be active during the heat of the day. Desert animals like the camel use the fur to keep cool. This sounds oxymoronic, but it is true. When the daily temperature and radiant energy of the sun combine, the temperature at the surface of the camel's fur can be as high as 70-80 degrees centigrade. The insulating effects of the fur allow the skin underneath to remain at around 40 degrees centigrade, a 30-40 degree gradient in a short distance.[23] However, camels, with their fur, like to find shade and remain still when the temperature is very hot. In fact they huddle together to keep cool.[24] Secondly, sweat evaporated directly from the skin removes more internal heat than does sweat evaporated from hair. The hair shaft has poor thermal conduction and so cooling the hair itself does not cool the body. Thirdly, hair reduces airflow between the external atmosphere and the skin by trapping a layer of air beneath the fur. This trapped air quickly evaporates all the water it can and becomes saturated with moisture and no further evaporation can take place.[25] Because of these factors, there would be strong selective pressure on any animal active during the heat of the day. Zihlman and Cohn relate:

> "How might early hominids have dissipated the heat load generated internally, as well as externally from the sun? One way is through the skin. The skin of modern humans contrasts with that of other, nonhuman primates in four features: 1) humans have a great density (over two million) of functioning eccrine sweat glands over the entire body surface; 2) loss of the apocrine sweat glands has been associated with hair loss, and has occurred except in the ano-genital and axillary regions; 3) hair follicles are diffuse and hair shafts are noticeably reduced in size; 4) skin pigment ranges from dark to light.
>
> "How might these features be interpreted in a functional and evolutionary way? There is the remarkable thermo-regulatory function of eccrine sweat glands. Sweating can deliver two litres of water to the skin surface in two hours and carry off almost 600 calories of heat. Hair tends to trap moisture, so that sweat evaporation is more effective with reduced hair. Interestingly, the number of hair follicles in humans is similar to that in chimpanzees and gorillas, but the much reduced size of hair shafts in humans gives a hairless appearance."[26]

Our lack of thick hair enhanced the evaporative and cooling potential of sweating.

But hairlessness causes other problems. Some of these problems make clothing desirable. The first difficulty concerns the effects of ultraviolet light on the naked skin. Ultraviolet light destroys skin cells, as anyone who has suffered a severe sunburn will know. In the tropical sun, this becomes quite a problem. Melanin, the dark skin pigment, absorbs ultraviolet light and protects the skin, meaning that a population will experience a large selective pressure toward dark skin in such an environment.[27] The thermal gradient from the fur surface to the skin, which was mentioned above in relation to the camel, also applies to men. Such a protective feature of hair is probably the reason that mankind has retained hair on the top of their heads. It is easier to stabilize the brain's temperature if you do not allow solar radiation to impinge on the skin itself, thus heating the scalp with the large, delicate brain inside.[28] Not only that, it is a fact that dark hair will perform this function better than light hair because the radiant energy is absorbed at the hair blanket's surface and then re-radiated to the air.[29]

Another problem which hairlessness causes is an inability to inhabit regions of the earth where it is cold. Naked humans do not live long outdoors during the winter in Minnesota, Germany, Russia or northern China. The earliest evidence of habitation in a region which would require clothing is the Dmanisi *Homo erectus*, which was found just east of the Black Sea in Dmanisi, Georgia, and is 1.6 million years old.[30] The extreme of temperature during the winter in the Georgia region is -10 to -20 degrees Fahrenheit.[31] Considering that 1.6 MYR ago was during the middle of the Donau glacial age in Europe, it might well have been colder than today. By 300,000 years ago man had successfully inhabited Diring Yuriakh, Siberia latitude 61 degrees north. Winter temperatures there today reach as low as -50 degrees Fahrenheit. These men needed clothing. The nakedness that resulted from our big brain causes our need for more clothing than a loin cloth.

From this we can see that a large brain brings about pain in childbirth, brings about the need for an efficient system to remove heat (the sweating system), which, in turn, caused the evolution of the emissary veins in the skull which allow the sweat of the brow to cool the brain when we work hard. A large brain also causes the need for hairlessness over most of the body except for the scalp, further requiring in tropical environments the need for dark hair and dark skin. Hairlessness also caused the need for clothing in many environments that man would eventually inhabit. When, in the archaeological record, do we first see evidence for these features? Amazingly it was around 1.6 million years ago.

As we have seen, *Homo erectus/Homo habilis* had spread quite widely by 1.4 million years go. He was found spread from Java at 1.8 Myr,[32] to China at 1.9 Myr[33], to Pakistan at 1.9 Myr[34], to Georgia at 1.6 Myr[35], to Jordan at 1.4 Myr[36] and of course, Africa by this time period. The widespread environmental conditions encountered would require fire and clothing. Fagan writes:

> "For *Homo erectus* to be able to adapt to the more temperate climate of Europe and Asia, it was necessary not only to tame fire but to have both effective shelter and clothing to protect against heat loss. *Homo erectus* probably survived the winters by maintaining permanent fires, and by storing dried meat and other foods for use in the lean months."[37]

I would argue that any being who wears clothing must be a descendant of Adam since clothing was part of the curse. Mankind was also given a unique sweating system and we first find evidence for this sweating system in *Homo erectus* 1.6 million years ago.

On August 22, 1984, Kamoya Kimeu, a member of Richard Leakey and Alan Walker's anthropological search team, walked to a hill across a small stream to take a break from the frustration of the past two weeks of fruitless searching for fossil hominids. The hill was the site where Turkana boys kept their goats. People and goats trampling the hillside would make the finding of a hominid at this site quite unlikely. The hill also was not eroded very severely, so fossils weathering out might disintegrate due to chemical dissolution prior to the time that they were exposed on the surface. As he got to the hill, he saw a matchbook sized pebble lying on the ground. Part of a skull, it belonged to an extinct species of man called *Homo erectus*.[38] Subsequent excavation of this unlikely site produced the most complete skeleton of *Homo erectus* ever found. It was given the official name KNM-WT15000. The Nariokotome boy (Figure 18) died at approximately 11 to 12 years of age. This is known from evidence of which teeth had erupted and which hadn't. From the fossil material it was possible to construct what the boy's skeleton looked like. And from that, it was possible to scale it up to adult size. The importance of this will become evident when we look at the erectus pattern of child birth. Estimates of the ultimate height of the Nariokotome boy, if he had lived, range to more than six foot one inch.[39] This is consistent with what is found among other *Homo erectus*. Studies of them show an average height of 5 feet 7 inches.

The boy presented some very interesting indirect evidence for having a sweating system like ours. People who live in cold environments have thick short bodies since that conserves heat during the cold. However, people who live in the tropics tend to have long thin bodies which enable the body to rid itself of extra heat. The Nariokotome boy, which is what the skeleton was informally named, had a body which clearly showed adaptation to an extremely hot environment. Walker and Shipman write:

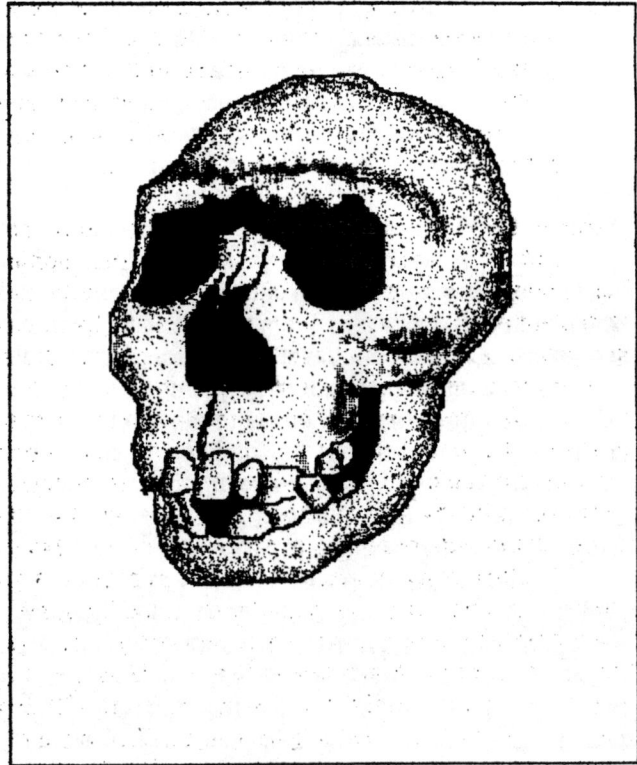

Figure 18 Nariokotome Boy (Homo erectus)

"We found that, compared to living African (tropical climate) and living European (temperate climate) populations, the boy was extraordinary. He was not simply tropical [in shape-GRM]; he was hypertropical, with ratios below the range of ratios recorded for living Africans.

"The antiquity of this bodily to heat stress told us something else too: the boy was probably running around in hot, open country and *sweating*. Although it is indirect evidence, the boy's body build suggests that he, and all *Homo erectuses*, had lost whatever body fur or hair our more ancient ancestors probably possessed. If he had been hairy or furry, then panting (and avoiding activity during the hottest hours of the day) would have served as his main mechanism for heat loss, as it does for the other animals of the African savanna. Because he had no furry protection from the harsh equatorial sun, the boy was probably also very darkly pigmented. Another fascinating feature reflects another aspect of the boy's adaptation to functioning in such a hot and arid environment: he had a nose, a real nose. Bob Franciscus and Erik Trinkaus of the University of New Mexico documented the fact that *erectus* was the earliest species to have a projecting, human-type nose. In contrast, australopithecines and *Homo habilis* had flat, apelike noses, so that (prior to *erectus*) the nostrils leading to the nasal aperture--the opening of the respiratory tract through the nostrils--were sunken into the surface of the face rather than being part of an external nose."[40]

The discovery of the Nariokotome boy sheds much light on the issue of human birth and when that pattern of birth and childhood was first seen in the fossil record. And it sheds light upon when men first began experiencing the curse of pain in childbirth.

Among mammals there are two patterns of birth and childhood. The first pattern is called altriciality. In this pattern the animal is born helpless and extremely immature. The brains of altricial animals are usually half the size of the adult's, and double in size by adulthood. Because of this it takes lots of parental effort to raise the young. Animals following this pattern usually have litters and perform this care for multiple offspring at once. If a mother must be burdened with the care of her offspring, there might as well be many them. Cats, with their blind and helpless kittens, are altricial. The other pattern is precocial. In this pattern the offspring are usually born single and from birth are able to get around quite well. Their brains are nearly adult size at birth. They are alert and all their organs are functioning. Precocial animals are able to locomote in the species pattern at birth and are able to use the species communication system immediately. This latter fact is because the brain of the precocial animal is essentially adult sized. Examples of this birth pattern are the horse, the wildebeest etc., where the young will run with the herds within minutes.

Now, according to Walker and Shipman,[41] altricial species almost never have bigger brains than precocial species. The reason is that for all mammals save one, the brain grows rapidly during gestation but then grows less rapidly after birth. There is a kink in the graph of brain size vs. time which occurs at birth. Altricial species, with their immature state at birth and their subsequent slowdown in the rate of brain growth, forever remain behind the more maturely born precocial species.

What humans seem to have accomplished is the trick of keeping the brain growing at the embryonic rate for one year after birth. Effectively, if humans were a fundamentally precocial species, our gestation period is (or should be) 21 months. It is at about 21 months after conception that the infant begins to walk like their parents and shortly thereafter begins to communicate in the human fashion with language. As we saw above, precocial species are born with an ability to move and communicate. So humans are precocial with an overlying pattern of altriciality. This birth pattern, unique among all mammals, is necessary for our survival. No mother could possibly pass a year old baby's head through the birth canal. And if human females were designed so that a year-old infant's head could pass through the birth canal, they would be unable to walk bipedally. Thus, human babies are born "early" to avoid the death of the mother. Walker and Shipman write:

"Humans are simply born too early in their development, at the time when their heads will still fit through their mothers' birth canals. As babies' brains grow, during this extrauterine year of fetal life, so do their bodies. About the time of the infant's first birthday, the period of fetal brain growth terminates, coinciding with the beginnings of speech and the mastery of erect posture and bipedal walking."[42]

This pattern of growth has huge implications. Every other primate doubles their brain weight from birth to adulthood. Humans, however, triple their brain's birth weight in that time. Our last 12 months of fetal growth rate of the brain occurs outside the sensually deprived womb. The vast quantity of sensory input during the first year of life affects the rate and nature of the neural connections. Because of this year of helplessness, parents must provide close physical and emotional support for the infant. Unlike chimp babies who can cling to their mother's fur, human infants cannot even hang on to mother in spite of having the hand reflex. The mother has no fur because she sweats and she sweats because of a big brain which is why she gives birth to her child early. This early birth then requires the mother to care for the infant and increases the bond between mother and child which partially makes us human. So, what is the birth pattern in Homo erectus? It is human.

Daryl Wilson, one of the reviewers of this manuscript suggested that this pattern of growth may have been designed by God to bring about the nuclear family. He cited two facts in support of this contention: the family unit is theologically a symbol of the God-head and in heaven we are told that families are no more. It is an interesting speculation.

Knowing the adult height, there are anatomical relationships between the average male pelvis size and the average female pelvis size. Using this relation, it is possible to examine the birth pattern of *Homo erectus*. Shipman and Walker point out that the adult Homo erectus cranial capacity was 950 cc.[43] If they followed the apelike pattern of doubling their brain size after birth, they would need to be born with a brain size of around 400 cc. Following the discovery of the Nariokotome skeleton, the approximate size of the erectus birth canal is known. A head with a 400 cc brain is too big to fit through the birth canal. Estimates place the maximum fetal brain size able to fit through the erectus birth canal at just 231 cc.[44] This means that *Homo erectus* tripled his brain size from birth to adulthood. *Homo erectus* had a human pattern of birth and must have endured similar pain in childbirth.

To close, it would appear that there is a single underlying cause of God's curse for the man and woman and it is an increase in brain size. This increase also caused the loss of hair requiring clothing when mankind eventually inhabited northern climes. *Homo erectus* is found in European Georgia 1.6 million years ago. Without fire or clothing, he would have been unlikely to survive the more severe winters in that area.

The fact that Homo erectus was saddled with the problems given to Adam and Eve after the fall has theological implications for the status of Homo erectus, the time during which Adam lived, and the issue of who is eligible for salvation. *Homo erectus* was subject to the three aspects of God's curse. He sweated and had emissary veins. She had pain in childbirth and raised her children as we do. They both wore clothing. I have long contended that humanity in the theological sense is much older than most Christians are willing to admit. If sweat, increased pain in childbirth and clothing are not signifying of humanity and the Fall, what then does theologically separate us from animals?

This view of the curse was criticized by one friend for making the Bible sound like a "just-so" story. A "just-so" story is one that purports to tell the listener the origin of various features of the world. Cassiopea, the queen placed into heaven by an angry god in Greco-Roman mythology is such a story to explain the pattern of stars in the sky. Myths about how the leopard got his spots would be another. The difference between the view advocated here and a "just-so" story is that the ancient Hebrew writer could not have known of emissary veins and their reversal of flow. Nor could he have known of the uniqueness of the human sweating system. Thus he could not have known that this was so.

To turn this criticism around, it is very intriguing to me that the ancient Hebrew writer would choose as a curse for man and woman, two different maledictions which can be caused by a single phenomenon--an increase in brain size. This single cause also would require the loss of hair and the subsequent need for clothing. Since there is no way that the Hebrew writer could have had the scientific knowledge to purposefully construct this tale, is this a fortuitous conjunction of statements or is it divine inspiration?

References

1. Alan Walker and Pat Shipman, <u>The Wisdom of the Bones</u>, (New York: Alfred Knopf, 1996), p. 218
2. Alan Walker and Pat Shipman, <u>The Wisdom of the Bones</u>, (New York: Alfred Knopf, 1996), p. 218
3. Glenn R. Morton, <u>Foundation, Fall and Flood</u>, (Dallas: DMD Publishing, 1995), pp 142-163
4. Donald Johanson and James Shreeve, 1989, Lucy's Child, (New York: William Morrow, 1989), p. 263
5. Steven Mithen, <u>The Prehistory of the Mind</u>, (New York: Thames & Hudson, 1996), p. 11
6. L. C. Aiello, "Human Body Size and Energy," in Steve Jones et al, editors, <u>The Cambridge Encyclopedia of Human Evolution</u>, (New York: Cambridge University Press, 1992), p. 45
7. C. R. Taylor, "The Eland and the Oryx", <u>Scientific American</u>, January, 1969, p.92.

8. Wolfgang Zenker and Stefan Kubik, "Brain Cooling in Human--Anatomical Considerations" <u>Anat. Embryol.</u>, (1996) 193:1-13, p. 1-2

9. Dean Falk, <u>Braindance</u>,(New York: Henry Holt and Co., 1992), p. 156

10. P. E. Wheeler, "The Evolution of Bipedality and Loss of Functional Body Hair in Hominids," <u>Journal of Human Evolution</u>, (1984), 13:91-98, p. 91

11. Arthur C. Custance, <u>The Meaning of Sweat as Part of the Curse</u>, Doorway Papers, 50, (Ottawa: Privately Published, 1962), p. 11

12. Bernard Campbell, <u>Human Evolution</u>, (Chicago: Aldine Publishing, 1974), p. 280-282.

13. "Skin, Human", Encyclopedia Britannica,16, (Chicago: Encyl. Brit., 1982), p. 843

14. "Skin, Human", Encyclopedia Britannica,16, (Chicago: Encyl. Brit., 1982), p. 843

15. Bernard Campbell, <u>Human Evolution</u>, (Chicago: Aldine Publishing, 1974), p. 280-282

16. Adrienne L. Zihlman and B. A. Cohn, "Responses of Hominid Skin to the Savanna," <u>South African Journal of Science</u>, 82:2, (1986), p. 307-308, p. 308

17. The values used in this calculation are from C.R. Taylor, "The Eland and the Oryx",Scientific American January 1969. The Eland uses 5.5 liters of water per 100 kg body weight when hot (p. 94-95). 60 % lost via evaporation. 20% via feces 20 percent via urine. (p. 89). This means that a 500 kg Eland will lose 5.5*5*.6=16.5 liters per day via evaporation. or 3.3 liters per day per 100 kg body weight. Compared to a human who weighs 70 kg,the Eland sweats 2.268 liters/day (a human weighs 70 kg so there are 7.14 humans per Eland.) On a body weight basis humans literally drip water off their body compared to an Eland.

Since the human can sweat 1 liter per hour, in a day a human can sweat 10 times more water in a day than can an Eland.

18. The values used in this calculation are from C.R. Taylor, "The Eland and the Oryx",Scientific American January 1969. The Oryx uses 3 liters of water per 100 kg body weight when hot (p. 95). 62% is lost via evaporation.(p. 89). This means that a 200 kg Oryx will lose 2.00*3*.62=3.72 liters per day via evaporation. or 1.86 liters per day per 100 kg body weight. Compared to a human who weighs 70 kg,the Oryx sweats 1.29 liters/day (a human weighs 70 kg so there are 2.88 humans per Oryx.) Since the human can sweat 1 liter per hour, in a day a human can sweat 18 times more water in a day than can an Oryx.

19. Wolfgang Zenker and Stefan Kubik, "Brain Cooling in Human--Anatomical Considerations" <u>Anat. Embryol.</u>, (1996) 193:1-13, p. 1-2.

20. Dean Falk, <u>Braindance</u>,(New York: Henry Holt and Co., 1992), p. 153

21. Dean Falk, <u>Braindance</u>,(New York: Henry Holt and Co., 1992), p. 159

22. Noel T. Boaz, <u>Quarry</u>, (New York: Free Press, 1993), p. 38

23. Hilde Gauthier-Pilters and Anne Innis Dagg, <u>The Camel</u>, (Chicago: University of Chicago Press, 1981), p. 73

24. Hilde Gauthier-Pilters and Anne Innis Dagg, <u>The Camel</u>, (Chicago: University of Chicago Press, 1981), Figure 25.

25. Adrienne L. Zihlman and B. A. Cohn, "Responses of Hominid Skin to the Savanna," <u>South African Journal of Science</u>, 82:2, (1986), p. 307-308, p. 308

26. Adrienne L. Zihlman and B. A. Cohn, "Responses of Hominid Skin to the Savanna," South African Journal of Science, 82:2, (1986), p. 307-308, p. 308

27. Alan Walker and Pat Shipman, <u>The Wisdom of the Bones</u>, (New York: Alfred Knopf, 1996), p. 196-197

28. A. L. Zihlman and B. A. Cohn, "The Adaptive Response of Human Skin to the Savanna" <u>Human Evolution</u>, 3:5(1988):397-409, p. 404

29. Boyce Rensberger, "Racial Odyssey," in Elvio Angeloni, Editor, <u>Annual Editions Physical Anthropology 94/95</u>,(Sluicedock,Guilford, Conn.: The Dushkin Publishing Group, Inc., 1994), p.40-45, p. 43

30. Roy Larick and Russell L. Ciochon, "The African Emergence and Early Asian Dispersals of the Genus Homo," <u>American Scientist</u>, 84(Nov/Dec, 1996), p.548-550 and Alan Walker and Pat Shipman, <u>The Wisdom of the Bones</u>, (New York: Alfred Knopf, 1996), p. 233

31. Howard J. Critchfield, <u>General Climatology</u>, (Englewood Cliffs: Prentice-Hall, Inc., 1966), p. 210

32. Vadim A. Ranov, Eudald Carbonell, and Xose Pedro Rodriguez, "Kuldara: Earliest Human Occupation in Central Asia in Its Afro-Asian Context," <u>Current Anthropology</u>, 36:2, April 1995, p. 337-346, p. 342

33. Roy Larick and Russell L. Ciochon, "The African Emergence and Early Asian Dispersals of the Genus Homo."<u>American Scientist</u>, 84(Nov/Dec, 1996), p. 540

34. Roy Larick and Russell L. Ciochon, "The African Emergence and Early Asian Dispersals of the Genus Homo."<u>American Scientist</u>, 84(Nov/Dec, 1996), p. 540

35. Alan Walker and Pat Shipman, The Wisdom of the Bones, (New York: Alfred Knopf, 1996), p. 233
36. Roy Larick and Russell L. Ciochon, "The African Emergence and Early Asian Dispersals of the Genus *Homo.*" American Scientist, 84(Nov/Dec, 1996), p: 544
37. Brian M. Fagan,. The Journey From Eden, (London: Thames and Hudson, 1990), p. 76
38. Alan Walker and Pat Shipman, The Wisdom of the Bones, (New York: Alfred Knopf, 1996), p. 7-13.
39. Alan Walker and Pat Shipman, The Wisdom of the Bones, (New York: Alfred Knopf, 1996), p. 194
40. Alan Walker and Pat Shipman, The Wisdom of the Bones, (New York: Alfred Knopf, 1996), p. 196-197.
41. Alan Walker and Pat Shipman, The Wisdom of the Bones, (New York: Alfred Knopf, 1996), p. 220-222
42. Alan Walker and Pat Shipman, The Wisdom of the Bones, (New York: Alfred Knopf, 1996), p. 222
43. P. Shipman and A. Walker, 1989. "The Costs of Becoming a Predator," Journal of Human Evolution, 18, 373-392, p. 388-389
44. Alan Walker and Pat Shipman, The Wisdom of the Bones, (New York: Alfred Knopf, 1996), p. 226-227

THE HARMONY OF EYE AND EAR

"His brother's name was Jubal; he was the father of all who play the harp and flute." Genesis 4:21 (NIV)

"When Laban had gone to shear his sheep, Rachel stole her father's household gods." Genesis 31:19 (NIV)

Genesis 4:21 is the first verse in the Bible that gives any hint of an artistic or esthetic sense among the descendants of Adam. Music is mentioned long before any form of image-making art. Laban's household gods are the first indication that statuary existed in the Bible. Rachel lived somewhere around 1800 BC. Yet in spite of this extremely late mention of representational art, many Christians have decided that if the people in a culture do not engage in art, then they cannot be human. In other words, the lack of art or artistic ability excludes you from spirituality. Hugh Ross, who believes that the Bible limits the creation of Adam to less than 60,000 years ago, speaks of art in the Upper Paleolithic French caves:

> "In the case of the cave drawings and pottery fragments, the degree of abstractness suggests the expression of something more than just intelligence. Certainly no animal species other than human beings has ever exhibited the capacity for such sophisticated expression. However, the dates for these finds are well within the biblically acceptable range for the appearance of Adam and Eve -- somewhere between 10,000 and 60,000 years ago according to Bible scholars who have carefully analyzed the genealogies. Since the oldest art and fabrics date between 25,000 and 30,000 years ago, no contradiction exists between anthropology and Scripture on this issue."[1]

Although, in other writings, Ross has stated that art does not define spiritual humanity, his statement above can be interpreted in no other way. If art defines spiritual man then its existence prior to 60,000 years ago would falsify Ross's views. If art doesn't define spiritual man, then why care that the oldest art is less than 60,000 years old? The last sentence of the above quotation makes no sense whatsoever, if not in the context of art demonstrating spirituality. The existence of breathing prior to 60,000 years ago is no contradiction because breathing does not define mankind. Ross clearly is using the lack of art in a people to exclude spirituality when the Bible does not.

Other Christians have also followed this direction. David Wilcox wrote:

> "The evidence for artistic or religious expression among the Neanderthals is almost nonexistent. There is debate over whether (and for what reasons) they may have occasionally buried their dead, over whether they used ochre as paint, and over their hunting methods... However, there is no evidence of art, no ornaments, no symbolism, no indication of graving tools or sewing, and no clear indication of permanent settlements or trade of raw materials."[2]

Then on the next page, Wilcox re-lists his reasons for rejecting Neanderthal as a spiritual child of God by arguing that the lack of art, record keeping and language excludes him. He goes so far as to say that Neanderthals are a very different species than us, although we looked a lot alike. We have seen (in Chapter 5) that more recent evidence than Wilcox used in his article shows that Neanderthal could speak, but where in the Scripture does it say that if you don't have artistic and record keeping abilities you are not a man? This would seem to be a clear addition by Wilcox to the Scriptural account and, as we shall see, would exclude many people who are clearly spiritual humans.

One final example, Davis and Kenyon, speaking of *Homo erectus*, state,

> "It also left no evidence that it buried its dead, no signs of art, or other recognizably human culture."[3]

This statement is absolutely wrong as we shall see.

This widespread belief among Christians that art is important to be human is one which deserves to be questioned because of the implications it has for evangelism. Would or should we refuse to evangelize a tribe which lacked art? To cite some of the other criteria used above, should we refuse to send a missionary to a tribe that lacked record keeping abilities? Lacked creativity? (One could advance an argument that modern stone-age peoples all lacked creativity or they wouldn't be stone-age people). Would we refuse to tell the gospel to a tribe in the Amazon that did not sew because they had no need of it? Or would we shun a Siberian tribe who even today do not bury their dead in graves? Must we check to see if a tribe has graving tools prior to supporting the Wycliffe translator that wants to give them the Bible? Since the nomadic Kalahari !Kung have no "permanent settlements", are we to withdraw our missionaries?

The entire concept of deciding whom to evangelize based upon these various criteria would be rejected by all of the above authors. Indeed, they would be horrified at the logical consequences of their view. Yet they insist upon using such a criterion to determine which fossil man was and was not a spiritual descendant of Adam. This is the wrong approach. One should not define humanity by what they don't have but by what they do have. If a being displays evidence of a uniquely human activity, they should be considered human. One must never forget that any particular activity, such as sewing or burial, may be culturally or environmentally constrained. A tribe living in a tropical jungle does not need sewing but they display lots of other evidence for humanity such as control of fire, making tools with which to make other tools, and so forth. Eskimos and Siberians, who live on frozen ground, have little inclination to dig into the hard frozen ground to dispose of a relative's body when the relative dies in the winter. Thus, they would not leave graves for the archeologist to find. The criteria that *must* be used to define humanity in our dealings both with primitive tribes and in the fossil record is merely the possession of *any* uniquely human attribute.

There are two reasons why one might not find art associated with either modern or ancient humans. First, there might be a social taboo or custom that prohibits the production of a particular form of art, or, indeed, all art. Secondly, there may be a technological or preservational reason. As we saw earlier, anything that is made out of wood is not likely to be preserved in the fossil record for even 10,000 years. Third, the culture may not have the technological inventions necessary to produce a given type of art. Let's examine the cultural aspect first.

While mankind does have an esthetic sense, the making of any given type of art is clearly not a universal characteristic of humans. Some cultures discourage the production of representational art and only encourage geometrical forms of art, i.e., designs. Moslems discourage the manufacture of representational art, especially in their religious art. The early Moslem theologians believed that the representation of living things was competition with God for God alone is capable of creating something that is alive. This is why almost no mosque (unless it is an old Christian Church) has representational art. There were some early Moslems who thought Christians were idolaters because they made statues and placed them in the churches. The prohibition against representational art, which entered the Moslem world in the 8th century, has not always been followed. The Encyclopedia Britannica states:

> "For practical purposes, representations are not found in religious art, although matters are quite different in secular art."[4]

Similarly, certain Jewish sects have abstained from representational art. The Hasidim (450-200 A.D.) argued for complete abstinence from any form of adornment or art.[5] Even today some orthodox Jews do not like pictures taken of them because of the prohibitions against making graven images. However, this is not limited to Jews. An old friend of mine, a Christian, firmly believes that Christians should not engage in any form of art. This has caused problems between him and a mutual friend, who happens to be an artist. Since my old friend does not engage in art himself, does this mean that he is not a descendant of Adam, as Wilcox above would suggest? The only difference between my friend and the Neanderthal, who produced only small quantities of art, is that we can observe and talk to my friend today; we can't do that with the Neanderthals and so they become an easy target.

Between 1934 and 1937, Meyer and Sonia Fortes studied the Tallensi of Northern Ghana. The Tallensi had only been under Ghanian administrative control for 25 years at the time the Meyers worked with them. The nearest missionary was 10 miles away and the nearest civil servant was 7 miles away. They had only been slightly influenced by the European culture. Fortes reports that the adult culture was "markedly lacking any form of art or decorative product."[6] The nearest thing to adult art was lozenge-like blobs, irregular chevrons drawn on home walls or a crudely drawn crocodile or chameleon sketched in the mud on an ancestor shrine. Other than these things, adult art was unknown. The children would model clay figures of humans but this was apparently not carried on into adulthood. Meyer thought such clay figurines were used not as art but as a three-dimensional diagram "with discriminatory anatomical features shown, rather than a portrait."[7] Fortes cited this as evidence that the Tallensi children had a good grasp of objects in space.

Other than the above, the Tallensi culture was lacking two-dimensional representational art. This is the type of art which so amazes people in the French caves. As Fortes found out, the Tallensi people were totally unable to produce such art, at least initially. The children had never been exposed to drawing on two-dimensional surfaces and the concept was entirely new. On March 18, 1934, Fortes had about a dozen Tallensi boys in his home. He drew a picture of an animal on a piece of paper and showed it to one of the older boys. He then gave the boy a pencil and told him to make lines. The boy covered the page with haphazard squiggles.[8] Fortes distributed the pencils and paper to the other boys who also began by covering the paper with random squiggles. When Fortes suggested that they draw a cow or a donkey, no one responded. One boy showed Fortes what he called a man (Figure 19). Fortes noted that this work appeared to have the quality that is normally ascribed to the pictures drawn by 3 or 4 year olds, though the Tallensi artist was 11 or 12 years old.[9] One item Fortes noted was that all the Tallensi children tried to fill the entire space of the drawing field. Unlike Western-educated children the Tallensi did not see the paper as a framed space upon which to draw, they saw it as an empty space needing to be filled up. As we will see, this is a cultural view of how space should be used and says nothing about the innate intelligence or humanity of the Tallensi.

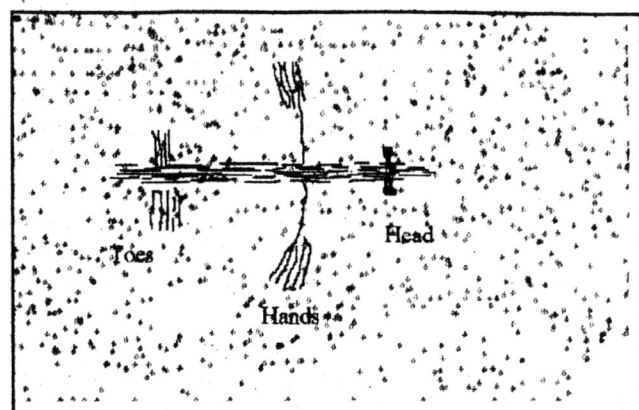

Figure 19 Redrawing of the first Tallensi two-dimensional artwork.

Fortes also had a test that was used by English psychologists to determine the mental condition of people. The test consisted of fifty-four pieces of wood cut into squares, lozenges and triangles each about an inch and a quarter on a side. Each shape had six colors. The object was to place the pieces on the board in any way one wanted. Normal English children used the board as a framed space and began in the center working outward. The patterns they made were connected.

The Tallensi, on the other hand, did not treat the board and shapes in such a fashion. The entire experience was foreign to them; a white man's pointless game that was stranger than the idea of drawing on a piece of paper. When the test was administered to a variety of people, an odd result was seen. Everybody tried to fill the entire space up with the pieces. This was not normally seen in European peoples. The young children laid the various colored shapes down in random patterns. The older children and adults and people made lines of shapes or alternations of colors but no patterns connecting the lines. The Tallensi culture's view of two-dimensional space was entirely different from that of Western culture. Fortes reports:

> "When I showed the Tallensi efforts to Dr. Lowenfeld she said they were exactly like the work of mental defectives! Needless to say none of the Tallensi subjects, least of all adults like Sinkawol, could by any criterion be regarded as mentally defective."[10]

So what is the problem with the Tallensi? There is nothing wrong with them. There is a problem with how Christians, and Westerners in general, have treated art. Art is culturally conditioned, and is not "innate" or part of the image of God, as has often been implied by Christian apologists. The Tallensi were initially unable to deal with two-dimensional art because their culture did not incorporate such art. In the 1970s similar drawing tests were conducted on the children and grandchildren of the initial Tallensi artists. The Tallensi of the 1970s were given a Western education. The progression of their two-dimensional drawings was absolutely identical with any European child. They no longer tried to fill the space, they drew men with body parts drawn in proper proportion, and, amazingly, people were most often drawn facing left, exactly as European children most often do. This data proves that the Tallensi are not genetically disposed to a different artistic use of space. Today they perform like Europeans, as indeed they should after a European education. This data clearly shows that art and the use of space is determined by the culture and the education.

The implications of the Tallensi data to the treatment of art among fossil men are quite clear. The Tallensi are clearly children of Adam, yet they would not have been able to create the art seen on the walls of the French caves. Neanderthals did not create the art on the walls of French caves. Applying the logic often used, if, as David Wilcox says, the Neanderthals are quite a different species from us because they do not engage in what he presumes to be the proper form of art, then one must conclude that the Tallensi are equally a very different species from the rest of humanity. This ridiculous conclusion follows from making the lack of a particular type of art in a culture as evidence of species difference.

This difficulty is also evidenced by the lack of representational art among some *Homo sapiens* cultures found in the archaeological record. The Magdalenians were the peoples who produced the best of the Upper Paleolithic art in the French caves. They lived between 18,000 and 12,000 years B.P. The Azilians succeeded the Magdalenians and were *Homo sapiens*, just like us, who lived between 12,500 and 10,000 years B.P. But they did not engage in the manufacture of much art. The only art objects that have survived to our times from these humans are small pebbles that have been painted with simple geometric figures such as crosses. This is not at all the sophisticated art of which Dr. Ross praises as having an abstractness that suggests "the expression of something more than just intelligence."

The widespread view that art is an evidence of spiritual man and that the lack of art implies the lack of this spirit must also deal with the difficult fact that modern *Homo sapiens* were on earth for at least 80,000 years prior to the "supposed" explosion of art around 40,000 years ago. If the lack of art is evidence of a lack of spirituality, then does this imply that for 80,000 years there were animals walking around in human bodies on the earth? It also raises many other difficult theological questions. If there were humans who were nothing more than non-spiritual animals, are any of their descendants still living on earth? If so, how do we recognize them? Do we send missionaries to them? Were the Tallensi such creatures? Obviously, the implications of this position must be rejected by Christians. Brian Hayden wrote of this view,

> "If the proponents of biological change are to be believed, art should be coterminous with the spread of the sapient gene. Yet, art does not really flourish until the latter half of the Upper Paleolithic when site density (and by implication population and resource extraction potential) most dramatically increases. Moreover, there are many examples of hunter/gatherers elsewhere in the Upper Paleolithic and even in modern times who did not create paintings. For example, the Bushmen have no recollection of rock painting and not all Australia Aborigines created rock art. Are these groups, then, somehow less than fully sapient?"[11]

One could ask of the Christian apologists, "Are these groups, then, somehow less than fully spiritual?"

The second reason that art might not be found among some ancient hominid cultures is that any art they created simply has not survived. Any art that was produced on perishable materials like skin or wood simply would not survive. The Gilyaks of Siberia carve wooden gods that they hang over their huts.[12] Such art will not survive for a long time. In western Australia this century there were aborigines who made extremely crude stone tools.[13] If this was all that we knew of their culture one could possibly conclude that they were less than human. But in point of fact, their wooden tools were extremely elaborate. They painted art on these wooden tools. However, in 10,000 years none of this will exist and these peoples, fully sapient and fully religious, will appear to the archaeologist as backwards. One simply cannot tie humanity to a given technology or technical achievement, be it in art or culture. In a previous chapter we saw the extent of religious art as it went back in time. The earliest possible object of religious representational art is the Berekhat Ram figurine found in the Golan Heights which dates to 300,000 years ago.[14] Art takes more forms than Venus figurines. There is rock art, geometric art, statues, jewelry, esthetic decorations on tools, body-painting and fossil collecting. We will look at how far back the history of each of these art types extends into the past.

Rock art is any art left on the surface of a rock. It can be inside or outside of a cave. If the art is left outside of a cave, where the wind, water, snow etc. is able to attack it, it will not last very long. Because of this, examples of truly ancient rock art found outside of caves are quite rare. This issue is extremely important as we go further back in time, because it provides a reason for the "sudden" appearance of cave art 30-40,000 years ago. The earliest cave art dates from the time man first entered caves to paint and the appearance of an "explosion" is merely due to the fact that rock art in caves is more likely to be preserved than art outside of the cave.[15]

Rock art is also extremely difficult to date. Usually the pigments do not contain charcoal or other materials which can be dated by C^{14}. Thus, they normally must be dated by indirect means. In a cave when a layer covering part of a picture also contains a bone, tooth or ash from a fire, that can be dated, the conclusion is inescapable that the wall art must be older than the age of the layer. Consider the cave shown in Figure 20. The circle art above the sediments is undatable but those circles covered by the 23,000 year old layer must be older than that sediment and the circles covered by and below the 116,000 year old layer must be older than 116,000 years. The only reason that the X art can be dated is because a piece of the cave roof containing an X fell into and was covered by the upper part of the 23,000 year old layer. These are the means by which rock art is dated. Only a couple of direct dating processes for pictures have been advanced. In Australia, the thickness of the rock varnish has been used to date some rock art directly. Also occasionally the art contains some charcoal that can be dated by radiocarbon. With these limitations let us look at rock art.

Almost all of humankind, both technologically primitive and advanced, engage in rock art. The primitive peoples have often painted pictures of animals, gods or the results of shamanistic visions on any available surface. Modern, technologically advanced cultures also engage in this activity. City gangs paint their logos on brick walls, bridges or other inappropriate places. As a boy in rural Oklahoma we had a local rock, known as Devil's Rock, upon which every child would carve his initials or the announcement of his latest love. Once in a while someone would paint on this sandstone rock. The writing of initials and names in wet cement is also an expression of this urge. Rock art is a universal tendency; we just do it differently today.

Going back into the past, American Indians were well renown for their petroglyphs (which is what an archaeologist calls graffiti after it has aged for a while). Such petroglyphs can be found throughout North America, Africa and Eurasia. Dating from the first millennium BC a petroglyph at Tanum Bohuslan, Sweden depicts men in horned helmets engaged in fighting.[16] In the central Sahara, at Tassili n'Ajjer, there is much rock art which shows herds of animals being tended by men. This rock art is believed to have been executed somewhere between 8000 and 3000 BC.[17] Prior to this we come to the rock art of the Upper Paleolithic.

Prior to 10,000 years ago, most of the rock art which is still preserved comes to us from caves. Bahn and Vertut write of these ancient Paleolithic artists,

"Since they spent almost their whole lives in the great outdoors, it has always been assumed that they must also have produced art outside the caves, but that it has not survived the millennia of weathering and erosion."[18]

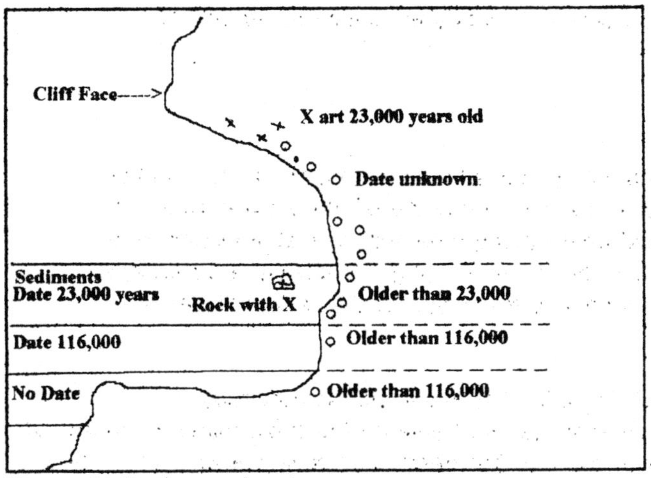

Figure 20 Dating rock art

The marvelous cave art of the Upper Paleolithic, which dates from 32,000 to 10,000 years ago, comes from these caves. There are only a very few sites that have been claimed as Paleolithic open-air art sites. For years Soviet archaeologists claimed that some Siberian open-air figurines were Paleolithic in age. With no means of dating them no one was prepared to believe their claims. The Siberian art which occurs near Chichkino, includes numerous drawings of animals over a 3 kilometer area. The pictures include horses and a wild bovid which are considered characteristic of Paleolithic times. Consistent with the Eurocentric bias of the anthropological art experts, credibility for the Soviet claims had to wait until such things were found in Europe in the outdoors. Such a discovery occurred at Domingo Garcia in Segovia, Spain. A horse figure was hammered into a rock. The only means of dating it is by its artistic style, but it is similar to the nearby horses carved at La Griega, which is of Upper Paleolithic age. Then in Portugal three animals, including a horse 37 centimeters tall and 62 centimeters long were found along the Douro River in a situation that protected the figures from the elements. This animal also was in the La Griega style. Finally, in the eastern Pyrenees, at Fornols-Haut, an ibex and an isard were found carved into an open-air rock 750 meters up the mountain.[19] Suddenly, the Russian claims of open-air Paleolithic art appeared reasonable.

But since all of these upper Paleolithic sites are mostly less than 20,000 years old, the question must be asked, "How much further back in time did mankind engage in open-air art?" If the survival time of such art is maximally 20,000 years, then the lack of such art in times prior to 20,000 years BP is understandable. What does this say about the preservation of rock art which occurred in caves? Many authorities suggest that this destruction of open-air art has a lot to say about why we find cave art when we do. Brian Hayden suggested that the explosion of art in the Upper Paleolithic was *not* due to the invention of art, but due to the decision to use deeply recessed caves as surfaces for the art. In this way art which was made for millennia upon millennia was now able to be preserved long enough for us to see it. He wrote,

> "However, first, I suggest that the feature that is most unusual about the Upper Paleolithic art tradition may *not* be the use of symbolic coloring materials nor even artistic expression, but *the use of deep recesses in caves* for painting (as opposed to surfaces more prone to weathering or decay), together with the *sophisticated* representational nature of parietal art and the creation of *sophisticated* portable art. Thus, rather than trying to explain the emergence of 'art' in the Upper Paleolithic, it may be more realistic to assume that there was a long tradition of art in Dissayanyake's sense, possibly including representational forms, extending back into the Middle or Lower Paleolithic. In this scenario, the phenomenon most in need of explanation is the use of deep caves for art in certain special regions during the Upper Paleolithic (overwhelmingly during the *late* Upper Paleolithic) and the sophisticated elaboration of realistic animal representations by specialists. Deep caves are, of course, ideal contexts for preserving wall paintings, and it is possible that it is above all the use of deep caves for painting beginning in the Upper Paleolithic which makes it *seem* that parietal art appeared suddenly in the Upper Paleolithic."[20]

Robert G. Bednarik agrees,

> "The petroglyphs in Auditorium Cave are the only rock art we have found so far from the Lower Palaeolithic, and empiricists might perceive them as evidence of the earliest rock art. However, taphonomic reasoning renders this most unlikely: if the earliest available evidence of a phenomenon is the most deterioration-resistant manifestation of it (e.g., deeply carved cupules, in the case of rock art), it most likely reflects taphonomic truncation of the record rather than historical reality."[21]

(Taphonomic reasoning concerns what happens to something after it is made and abandoned).

If it is true that cave art was the result of the time that man first entered the deep caves to create art, one invention above all others was required to make their trip into the deep dark cave possible. This item is nothing less than the oil lamp. Caves are dangerous places, and, without a light, a traveler is liable to fall into a hole or off a cliff. Thus, cave art could not be produced until a light was available! Art is not created in a vacuum but requires other inventions. Spirituality is not the sole cause of the supposed 'sudden' appearance of art. Preservation and the invention of the lamp allowing entree to the dark parts of caves was necessary.

An examination of the archaeological record of lamps reveals a very interesting correlation. There is only one Middle Paleolithic oil lamp. The Middle Paleolithic is generally defined as the time prior to 40,000 years. There are 34 lamps found in the earliest Upper Paleolithic sites. This explosion of lamps occurs exactly at the same time that Christian apologists say that art and spirituality are supposed to have exploded.[22] I would contend that the invention of lamps allowed man for the first time to enter dark caves and paint his art there. Art had been in existence for millennia prior but had always been done outdoors and was destroyed by the elements.

Recently, the first cave art ever found in central Europe was discovered. Were the Central European men less artistic than Western Europeans? No. The reason for the lack of central European, Upper Paleolithic cave art lies in the lack of preservation. Bednarik states,

> "So-called 'cave art' was not endemic to caves, it is found in caves not because it was only produced there, but because it only survived there. This is well illustrated by the first Pleistocene rock art ever discovered in central Europe. At the sites Hohler Fels and Geissenklosterle, southern Germany, rock art was recently found on exfoliated cryoclasts, which confirms again that the apparent absence of cave art in central Europe is a function of speleoclimate and speleogenesis, not of cultural factors. It is a *taphonomic* phenomenon."[23]

G.A. Clark goes further and questions whether the entire model of the origin of art in European caves 35,000 years ago is due to the invention of art or is due instead to the destruction of the earliest art. He says,

"Early Upper Paleolithic art, therefore, probably just represents the deepest point in time from which we have a relatively large corpus of surviving examples. Although I have not compiled a data base like Bednarik's, I do monitor this literature, and my construal of pattern in the evidence for symbolic behaviour is in broad agreement with his. I think that CA [Current Anthropology--GRM] readers need to know that the dominant paradigm for the appearance of evidence for symbolism [what I call 'the standard model'- . . .], based on decades of European research, can in fact be called into question on empirical grounds."[24]

Thus, the "explosion of art," which some Christian writers are using to date the creation of Adam, may represent nothing more than the earliest time that art has been preserved. Is there evidence that this is true? Yes. While many will cite the 32,000 year old Chauvet Cave as being the oldest cave art in the world, the painted art at Carpenter's Gap cave in Australia has been dated to 39,000 years old.[25] The rock art of western Europe represents only .03% of the world's rock art yet it seems to be the only art to which Christians pay attention.[26] The idea that cave art began in Europe is most decidedly wrong. Although still controversial, at Jinmium, Australia, rock art has been found which dates 116,000 years old.[27] This is prior to the time that modern man was in Australia and part of the art here may go back as far as 176,000 years ago. The art consists of thousands of cupules, small circles chipped into the rock face. One wall has 3500, 1.5-inch circles carved into it and another has 3200.[28] Considering these numbers, it is quite reasonable to think that there may have been some religious significance to this site. If so, it would be the oldest religious site in the world.

An even older example of art is found at Auditorium Cave in Bhimbetka, India. This art work consists of some large circular holes, chipped into the rock, and a meandering line. As we shall see, there may be religious significance to the meandering line. These were found covered by a layer with Acheulean artifacts in it. Acheulean tools were the tools made by *Homo erectus*. This fact alone makes it likely that the date of this artwork is several hundred thousand years old.[29] Thus if Christians want to talk about the origin of cave art as indicating the time when Adam and spirituality were created, then we need to look several hundred thousand years ago, to *Homo erectus*.

Forms of geometric art are found carved on bones, walls and other objects for thousands of years into the past. The meander or zigzag symbol, formed from parallel meandering lines, is a symbol which has been found quite far back in time. This symbol, which is in the form of a snake, is closely associated with an Australian aboriginal religion. In Arnhem Land, the earliest occupied part of Australia, Europeans discovered the aborigines worshipping the Rainbow Serpent.[30] Rock paintings of this sinuous snake have been found dated to 9000 years ago. Alexander Marshack traced this symbol through the Upper Paleolithic. At Romanelli, Italy, a meander symbol was found on a block of limestone dated to 8500 B.C. Marshack says of the meander symbol that "It is perhaps the most prevalent motif of the Upper Paleolithic, yet it has not until now been systematically studied."[31] He further cites the meander symbol at Grotte de Goyet, Belgium at 12,000 B.C. and in Koonalda Cave, Australia dated at 20,000 years B.C.

Even Neanderthals scratched geometric art onto objects. At Bacho Kiro, 41,000 B.C., a bone fragment was found with the zigzag motif on it.[32] Appealing to our rule that there are no supergeniuses, we can say that earlier Neanderthals would have also been able to engage in this behavior. The only reason they didn't was because the technology had not been invented. An even more fascinating find was that reported recently from Eastern Europe. At Vindija Cave, north of Zagreb, Yugoslavia, a layer called complex G was found to contain the remains of Neanderthals. These Neanderthals were among the last to be found dating from 33,000 years ago. The tools that were found with them were Upper Paleolithic Aurignacian tools, not the typical Mousterian tools.[33] This incredible discovery has garnered very little attention in the Western World. Many anthropologists and Christians have said that Neanderthal was unable to manufacture Aurignacian tools. This obviously is not true. But even more amazing is the fact that geometric art was also found in this fully Upper Paleolithic site. A bear bone with a circle carved on it was also found. This is *Neanderthal art!* Considering that many Christian writers have made a major stand on the "fact" that the Neanderthals did not produce art, and thus must be ejected from the human family, this data shows that these writers are absolutely wrong.

A meandering symbol was pecked into the wall of Auditorium Cave at Bhimbetka, India, and is probably several hundred thousand years old.[34] The earliest dated use of this meandering geometric motif dates from 300,000 years ago and is found on a rib found at Pech de L'Aze.[35] The age of this meander means that it was probably created by *Homo erectus*. We have a non-iconic symbol whose use stretches from modern man back to *Homo erectus*.

The carving of a statue is not something that occurs in a vacuum as most Christian apologists seem to believe. Most seem to think that all that is needed is spiritual awareness and voila, beautiful statues would arise. To carve a statue and produce the wonderful art of the Upper Paleolithic, one must have the proper tools and the proper planning abilities. Carving a statue requires the imagining of the shape and extracting that shape from the unformed rock. These are precisely the skills required to make a beautifully symmetric Acheulean bifacial hand-axe. The shape must be imagined and then the artificer removes the appropriate parts of the rock. In the case of representational statues, fine carving tools must be available or it is impossible to carve an image. The two required technologies are blade manufacture and then the manufacture from a blade of the burin. A burin was a stone tool an example of which is seen in Figure 21. The sharp point makes an excellent engraving tool which is necessary for carving the art work out of bone or soft stone. Anyone who has worked a lathe, as I did in college, knows that the burin is still in use today. It is now made of tempered steel and serves as the cutting tool on a lathe. The sharp triangular point digs into the metal or wood as it turns in the lathe.

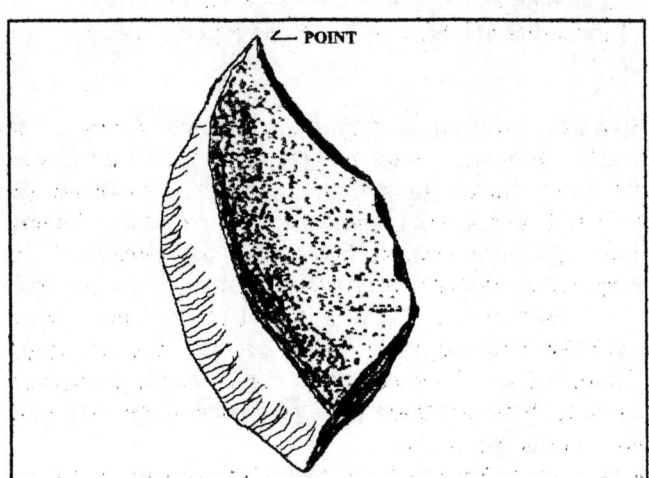

Figure 21 The Burin

The earliest burins I have been able to learn of are found in Chatelperronian sites dated around 36,000 years ago.[36] This of course does not mean that there were none prior to this time. The burin's manufacture depends upon another, earlier technological improvement. This is the blade making technology of the Upper Paleolithic, such as that made by the Magdalenians. A blade is defined as a flake of rock which is more than twice as long as it is wide. As time went by, the Stone Age men were able to make the flakes thinner and thinner. This can be seen in Table 3 which shows how much cutting surface could be extracted from a pound of flint by various paleolithic technologies.[37]

This technological improvement in stone-cutting meant that it was not possible to make a burin with the necessary thinness until the blade technologies were developed. Without a burin, no bone or ivory carving was possible. Without the bone and stone carving, statues could not be made out of nonperishable material. If prior to the invention of the burin, statues were made of perishable material, there would be very little evidence for mankind's artistic nature in the fossil record. This is exactly what we find. Without the proper tools any statue created out of stone would be very crude, like the Berekhat Ram figurine mentioned earlier.

A burin may also have been needed to produce bone tools which Hugh Ross thinks is a sign of spirituality. Speaking of the finding of complex (Upper Paleolithic-like) barbed points found in 80,000 year old deposits Ross said,

Tool Tradition	Approximate Dates	Length of Cutting Surface
Abbevillean	1.0-2.6 MYR	2 inches
Acheulean	.25-1.4 MYR	8 inches
Mousterian	40-250 KYR	40 inches
Magdalenian	11-20 KYR	40 feet

KYR=Thousand years; MYR=Million years

Table 3 Improvements in Stone Cutting

How does this archeological data square with the biblical record of humankind? As I discuss in my book Creation and Time, the design and use of tools is a function of intelligence, perhaps emotions and will, too, but not necessarily of the spiritual dimension of a creature. Therefore tool design and use is something all birds and mammals could exhibit, given adequate intelligence."[38]

In the above quotation, Hugh Ross is responding to the discovery of barbed bone points from ancient (80,000+ year) deposits which look identical to the points made by men Ross would consider as having the Image of God. These bone points would have required a burin to have been produced, and are described as follows:

"Stone Age humans first learned to make tools in Africa more than 80,000 years ago and not, as widely believed, in Europe many thousands of years later, researchers say.
"Alison S. Brooks, a George Washington University Archaeologist, said that bone tools with barbed points and blades dug up from an ancient lakeshore in Zaire predate similar tools found in Europe from some 60,000 years later.

. . .

"The finding shows that early humans in Africa invented sophisticated toolmaking long before their European counterparts. Barbed points like these appeared in Europe only 14,000 years ago."

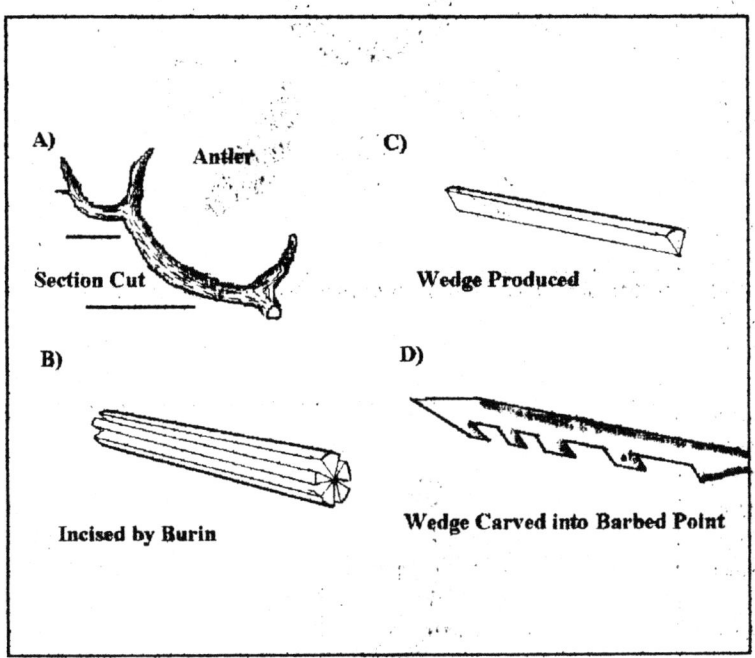

Figure 22 Barbed point manufacture.

The age estimates ranged from just under 80,000 to more than 170,000 years and Dr. Brooks, for convenience, referred to the ages as 'more than 80,000 years.'"[39]

The burin is necessary to make these tools because of the necessity of carving grooves. A section of bone or antler is cut, giving a cylinder (Figure 22 A). Then grooves are cut into the antler lengthwise and the groves are cut to the center (Figure 22 B). Looking at the bone end on, the grooves create wedges. The wedges are broken out by incising the antler to the center. (Figure 22 C). Each wedge of bone then becomes the blank which is further carved for the barbs. (Figure 22 D) Fagan writes:

"Reindeer antler is soft and easy to work when it is fresh from the beast, or collected when recently shed in the wild. The Cro-Magnons were able to rely on regular supplies of fresh antler, but the large beams and tines were of somewhat limited use as tools. They could serve as crude picks or levers for digging up plants, for excavating soil for semi-subterranean dwellings, or could be used for levering out lumps of red ocher. *It was not until the burin and the sharp-edged blade came along that antler as a material for tools came into its own, being ideal for fashioning lightweight spearheads and barbed harpoons for hunting all manner of game, and many other fine artifacts."*

"The blade and burin made possible the so-called 'groove and splinter techniques' that produced the necessary blanks. Armed with a sharp burin or blade, the hunter would work two deep, longitudinal grooves into the beam of a fresh antler, grooving carefully until the burin penetrated the soft, spongy tissue that formed the antler core. The grooves formed a V-shape through the tough outer layer of antler, and the long splinter could be undermined, then levered out of the beam, thereby providing a rough blank that could be cut and shaped very readily into any number of specialized artifacts." [40](emphasis mine)

To make the fine grooves in the bone, a burin is required. To make the burin, blade technology is required. And guess what, the first blade technologies were found in Africa, just prior to the time that these first barbed points are found. Fagan writes:

"The toolkits used by Stone Age Africans after 100,000 years ago display considerable variety, and include occasional collections of much thinner and finer 'blades' of far more standardized size and shape, struck from carefully prepared cores that were fashioned to produce considerable numbers of versatile artifact blanks. Such blades (fine, thin, and often parallel-sided blanks) often occur where local raw materials are of the finest quality, such as the fine quartzites of the Sahara and of coastal southeastern Africa." [41]

Prior to this time, the toolmaking technology was quite different. Before a blade tool can be made, the rock must be carefully prepared by the careful shaping of the rock, called a core. Once the core is properly prepared, blades can be removed with relative ease. Prior to this time, rocks from which flakes were struck were not properly shaped in order to make blades. This blade making technology did not appear in Europe until sometime after 50,000 years ago. The latest Neanderthals were the most eager users of blade technology in Europe.

It might be pointed out that it was not until the invention of the stone blade technologies that sewing needles were able to be fashioned out of bone, although wooden needles might have existed long before. A burin with an extremely fine point is called an awl, and these awls were used to drill the eye in the bone needle.

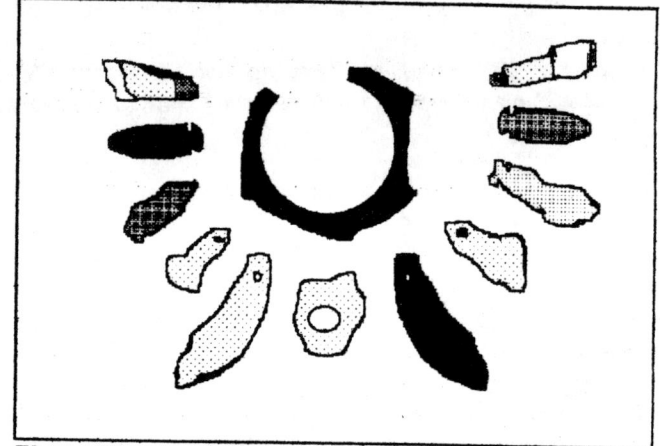

Figure 23 Neanderthal necklace from Arcy-sur-Cure

Thus, what is often interpreted as a major spiritual change point for humanity is probably nothing more than the invention of an extremely important tool—the burin. What Christian apologists have done is to confuse the material effects of the burin's invention as the effect of spirituality. Nothing could be further from the truth.

Unless and until Christian apologists are willing to dig into the smallest details of a science to examine what data actually exist in that science, we will continue to proliferate views, such as Ross has proposed, which can be knocked down by a study of the details. Unfortunately, I fear that Christianity has depended upon the easy answers to the problem science presents. By relying on the easy, non-detailed answer, we succeed in propounding easily destroyed theories, and we succeed in appearing uninformed. This is not meant to be harsh, but Christians should not expect our adversaries to be charitable to our failings, intellectual or otherwise. We should not accept anything less than the best answer we can find.

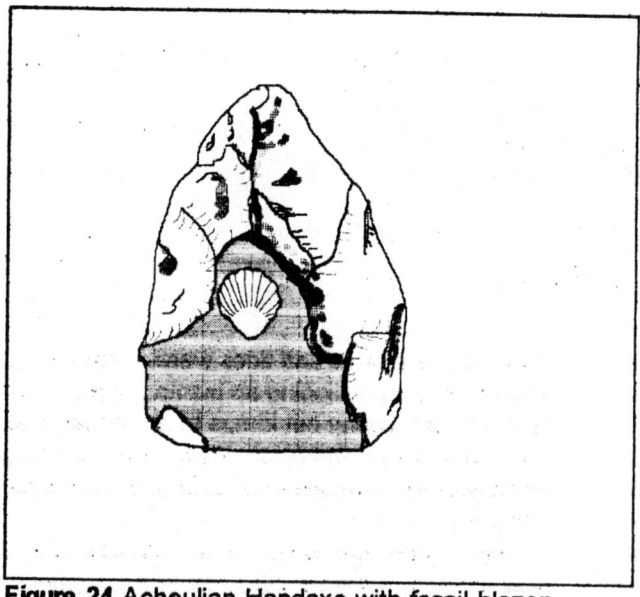

Figure 24 Acheulian Handaxe with fossil blazon. (250,000 years old)

Jewelry is another form of art with which all men are familiar. Jewelry hung from the neck can be identified by a bead or shell having a hole drilled through it so that the string can be threaded through it. Such items are found very far back in time. Bednarik notes,

> "Similarly, the idea that the advent of personal ornamentation coincides with that of the Aurignacian is attributable only to insufficient knowledge of the relevant material. Drilled animal teeth and other objects that are several times as old as that 'transition' (up to 300,000 years) have been known to exist for many decades."[42]

A beautiful Neanderthal necklace, 35,000 years old, was found at Arcy-sur-Cure(Yonne), France (Figure 23).[43] At the Neanderthal site of Bacho Kiro in Bulgaria perforated animal-teeth pendants were found which date to 41,000 years. Fox canines with an abandoned attempt to form a perforation were found at the Neanderthal site of La Quina and a wolf foot bone and a swan vertebra had holes bored through them at Bocksteinschmiede, Germany dating to 110,000 years ago.[44] Jewelry was invented by *Homo erectus*. *Homo sapiens* merely followed the older technology.

Why is there an explosion of art in the Upper Paleolithic, 30,000 years ago? It is due to the relative wealth of Upper Paleolithic peoples. At Sungir, 17,000 years ago, three people, a man and two children, were buried with a total of more than 13,000 ivory beads sewn to their clothing. The man was buried with 2936 beads,[45] the boy had 4903[46] and the girl had 5274.[47] The time spent making these beads was time *not spent obtaining food!* . The people at Sungir had to have sufficient sources of food to allow the expenditure of the time required for the manufacture of such items. Semenov demonstrated that each bead took about 45 minutes, but White thinks an hour is more appropriate.[48] This means that the 13,113 beads required between 9834 and 13,113 hours to make. Put in modern terms, the preparations for this burial required the output of 1639 eight-hour work days, or, since there are 240 work days per year for most Americans, this represents the output of nearly seven years of labor! That society had to be wealthy in order to support this magnitude of labor investment in objects that were to be buried! Art is something one does when one is not worried about starving to death.

There are a few cases of fossil collecting by ancient man. Thirty-two thousand years ago, fossils were pierced, apparently hung on a string and used as personal ornaments at Kostenki 17. There were three fossil coral polyps and four belemnite fossil beads.[49] At Arcy-sur-Cure, (35,000 BP) Neanderthals collected crystals of iron pyrites—fool's gold—and sometimes even engraved them.[50] Iron pyrite forms a gold colored cubic crystal and is collected avidly by some mineral collectors. At Combe Sauniere, Neanderthals carried chunks of iron pyrite 2-3 kilograms in weight from distances of up to 50 kilometers. At La Ferrassie, dating prior to 68,000 years ago,[51] Neanderthal man collected the unusual components of a geological collection: stones, fossil shells and minerals.[52] The Neanderthal geologist collected a Mesozoic fossil shell and coral and a piece of iron pyrite. Considering that a colleague I work with was collecting iron pyrite two weeks prior to my writing this sentence, the collection of minerals is a very human thing to do. Around 250,000 years ago, at West Tofts, England, a human picked up a Cretaceous flint which contained a partially eroded mollusc, *Spondylus spinosus*. The person saw something aesthetic in this piece of flint. He sat down and carefully began knapping the piece always careful not to hit the fossil which remained on the surface of his tool. After a while, the he had fashioned a handaxe with the fossil on the face, like a blazon. This must have been a highly prized object (Figure 24) But one cannot forget the oldest mineral collections in the world, which was collected by *Homo erectus* at Zhoukoudian Cave near Beijing, China. Twenty pieces of quartz crystal were found.[53] This is dated 500,000 years ago. And at Gesher Benot Ya'aqov, Israel, Quartz crystals were also found, dated between 240- 750 thousand years ago, which must have been collected by *H. erectus*.

Body painting has probably been going on from the very beginning. One of the earliest sites, containing the genus *Homo*, yielded a material which when rubbed created a red powder.[54] Bruce Dickson writes:

> "...if these ochre lumps can be taken as evidence of early body ornamentation and the use of cosmetics, they are particularly significant. Lewis Mumford notes the human universality of what he calls 'technical narcissism.' By this he means body and facial decoration, the use of masks, costumes, wigs, tattooing, scarification, and so forth. He suggests that all of these are part of humankind's 'effort to establish a human identity, a human significance, a human purpose. Without that, all other acts and labors would be performed in vain.'"

> "Even more striking is Mumford's assertion that such technical narcissism (of which body and face painting are essential parts) indicates that:

'Primitive man's first attack upon his 'environment' was probably an 'attack' upon his own body; and that his first efforts at magical control were visited upon himself.....'

"The presence of worked ochre in Bed II at Olduvai Gorge suggests that the beginning of this 'attack' may even predate the appearance of *Homo erectus* and begin instead with *Homo habilis* or the australopithecines more than 1.5 million years ago."[55]

The site of Bed I at Olduvai Gorge, 1.6 million years old, has brought forth the oldest manmade piece of art ever found. This is art from long before any *Homo sapiens* walked the earth. Mary Leakey writes:

"In concluding this review of the lithic material from Oldowan and Developed Oldowan Sites the grooved and pecked phonolite cobble found in Upper Bed I at FLK North must be mentioned. This stone has unquestionably been artificially shaped. But it seems unlikely that it could have served as a tool or for any practical purpose. It is conceivable that a parallel exists in the quartzite cobble found at Makapansgat in which natural weathering has simulated the carving of two sets of hominid--or more strictly primate-- features on parts of the surface. The resemblance to primate faces is immediately obvious in this specimen, although it is entirely natural, whereas in the case of the Olduvai stone a great deal of imagination is required in order to see any pattern or significance in the form. With oblique lighting, however, there is a suggestion of an elongate, baboon-like muzzle with faint indications of a mouth and nostrils. By what is probably no more than a coincidence, the pecked groove on the Olduvai stone is reproduced on the Makapansgat specimen by a similar but natural groove and in both specimens the positions of the grooves correspond to what would be the base of the hair line if an anthropomorphic interpretation is considered. This is open to question, but nevertheless the occurrence of such stones at hominid sites in such remote periods is of considerable interest."[56]

One final, very interesting example of an aesthetic sense comes from an amazing 3 million years ago. It is the Makapansgat pebble mentioned by Leakey above. In 1925, W. I. Eitzman found a reddish cobble at an archeological site which eventually became the Limeworks Quarry at Makapansgat, Transvaal, South Africa. The pink breccia which was attached to the pebble was characteristic of the three million-year-old strata called Member 4! This cobble, weighing 260 grams, is made out of banded iron. The nearest outcrop of banded iron formation was three miles from where the pebble was found.[57] What would possess an *Australopithecus* to carry this pebble for three miles back to his rock shelter? The pebble carries two faces on it, faces which appear to be hominid. One face looks human and is the one which initially attracted the attention of Eitzman. The other face, which was not noticed for several decades, appears like what all reconstructions say the australopithecines would have looked like. Raymond Dart says,

"The 'facial proportions' from this new aspect are thus in excellent general agreement with those that reconstructional efforts have caused each modern artist, with minor variations, to produce for *Australopithecus*. This concordance of itself is sufficient justification of the inference that conceptual processes of a similar nature caused an australopithecine to transport the pebble to the cave at Makapansgat. In addition, the curious and to some extent corroborative fact is that once one admits the possibility that an *Australopithecus* had the intellectual ability to detect the presence of a face on this alien natural stone, then the social responses that capacity evoked, follow. The pebble would have had no point without an ability on his associate's part to comprehend and share the emotional reactions, the puzzlement and amusement, that the discoverer had had. And from this it may also be deduced that he and his fellows at the australopithecine phase of human evolution had already reached a humanoid level of self-realization and self-awareness."[58]

Exactly what this means for the humanity of *Australopithecus*, I am not sure. But it does mean that we should not categorically exclude *Australopithecus* from the human family. One must remember that there is very little evidence preserved of the behaviors of these beings and that the anatomical differences between them and us are not great. Ernst Mayr, the great taxonomist, wanted to include *Australopithecus* into the genus *Homo*.

In conclusion, art and art-making have been a part of the human scene for millions of years. For Christian apologists to continually proclaim that there is no evidence of art prior to 40,000 years ago, and that this "sudden explosion" of art is an evidence of the creation of spirituality, ignores a tremendous quantity of archeological evidence. Christians, who should desire truth above all, need to be very careful in what they say about the fossil record.

Music Man

Music has always been extremely important to man. It is so important that all known human societies have some form of music.[59] Bruno Netti states,

> "The most ubiquitous use of music, however, is as part of religious ritual. In some tribal societies, music appears to serve as a special form of communication with supernatural beings, and its prominent use in modern Christian and Jewish services may be a remnant of just such an original purpose."[60]

This is an important fact to remember as we go back through the history of musical instruments. If music is most often used for religious purposes, the fact that a culture made music is probably indicative of religion. The most difficult thing about the study of music in the past is that things like reed pipes, wooden instruments and stretch drums will long since have disintegrated.

Ancient Musical Instruments Made by Modern Man

The history of music and musical instruments goes a long way back into the past. One musical score was found carved on a rock and dated to 800 B.C. It is undecipherable.[61] One earlier musical score has been translated by Anne D. Kilmer, who is head of the Assyriology Department and the University of California, Berkeley.[62] The earliest historical references to music come from China and Mesopotamia. In 2697 B.C. the report says that the Emperor *Huang Ti* sent Ling Lun to make bamboo flutes.[63] (Since *Huang Ti* means Emperor in Mandarin whoever wrote the Encyclopedia Britannica article failed to get the emperor's name.) The Sumerians had stringed instruments, reed instruments and drums.[64]

Evidence of music is found much earlier than that, although the number of instruments found becomes much fewer. The reason for this is the durability of wood and skin artefacts. The only objects which appear from much earlier than the above musical instruments are those made of very durable material, such as bone, although bone is not as durable as many would surmise. Because of the progressive destruction of perishable musical instruments, the bone flute and bone whistles become the major survivors from earlier periods.

From layers dated 13-15,000 years ago, a beautiful, gray eagle bone flute was found. This flute now resides in cabinet number one at the national Museum of Antiquities in France. It is 4½ inches long. One end was intentionally cut by a flint knife; the other end was broken. This middle Magdalenian flute was found at Le Placard Cave. Such worked and decorated bird bones are not uncommon in Upper Paleolithic sites. The unbroken examples can still be played today.[65] This handmade flute is engraved on the outside by two linear sequences of parallel lines, and six sets of nested chevrons. The flute, as a flute, is very simple and could only make one sound. It had no finger holes to alter the pitch. Thus, technically this was a whistle.

The oldest picture of a flute may be from an 18,000 year old French site. Coles and Higgs observe,

> At Les Trois Freres (Ariege), a semi-human figure seems to be playing either a musical bow (although musically this is not in the correct position) or a flute. The association of the semi-humans at this site, with grouped animals, seems to indicate some ceremonial activity, whether it be sympathetic magic or not and music by this time had been in existence for some thousands of years."[66]

Another type of whistle used in ancient times was a reindeer phalange that contained a hole which had been drilled through it.. When blown, it whistles. J. V. S. Megaw observed of these that the phalanges, when bored into on one surface, will emit a high-pitched shrill whistle if one blows between the ends of the articular condyles. Based upon modern examples, these objects are often regarded as decoy whistles and they play like a cross-flute. Such items have been found at Dolni Vestonice and Pekama Cave in Central Europe, and at La Madeleine and Solutre in France. All of these are Upper Paleolithic modern human sites. There are examples of similar whistles made by modern American Indians at archaeological sites in North America.[67]

Megaw's description of the phalanges is accurate, but phalanges are not the earliest evidence of blown instruments, in spite of the constant claim each author makes for having found the earliest item. The claim for the "earliest" is one that is found quite often, and is usually wrong. Megaw catalogues lots of flutes and other musical instruments. The Hungarian cave named Istallosko surrendered two reindeer phalange whistles and a cave bear femur that had three holes bored into it, one hole on the posterior surface and two on the anterior. This bone was found in association with stone tools that are of a type labeled Aurignacian II, and subsequent dating has yielded an age of 31,000 years old for the flute.[68] The largest hole was near the proximal epiphysis and was 11 millimeters in diameter, which is close to the size of a lip-hole on a modern cross-blown flute. Since the epiphysis is positioned in relation to the large hole in such a manner as to prevent the lips from covering the hole, Megaw concluded that this instrument was a forerunner of the modern notch flutes used by primitive tribes.[69]

Many other sites have yielded possible musical instruments. At Lokve another cave bear bone was found with three 'finger holes' drilled into one side. The bird's ulna from Drachen, near Mixnitz, Austria, had three large holes and several smaller ones. While it is doubted by some that this was an instrument, others believe that it was. Another flute from Salzhofen, Austria bears much resemblance to the flute recovered at Istallosko. This bone had three holes bored on one side and two on the other. A reindeer radius was made into an instrument 20,000 years ago at Badegoule. It was associated with Solutrean leaf-shaped stone points. This instrument had a damaged distal end but that end still contained a large hole, with a smaller one on the opposite side. A second small hole was found on the proximal end of the bone.[70]

From the 27,000 year old site of Isturitz, Basses-Pyrenees,[71] one of the flutes found was made from the cubitus of a vulture (Figure 25). The broken end preserves a rectangular hole and below it are two other complete holes.[72] It is decorated with simple notches along the sides. Further excavations in this cave recovered 14 bird-bone flutes.[73] Of the Isturitz find, the original report, written in French, describes it thusly,

> "Enfin, j'ai decouvert en 1921, une piece qui est sans doute unique, un gros os d'Oiseau, malheureusement brise a une de ses extremites, mais qui porte encore sur une seule rangee trois larges trous, comme dans une sorte de flute (pl. VII). C'est, sans doute, le plus ancien instrument de musique connu."[74]

Translation:

> "At last, I discovered in 1921 a piece which is without doubt, unique, a big bird bone, unfortunately broken at one of the ends, but which still carried three holes, like that of some sort of flute. It is without doubt the most ancient musical instrument found." [trans. by David Morton]

Gravettian sites in eastern Europe have also yielded several flutes. Coles and Higgs report,

> "Also in Moravia are the important Gravettian sites of Predmost, Pavlov and Brno. At Pavlov a large number of hut plans have been identified, oval, round and five-sided in shape, with some postholes and hearths. The associated industry included decorated bone and ivory objects including animals and human figures, and a number of phalange whistles; the occupation has been radiocarbon dated to c. 25,000 B.P."[75]

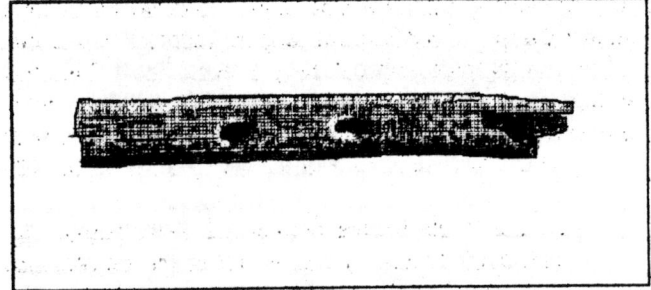

Figure 25 Isturitz Vulture bone Flute

At Dolni Vestonice, Czechoslovakia, flutes are found. This site is approximately 27,000 years old. Coles and Higgs relate,

> "Decorative objects include perforated shells and other pendants, and tubular beads; bone tubes, one with a plug of resin, probably were panpipes."[76]

The oldest flute made by *Homo sapiens* comes from Abri Blanchard about 30,000 years ago.[77] But remember, anatomically modern humans have been on earth for 120,000 years. Were they making flutes earlier than this? Most likely, but they may very well have been made of reeds which would not be preserved. It is highly unlikely that the very first flute was made of bone.

Lithophones were not made but were used by ancient man. These are draperied stalactites which emit musical tones when struck with something hard. Such naturally made but intentionally used musical instruments are found at many sites. There is one each at Nerja, Spain, Roucadour, Pech-Merle, and Les Fieux in France, and Escoural, Portugal. There are three lithophones at Cougnac, France.[78] Since these are natural objects, it is difficult to date when they were used.

Other instruments include bull-roarers, xylophones, drums and rattles. Several oval or elongate objects with a small hole at one end are believed to be "bull-roarers." Objects like these make a loud, low-pitched humming when swung around on a string at a rapid speed.[79] Bahn and Vertut show a picture of a beautiful example which dates to around 15-18,000 years ago. At Mezin, Russia, near Kiev, Soviet archaeologists have cut a record using a mammoth-hip-bone xylophone, shoulder blade drums and jaw-bone rattles they found.[80]

Neanderthal Musical Instruments

Up to this point all musical instruments have been younger than 30,000 years B.P. I wanted to establish above which instruments have been preserved that were made by modern man between 15,000 and 30,000 years ago. There are two kinds of instruments, phalange whistles and flutes. Amazingly, these same instruments are found at Neanderthal sites but in spite of this, statements continue to be made that the oldest flute is 30,000 years old made by modern man. These statements are simply not true. Examples of these kinds of statements are,

Bowers:
> "Music assumed an important role; the first known instrument, a bone flute found in France, dates to around 30,000 years ago."[81]

Ramm:
> "In the fourth and fifth chapters of Genesis we have lists of names, ages of people, towns, agriculture, metallurgy, and music. This implies the ability to write, to count, to build, to farm, to smelt, and to compose. Further, this is done by the immediate descendants of Adam. Civilization does not reveal any evidence of its existence till about 8000 B.C. or, to some 16,000 B.C. We can hardly push it back to 500,000 B.C. It is problematic to interpret Adam as having been created at 200,000 B.C. or earlier, with civilization not coming into existence till say 8000 B.C."[82]

Tattersall:
> "The subject of behavior is complicated by the fact that whereas in Europe Upper Paleolithic stone and bone tools were associated from the beginning with evidence of 'creativity' in the form of engravings, sculpture, notation, musical instruments, and so forth, this was not the case in the Levant. What's more, the earliest Upper Paleolithic tools from Boker Tachtit, while fully Upper Paleolithic in concept, were made using techniques that had been current in the Middle Paleolithic. However, since anatomically modern humans had made Middle Paleolithic tools for the first 50 kyr of their existence, we probably shouldn't find this too surprising."[83]

Other christian writers have uncritically accepted these statements and used them as support for apologetical positions. Hugh Ross goes so far as to equate the making of musical instruments with the singing abilities of birds. This equivocation makes it sound as if the manufacture of musical instruments is not a human activity.[84] I have found it curious that apologists would so quickly grab hold of the Upper Paleolithic European "artistic explosion" as evidence of a recent creation of Adam. As Tattersall claims, the "explosion" was in Europe not the Middle East; and indeed, the advent of art in the Middle East was later than in Europe. Do we really think that Adam was created in SW France? The French probably do, but that is an uncertain apologetic for Christianity. So what are Christians to think when they find out that Neanderthals were composing music and making musical instruments 90-100,000 years ago?

While Bowers (a journalist), Ross (an astronomer) and Ramm (a theologian writing before these discoveries) might be forgiven for not being aware of even more ancient Neanderthal-made musical instruments, Tattersall, an anthropologist, should be familiar with the literature of his profession. (And a Christian apologist should be extremely thorough.) These objects were found as early as in 1955 and reported in the scientific literature in 1967 by McBurney. What this illustrates is what Bednarik, an Australian anthropologist, has called the Eurocentric bias of anthropology, the belief that all things anthropological started in Europe and in the Upper Paleolithic. In point of fact, neither art (as we saw above) nor music began in Europe.

There are many more examples of musical instruments that were made by Neanderthal. The most recent find is also the youngest. It is a flute which is made in the same fashion as the Upper Paleolithic flutes made by modern men noted above. Thus, the tradition of flute making continues unaltered across the Neanderthal/Modern man transition. David Keys writes,

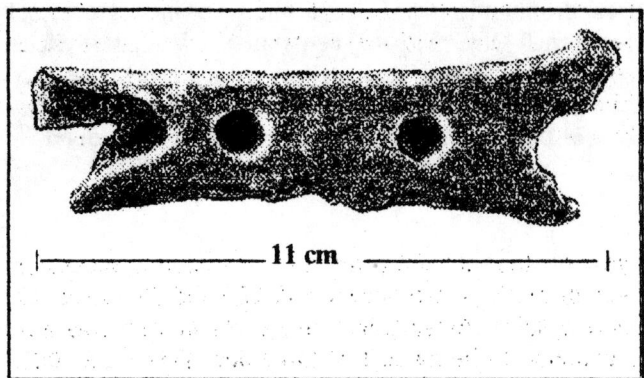

Figure 26 43,000 year-old Neanderthal flute from Slovenia

"Deep inside a cave in Slovenia, in the north of former Yugoslavia, archaeologists have unearthed the world's oldest true musical instrument - a flute which appears to have been made by Neanderthals around 45,000 years ago."[85]

The place where the flute (Figure 26) was found in Divje Babe I in the foothills of the Julian Alps. The Upper Pleistocene sediments at the site are over 12 meters thick and excavations have been underway since 1980.[86]
The flute is broken at each end and has two completely preserved holes in the center and two broken holes at either end. The spacing of the holes is nonlinear and the bone is hollow. It was found with Mousterian tools, the tools which are generally associated with Neanderthal man. Radioactive dating of the layers above and below the flute show that it is between 67,000 and 43,000 years old.[87]

Christian apologists have generally ignored the flute and its implications. One who hasn't is Hugh Ross. Ross claimed that this was not a flute but was a brace for twirling sticks with a bow and starting fires. He wrote:

"Three Slovenian archaeologists who made the discovery addressed, and reasonably dismissed, the idea that the holes might have been bored by the teeth of a large carnivore rather than by a bipedal primate. However, they seem to overlook some more obvious considerations. The bone was found near a hearth with charcoal and many burnt fragments of animal bones. One of these holes goes all the way through the bone and the other does not. These facts suggest at least some likelihood that the bone was an instrument for lighting fires (by twirling a twig in or through one of the holes with a bow). The holes may result from the bone's use as a hammer head or an ax head. Other possibilities abound."[88]

Ross continues by pointing out that the researchers could not be sure that this was a flute because they had not made a model of it.[89] As we shall see, this is not the case, a Canadian musicologist did just that.
There are three things seriously wrong with this approach. First, the burnt bones which are found near the flute are known to be burnt because they are scorched and blackened. If the bone had been used as a brace for a twirled stick, the friction between the stick and the bone would have been intense enough to char the bone. There is none of that on the flute and there should be if it was used with fire. None of the holes "go all the way through the bone" as Ross contends. The back part of the bone is totally missing behind the hole Ross claims is bored all the way through. Ross eventually acknowledged that the hole did not go all the way through and that the back of the flute had been broken.[90] Third, a hole made by hammering the bone would be entirely different

than one intentionally drilled through the bone. No one could strike the bone in exactly the same place with each hammer blow and produce a circular hole. The bone if hammered would fracture in an entirely different way.

Finally, the attempt to avoid the theological consequences of having Neanderthal make flutes has been dealt a serious blow by the researches of Bob Fink, a musicologist from Saskatoon, Canada. He studied the structure of the flute and summarized his findings by writing,

> "Holes 2, 3 & 4 on the bone (as shown, from left to right) stand in a significant relationship to each other: The distance between holes 2 and 3 is virtually twice that between holes 3 and 4. The line-up of the holes indicate that it is a flute.
> This means we are looking at a whole-tone and a half-tone somewhere within a scale. Such a combination of whole-tone and half-tone is the heart and soul of what makes up 7-note diatonic scales. Without making even one more measurement beyond this, we can already conclude: These three notes on the Neanderthal bone flute are inescapably diatonic and will sound like a near-perfect fit within ANY kind of standard diatonic scale, modern or antique. We simply cannot conceive of it being otherwise, unless we deny it is a flute at all."[91]

The significance of Fink's work is that the Neanderthals may have invented the musical scale used by *all* western European musical systems. When you listen to Beethoven or Brahms, their music is possibly based upon the Neanderthal musical scale![92] Is this scale a human universal? No. Other cultures use scales of as few as 2 or 3 notes as do many African cultures. Some cultures have as many as 12 notes in their scale.[93] The Chinese and much of eastern Asia use a 5-note scale for their music. What this does indicate is the antiquity of the European musical scale.

Now if the bone is not a flute, as Hugh Ross suggests, and was used to start fires, the fire-maker must have accidentally spaced the distance between his holes so that they would be spaced just so to appear as if they are holes for a flute. Fink calculated the probability of having the spacing of the holes by pure chance. It was 1 chance out of 680. But his calculation did not calculate the chance for all the holes to align on the same face of the bone. But if you add the chance that the holes would all be put on the same face by chance, the odds of this happening in addition to the proper spacing, the probability can be calculated as one chance out of a billion.[94] This would require that the "firestarter" was an exceedingly lucky fire starter. In reality the whole suggestion is untenable.

Like many claims for being the oldest musical instrument, the Neanderthal flute isn't. Neanderthals made phalange whistles (just like anatomically modern man and North American Indians). One was found at La Quina[95], which dates to 64,000 years ago.[96] However, this is not the end of the Neanderthal musical instruments. They extend much further into the past.

The oldest flute I am aware of is one from Haua Fteah in Libya. It had at least two perforations and thus was much more complex than the first flute I mentioned above, the *Homo sapiens*-made Le Placard Eagle bone flute. McBurney notes,

> "To these may be added a remarkable bone object most plausibly explained as a fragment of a vertical 'flute' or multiple pitch whistle, from spit 1955/64. In this position although directly associated only with a few non-diagnostic chips, splinters and splinters of bone it is none the less attributable to the Pre-Aurignacian owing to the clear indications provided by the overlying spits 1955/61-58, to be discussed in the next chapter. These last show every affinity with the material culture as described and certainly indicate the continued existence of the tradition in the area.
> "In all important respects preserved the bone tube reproduces the features of known paleolithic flutes from the European Gravettian both in the East and West, although older by a factor of at least 2 than any other specimen known."[97]

This object was recovered from the earliest, deepest occupation level at Haua Fteah, Libya. Glynn Isaac describes the dating of the layer in which the flute was found. (Mousterian levels are the Neanderthal layers and these were the layers that yielded two Neanderthal mandibles) He relates,

> "The stratigraphy at this cave site in Cyrenaica appears to span an unusually large segment of Late Pleistocene time and consequently deserves mention apart from its representation in the

frequency distribution patterns of C^{14} dates. About 5 metres of deposits were excavated below 'Mousterian' levels which have been C^{14} dated as follows:

W 85	Layer XXVIII(Mousterian)	$0.034.000 +/- 0.0028 \times 10^6$
GrN 2564	XXVIII(Mousterian)	$0.0434 +/- 0.0013 \times 10^6$
GrN 2022	XXVIII(rest fraction)	$0.04 +/- 0.0015 \times 10^6$
GrN 2023	XXVIII(bone fraction)	$0.47 +/- 0.032 \times 10^6$

"Extrapolation of the sedimentation rate down through the underlying strata gives a reasonable geochronometric estimate of at least 70 to 80,000 years for the base of the excavation. The small artefact sample from the lowest levels represents an idiosyncratic industry which includes fairly numerous blades (McBurney. 1967:91), burins, Acheulian elements (*ibid.*:Fig.IV,7:1,2,6), Mousterian elements (*ibid.*:Fig.IV, 1:7:Fig.IV,5:4:Fig. Iv, 7:3), the oldest known fossil musical instrument (*ibid.*:90: A.IV), and perhaps the oldest shell midden (*ibid.*: 99)."[98]

While 70-80,000 years seems old, Neanderthals were making music 10-20,000 years earlier than this. Prolom II is a Neanderthal site from the Crimea and it was probably the whistle capital of the ancient world. Forty-one phalange whistles made from *Saiga tatarica* were found there.[99] This is an early Wurm site, which means it is 90-100,000 years old.

The evidence that music and musical instruments extend back to at least 100,000 years ago should cause Christians to ponder the ability of our current apologetical schemes to handle the observational evidence. Only man manufactures complex instruments of music. And the earliest Neanderthal flute is more complex than many later examples made by anatomically modern men. Remember the comment made by Netti concerning the use of music as a part of religious activities. Only fallen man engages in religion. Non-spiritual animals do not worship. The concept that music is part of religious ritual is supported by the fact that the earliest known underground mines dating from around 125,000 years ago, were mining pigment which is used by primitive man for body painting. Music and art are found together at least as long ago as 100,000 years ago, and were carried out by both Neanderthals and archaic homo sapiens. It would seem difficult to reject a flute making human-like being from the human race. This data is strong evidence that Neanderthals and archaic *Homo sapiens* were human in a biblical sense of the word.

If true, this implies a change in human morphology from that time until now. A change of morphology *is* evolutionary change. At the very least these facts require that Neanderthal and archaic *Homo sapiens* were spiritual beings. But Neanderthals first appear on earth 230,000 years ago and archaic homo sapiens appear around 500,000 years ago. Even Ramm was unwilling to go that far back for human creation, feeling that that was too much to stretch the genealogies.[100]

These facts also present a tremendous problem for Christian apologetics. For the old-earth creationists, there is the problem of the place of Adam in the human race. Many of these views hold to a recent creation of Adam in the Upper Paleolithic. The problem is that at the same time the Neanderthals were making whistles in Prolom II, anatomically modern men were just leaving Africa. Were there two Adams? A Neanderthal Adam and an anatomically modern Adam? The Bible would support only one, meaning that the one Adam must be considerably prior to this time. For the young-earth creationist the response to these issues is equally bad. Morris says that all fossil men are descendants of Adam and that they lived after the Great Flood,[101] and yet, without these fossil men, he is left with no evidence of a single fossil man that he can point to and say that is a victim of the Great Flood.

But all of this does not take into account the evidence that Homo erectus was a carpenter, a manufacturer of water receptacles, a builder of pavement and huts, a maker of clothing (which is characteristic of fallen man) and a user of ochre for body painting. Some of this evidence goes back as far as 1.7 million years ago. These activities are quite like the activities of any modern primitive group. Any apologetical view which holds to an old earth and a recent creation of Adam ignores the clear evidence for spirituality among men who are morphologically archaic.

In light of the antiquity of music, one of the questions Christians should be asking concerns what musical instruments were they making out of wood? Wooden objects from times of that antiquity are extremely rare. Wood, skin and vegetable matter decay very rapidly leaving no trace in the fossil record. If Neanderthal was capable of making carved bone flutes, then he was certainly capable of carving wooden musical instruments which have not survived. Unless we are willing to believe that the whistles and flutes found in 90,000-year-old strata are the first instruments, the conclusion is inescapable that earlier flutes remain to be found. But since it is easier to make a flute from bamboo or other perishable material, it is quite likely that the first flutes were probably accidentally discovered by blowing through a hollow reed or bamboo in which a slit existed. This technology was probably transferred to bone much later.

The evidence seems clear. We have three choices: We can either ignore the evidence; we can conclude that the Bible is wrong; or we can develop a new apologetic which incorporates these facts. To attempt to say, as some have done, that Neanderthals are animals and that animals are capable of making musical instruments makes a mockery of what we call human. Ross, in order to avoid the consequences of an old humanity, suggests that music "may simply express the soulishness we share with bird and mammal species."[102] He then says that "bipedal primates might simply have imitated birds in producing various musical tones."[103] This, of course, is an entirely inappropriate response since *no* bird or other animal makes musical *instruments*. Ross has chosen to believe that mankind can be no older than 60,000 years and so he must do to any contradictory data what the young-earth creationists do. Deny it exists.

As we are seeing in our survey, evidence for human activity goes way back in the fossil record. Most demonstrably human behavior extends back at least a million years. Christians must incorporate such data into their apologetics. What I have been suggesting[104] (that the creation of man was several million years ago) is within the framework that Ramm says is an acceptable harmonization. Ramm writes:

> "The Bible itself offers no dates for the creation of man. We mean by this that there is no such statement in the text of the Bible at any place. We may feel that 4000 B.C. or 15,000 B.C. is more consonant with the Bible than a date of 500,000 B.C. But we must admit that any date of the antiquity of man is an inference from Scripture, not a plain declaration of Scripture.
>
> "If the anthropologists are generally correct in their dating of man (and we believe they are), and if the Bible contains no specific data as to the origin of man, we are then free to try to work out a theory of the relationship between the two, respecting both the inspiration of Scripture and the facts of science."[105]

References

1. Hugh Ross, "Art and Fabric Shed New Light on Human History," Facts & Faith, 9:3(1995)p. 2
2. David L. Wilcox, "Adam, Where Are You? Changing Paradigms in Paleoanthropology," Perspectives on Science and Christian Faith , 48:2(June 1996), p. 92
3. Percival Davis and Dean H. Kenyon, Of Panda's and People, 2nd Edition, (Richardson, Texas: Foundation for Thought and Ethics, 1993), p. 110
4. "Islamic Peoples, Arts of," Encyclopedia Britannica, 9, (Chicago: Encyclopedia Britannica, 1982), p. 984
5. "Jewish Peoples, Arts of",Encyclopedia Britannica, 10, (Chicago: Encyclopedia Britannica, 1982), p. 202
6. Meyer Fortes, "Tallensi Children's Drawings," in Barbara Lloyd and John Gay, eds. Universals of Human Thought (Cambridge: Cambridge University Press, 1981),pp. 46-70, p. 46-47
7. Meyer Fortes, "Tallensi Children's Drawings," in Barbara Lloyd and John Gay, eds. Universals of Human Thought (Cambridge: Cambridge University Press, 1981),pp. 46-70, p. 47
8. One can see this drawing in Meyer Fortes, "Tallensi Children's Drawings," in Barbara Lloyd and John Gay, eds. Universals of Human Thought (Cambridge: Cambridge University Press, 1981),pp. 46-70, p. 49
9. Meyer Fortes, "Tallensi Children's Drawings," in Barbara Lloyd and John Gay, eds. Universals of Human Thought (Cambridge: Cambridge University Press, 1981),pp. 46-70, p. 50-51
10. Meyer Fortes, "Tallensi Children's Drawings," in Barbara Lloyd and John Gay, eds. Universals of Human Thought (Cambridge: Cambridge University Press, 1981),pp. 46-70, p. 68
11. Brian Hayden "The Cultural Capacities of Neandertals," Journal of Human Evolution 1993, 24:113-146, p. 138
12. Ivar Lissner, Man, God and Magic, translated by J. Maxwell Brownjohn, (New York: G. P. Putnam's Sons, 1961), Figures 91-93.
13. James R. Shreeve, The Neandertal Enigma, (New York: William Morrow and Co., 1995), p. 249
14. Alexander Marshack in an extensive study of this object re-confirms that it is made by human hands. Alexander Marshack, "The Berekhat Ram Figurine: A Late Acheulian Carving from the Middle East," Antiquity, 71(1997):327-337.

15. Paul G. Bahn and Jean Vertut, Images in the Ice, (Leichester: Windward, 1988), p. 113
16. Anthony Harding, "Bronze Age Chiefdoms and the End of Stone Age Europe," in Goran Burenhult, editor, People of the Stone Age, (San Francisco: HarperSanFrancisco, 1993), p. 119.
17. Richard G. Klein, "Hunter-Gatherers and Farmers in Africa," in Goran Burenhult, editor, People of the Stone Age, (San Francisco: HarperSanFrancisco, 1993), p. 55
18. Paul G. Bahn and Jean Vertut, Images in the Ice, (Leichester: Windward, 1988), p. 113
19. Paul G. Bahn and Jean Vertut, Images in the Ice, (Leichester: Windward, 1988), p. 110-113
20. Brian Hayden "The Cultural Capacities of Neandertals," Journal of Human Evolution, 24(1993):113-146, p. 125
21. Robert G. Bednarik, "Concept-mediated Marking in the Lower Palaeolithic," Current Anthropology, 36:4(1995), pp. 605-634, p. 611
22. Sophie A. De Beaune, "Palaeolithic Lamps and their Specialization: A Hypothesis," Current Anthropology, 28(Oct. 1987):4: 569-577, p. 570
23. Robert G. Bednarik,"Who're We Gonna Call? The Bias Busters!" in Michel Lorblanchet and Paul G. Bahn, Rock Art Studies: The Post-Stylistic Era or Where do We Go from Here? (Oxbow Monograph 35, 1993), pp.207-211, p. 210
24. G.A. Clark, "Comments" Current Anthropology, 36:4(1995), pp. 605-634, p. 617
25. Paul G. Bahn, "Further Back Down Under," Nature, Oct 17, 1996, p. 577-578, p. 578
26. Robert G. Bednarik, "Comments", Current Anthropology, 29:2 (April 1988), p. 218
27. Paul G. Bahn, "Further Back Down Under," Nature, Oct 17, 1996, p. 577; R. L. K. Fullagar et al., "Early Human Occupation of Northern Australia: Archaeology and Thermoluminescence Dating of Jinmium Rock-Shelter, Northern Territory," Antiquity,70(1996):751-773.
28. "Engravings could be World's Oldest Artwork" Dallas Morning News, Saturday, September 21, 1996, p. 14A
29. Robert G. Bednarik, "Concept-mediated Marking in the Lower Palaeolithic," Current Anthropology, 36:4(1995), pp. 605-634, p. 611; and Paul G. Bahn, "Further Back Down Under," Nature, Oct 17, 1996, p. 577
30. Christopher Wills, The Runaway Brain, (New York: Harper Collins, 1993), p.147
31. Alexander Marshack, "Some Implications of the Paleolithic Symbolic Evidence for the Origin of Language," Current Anthropology, 17(1976):2:274-282, p. 278
32. Paul G. Bahn and Jean Vertut, Images in the Ice, (Leichester: Windward, 1988), p. 72, 80
33. Robert G. Bednarik, "Aurignacian Neanderthals?" The Artefact, 1995, 18:92
34. Paul G. Bahn, "Further Back Down Under," Nature, Oct 17, 1996, p. 577-578, p. 578;Robert G. Bednarik, "Concept-mediated Marking in the Lower Palaeolithic," Current Anthropology, 36:4(1995), pp. 605-634, p. 611
35. Alexander Marshack, "Some Implications of the Paleolithic Symbolic Evidence for the Origin of Language," Current Anthropology, 17(1976):2:274-282, p. 278
36. Brian Fagan, The Journey From Eden, (London: Thames and Hudson, 1990), p. 148
37. Chris Stringer and Clive Gamble, In Search of the Neanderthals, (New York: Thames and Hudson, 1993), p. 55.
38. Hugh Ross, "Art and Fabric Shed New Light on Human History," Facts & Faith, p.2
39. Associated Press, Dallas Morning News, May 1, 1995, p. 9d
40. Brian Fagan, The Journey From Eden, (London: Thames and Hudson, 1990), p. 156.
41. Brian Fagan, The Journey From Eden, (London: Thames and Hudson, 1990), p. 46.
42. Robert G. Bednarik, "Concept-mediated Marking in the Lower Palaeolithic," Current Anthropology, 36:4(1995), pp. 605-634, p. 606
43. see the cover of Nature May 16, 1996; for the date see Paul G. Bahn and Jean Vertut, Images in the Ice, (Leichester: Windward, 1988), p. 72
44. Paul G. Bahn and Jean Vertut, Images in the Ice, (Leichester: Windward, 1988), p. 72
45. Randall White, "Technological and Social Dimensions of 'Aurignacian-Age' Body Ornaments across Europe," in H. Knecht et al, editors, Before Lascaux, (Boca Raton: CRC Press, 1993), p. 286
46. Randall White, "Technological and Social Dimensions of 'Aurignacian-Age' Body Ornaments across Europe," in H. Knecht et al, editors, Before Lascaux, (Boca Raton: CRC Press, 1993), p. 291
47. Randall White, "Technological and Social Dimensions of 'Aurignacian-Age' Body Ornaments across Europe," in H. Knecht et al, editors, Before Lascaux, (Boca Raton: CRC Press, 1993), p. 292
48. Randall White, "Technological and Social Dimensions of 'Aurignacian-Age' Body Ornaments across Europe," in H. Knecht et al, editors, Before Lascaux, (Boca Raton: CRC Press, 1993), p. 282, 294

49. R. White, 1993 "Technological and Social Dimensions of 'Aurignacian-age' Body Ornaments across Europe," in H. Knecht et al ed. Before Lascaux: the Complex Record of the Early Upper Palaeolithic pp 247-299, p. 286-287
50. Brian Hayden "The Cultural Capacities of Neandertals ", Journal of Human Evolution, 24(1993):113-146, p. 123-124
51. Yuri Smirnov "Intentional Human Burial: Middle Paleolithic (Last Glaciation) Beginnings," Journal of World Prehistory, 3:2(1989), pp 199-233, p. 219
52. Andre Leroi Gourhan, The Hunters of Prehistory, transl. Claire Jacobson, (New York: Atheneum, 1989), p. 92-93
53. W.C. Pei, "Notice of the Discovery of Quartz and Other Stone artifacts in the Lower Pleistocene Hominid-Bearing Sediments of the Choukoutien Cave Deposit," Bulletin of the Geological Society of China, 11:2:1931:109-146, p.120
54. D. Bruce Dickson, The Dawn of Belief, (Tuscon: The University of Arizona Press, 1990), p. 42-43
55. D. Bruce Dickson, The Dawn of Belief, (Tuscon: The University of Arizona Press, 1990), p. 44
56. M.D. Leakey, Olduvai Gorge 3, Excavations in Beds I and II, 1960-1693, (Cambridge: Cambridge University Press, 1971), p. 269
57. K. P. Oakley, "Emergence of Higher Thought 3.0-0.2 Ma B.P.", Phil. Trans. R. Soc. Lond. B, 292, 205-211 (1981), p. 205-206
58. R.A. Dart, "The Waterworn Australopithecine Pebble of Many Faces from Makapansgat," South African Journal of Science, 70(June 1974), pp 167-169, p. 168
59. Bruno Netti, "Music" 1994 Microsoft Encarta.
60. Bruno Netti, "Music" 1994 Microsoft Encarta.
61. "Music, Western" Encyclopedia Britannica, 12, (Chicago: Encyclopedia Britannica, 1982), p 704
62. http://www.webster.sk.ca/greenwich/fl-compl.htm
63. "Music, East Asian", Encyclopedia Britannica, 12 (Chicago: Encyclopedia Britannica, 1982), p 671
64. "Music, Western" Encyclopedia Britannica, 12, (Chicago: Encyclopedia Britannica, 1982), p 704
65. Alexander Marshack, The Roots of Civilization, (New York: McGraw-Hill, 1972), p. 147.
66. J.M. Coles and E. S. Higgs, The Archaeology of Early Man, (New York: Frederick A. Praeger, 1969), p. 226-227
67. J.V.S. Megaw, "Penny Whistles and Prehistory," Antiquity XXXIV, 1960, pp 6-13, p. 6-7
68. J.V.S. Megaw, "Penny Whistles and Prehistory," Antiquity XXXIV, 1960, pp 6-13, p. 6-7
69. J.V.S. Megaw, "Penny Whistles and Prehistory," Antiquity XXXIV, 1960, pp 6-13, p. 7-8
70. J.V.S. Megaw, "Penny Whistles and Prehistory," Antiquity XXXIV, 1960, pp 6-13, p. 7-8
71. Alexander Marshack, The Roots of Civilization, (New York: McGraw-Hill, 1972), p. 96-97.
72. J.V.S. Megaw, "Penny Whistles and Prehistory," Antiquity XXXIV, 1960, pp 6-13, p. 7-8
73. Paul G. Bahn and Jean Vertut, Images in the Ice, (Leichester: Windward, 1988), p. 68-69
74. E. Passemard, 1944, "La Caverne d'Isturitz en Pays Basque," Prehisoire 9:1-84, p. 24.
75. J.M. Coles and E. S. Higgs, The Archaeology of Early Man, (New York: Frederick A. Praeger, 1969), p. 298
76. J.M. Coles and E. S. Higgs, The Archaeology of Early Man, (New York: Frederick A. Praeger, 1969), p. 298
77. Goran Burenhult, editor,American Museum of Natural History The First Humans, (San Francisco: Harper,1993), p. 103 and Richard Leakey and Roger Lewin, Origins Reconsidered, (New York: Doubleday, 1992), p. 322
78. Lya Dams, "Palaeolithic Lithophones: Descriptions and Comparisons," Oxford Journal of Archaeology, 4(1) 1985, p. 31-46
79. Paul G. Bahn and Jean Vertut, Images of the Ice Age, (Leicester, England, Windward, 1988), p. 69
80. Paul G. Bahn and Jean Vertut, Images of the Ice Age, (Leicester, England, Windward, 1988), p. 69
81. Bruce Bower, "When the Human Spirit Soared," Science News, 130, Dec. 13, 1986, p. 378
82. Bernard Ramm, The Christian View of Science and Scripture, (Grand Rapids: Eerdmans Publishing Co., 1954), p. 228
83. Ian Tattersall, The Fossil Trail (New York: Oxford University Press, 1995), p.225
84. Hugh Ross, "The Meaning of Art and Music," Facts & Faith, 10(1996):4, p. 11. See the sudden switch from discussing music to the discussion of singing birds. By doing this he evades the implication that making musical instruments are clearly only accomplished by humans.
85. David Keys, Archaeology Correspondent, Independent Sunday 2/25/96, p. 15 Manchester England.
86. A report on internet has been posted at http://www.zrc-sazu.si/www/iza/piscal.html

See also, Ivan Turk, Janez Dirjec and Boris Kavur, "Ali so v Sloveniji Nasli Najstarejse glasbilo v Evropi?" Razprave IV, razreda SAZU, XXXVI(1995, pp 288-293

87. See the internet report at http://www.apnet.com/inscight/04031997/grapha.htm
88. Hugh Ross, "The Meaning of Music and Art", Facts & Faith, 10:4,4th qtr, 1996, p. 11.
89. Hugh Ross, "The Meaning of Music and Art", Facts & Faith, 10:4,4th qtr, 1996, p. 11.
90. Hugh Ross, "Response To Glenn Morton's Critique," Facts & Faith, 11(1997):1, p. 6-7
91. A copy of this report can be ordered from
(including postage & hndlg) for $5.00 from Greenwich 516 Ave K South, Saskatoon, SK, Canada S7M 2E2
The report can be found on the internet at http://www.webster.sk.ca/greenwich/fl-compl.htm
92. See the internet report at http://www.apnet.com/inscight/04031997/grapha.htm and http://www.webster.sk.ca/greenwich/fl-compl.htm. See also Kate Wong, "Neanderthal Notes," Scientific American, Sept. 1997, pp28-30. One other possible musical scale which would fit the flute is the South Indian musical system.
93. Copyright - 1992 Grolier Electronic Publishing, Inc."Music" The Software Toolworks Multimedia Encyclopedia.
94. Fink calculates that the odds of the spacing is 1/680 to keep the tuning. But if the bone is divided into 10 degree units any hole could be found in any of the 10 degree segments, then each hole has a 1/36 chance of being in a given segment. with 4 holes the probability of the alignment becomes $1/36^4$= 1.6 million. Multiplied by the spacing probability the odds become 1 out of a billion.
95. Paul Mellars, The Neanderthal Legacy, (Princeton: University Press, 1996), p. 373
96. Paul Mellars, The Neanderthal Legacy, (Princeton: University Press, 1996), p. 404
97. C.B.M. McBurney, Haua Fteah (Cyrenaica),(Cambridge: Cambridge University Press, 1967), p. 90
98. Glynn Isaac,"Chronology and the Tempo of Cultural Change during the Pleistocene." in Calibration of Hominid Evolution, ed. W.W. Biship and J. Miller, Edinburgh: Scottish Academic press (1972), p. 381-430 reprinted in Barbara Isaac, editor, The Archaeology of Human Origins, (Cambridge: Cambridge University Press, 1989), p. 71
99. Vadim N. Stpanchuk, "Prolom II, A Middle Palaeolithic Cave Site in the Eastern Crimea with Non-Utilitarian Bone Artefacts," Proceedings of the Prehistoric Society 59(1993), pp 17-37, p. 33-34.
100. Bernard Ramm, The Christian View of Science and Scripture, (Grand Rapids: Eerdmans Publishing Co., 1954), p. 228
101. Henry M. Morris, The Remarkable Birth of Planet Earth, (Minneapolis: Dimension, 1972), p. 46-47
102. Hugh Ross, "The Meaning of Music and Art", Facts & Faith, 10:4,4th qtr, 1996, p. 11.
103. Hugh Ross, "The Meaning of Music and Art", Facts & Faith, 10:4,4th qtr, 1996, p. 11.
104. Glenn R. Morton, Foundation, Fall and Flood, (Dallas: DMD Publishing Co., 1995), (16075 Longvista Dr., Dallas 75248)
105. Bernard Ramm, The Christian View of Science and Scripture, (Grand Rapids: Eerdmans Publishing Co., 1954), p. 220

THE PHILOSOPHER'S STONE: THE TECHNOLOGY OF FOSSIL MAN

Alchemists of the Middle Ages searched for the philosopher's stone. It was a substance supposed to possess the ability transform lead into gold, cure ills and prolong life. Various alchemists described it differently. Some felt that it was a substance or trait found everywhere but unrecognized and unappreciated. If one could identify this ubiquitous material one could produce all these beneficial effects. The alchemists were correct; there was a philosopher's stone. It is human technology. While the alchemists did not live to see it, in the middle of the 20th century human technology, by means of carefully controlled nuclear reactions lead was literally turned into gold. Unfortunately, the cost was vastly higher than it could be obtained by means of mining and refining. By means of technology illness has been cured and life prolonged. These amazing fulfillments of the alchemist's dream were ultimately based upon the development of a primitive technology by ancient man. All of our modern technology descends from the inventions of men who lived long before civilization.

In looking at the technology of ancient man we are going to set up one rule. There are no supergeniuses in a population. If one individual in a population is capable of performing a certain feat, then it must be assumed that most of the population are capable of that feat. This can be illustrated by Albert Einstein. He used Riemannian geometry to describe the way light moves in the universe and to reformulate the laws of physics. He is correctly honored for having the ingenuity to put relativity together, but today, thousands of physicists are capable of doing relativistic calculations. There is no reason to doubt that Claudius Ptolemy, the ancient astronomer, would have been capable of performing relativistic calculations had he been given the proper education. Great-grandchildren of plains Indians, who lived a stone-age lifestyle, are not only quite capable of understanding relativity, many of them work as physicists performing such calculations. Today, physicists come from all countries, China, Japan, Uganda, etc. An invention, like relativity, is just that, an invention, not a speciation event. Because of this clear lesson of modern times, the conclusion is inescapable--if one member of a species could perform a particular task, then many were capable of that task--given the proper education. This lesson must be applied to the fossil record of man. We can no more believe that the invention of the sewing needle was somehow correlated with the evolution of a new level of intelligence than we can believe the same of the invention of the television.

One other issue must be addressed. How do you know that I am human and not merely a robot? I can construct English sentences and verbalize them. I can respond to your questions in ways that make you believe that I am like you. I have a wife and children. I go to work, have hobbies and in all respects behave as you do. From this you conclude that I am human. If an individual lacks the behavior pattern characteristic of humans, then we become nervous around him. Schizophrenics make us uncomfortable because they look like us but do not act like us. This discomfort extends to other situations.

When my oldest son was born, we were thrilled to allow other people to hold him. However, my wife has a seriously retarded uncle who has the intelligence of a 3-5 year old. He is the most loving of individuals, but he does not act quite like most adult men. Because of this, and my lack of experience with him, I was very hesitant to allow him to hold the baby. My wife had no such concerns, but then she had known her uncle since she was a child. Her uncle did fine with my son and he didn't notice the nervous father standing there ready to grab the baby back at the first hint of danger. I relate this because it illustrates my point that I had no difficulty allowing anyone who behaved as a normal adult to hold my son. Why did I have difficulty with her uncle? Because, ashamedly now, but at that time, I judged him as something other than human. He didn't behave like most humans. The point of this is that we judge modern people as human because they behave like humans. We should extend that criteria to the various forms of fossil man. If they behave humanly, then we should consider them *human!* With this as a preface, let us examine the technology of the ancients and see what it says about their abilities.

No Shortage of Housing.

"Adah gave birth to Jabal; he was the father of those who live in tents and raise livestock." Genesis 4:20 (NIV)

As the Scripture says, Jabal lived in tents. We know from this that tents were an invention of spiritual man. Modern men live in several characteristic forms of housing. Those who live in the developed world live in apartments or houses. Those in the less developed parts of the world live in tents made of animal skin, wood and grass huts, or other similar habitations. The point is that it is a characteristic behavior of humanity to build a habitation and organize the space in which they lives. Mankind uses space differently than any other being. I am writing this in a room that is set aside as my office. The adjacent room is the kitchen where cooking is carried out. No cooking is done in my office. A few rooms away are the bedrooms. Out back is a work room where some occasional woodworking goes on, and a garden of tomatoes is planted out back. My fence and the walls of my house separate my family from other families. Metaphorically, it separates *us* from *them*. Next door my neighbor has a similar situation and a wall separates my space from his space. Down the road is a park, a commons, where community activities take place such as art and craft fairs.

This social use of space is a hallmark of humans. None of the great apes construct their environment so that walls separate them and the areas of their various activities. The only building chimpanzees engage in are their nests up in a tree, which consists of a great interweaving of branches.[1] They do not prepare various parts of the tree into regions for differing behavior.

As we go back in time, the separate use of space gets more difficult to discern in the archeological record but it still exits. We will see, as we go from modern times to more ancient times, the type of activity involved in the construction of living spaces does not change greatly. This implies that Neanderthals, archaic *Homo sapiens*, and *Homo erectus* engaged in similar behaviors to us.

Throughout agricultural times, back to around 12,000 years ago, neolithic farmers have lived much like us. They had a house or hut, storage facilities for the grain, threshing areas, fields for the plants. The oldest protective wall is found at Jericho in the 9th millennium B.C.[2] Four hectares of round huts were uncovered; they were apparently surrounded by a stone wall 10 feet thick. One stone tower was found which was 30 feet high and 33 feet thick.[3] The city wall separates the world into two classes: us and them. Such walls were needed for the protection of the yearly harvest. The oldest city, which is the oldest evidence of a modern lifestyle, occurs at Mallaha in Palestine. This is a 10th millennium city and consists of a group of round huts ten to thirteen feet in diameter. Occasionally there is a hut that is 30 feet across. Huts were dug into the earth and surrounded by walls.[4] At Mureybet, Syria, similar circular huts were found, also dug into the ground. Once again, the house was surrounded by a low wall of wood.

As one goes further back into the primitive past, the habitations become more and more primitive. What should a habitation look like? A picture of a modern day primitive habitation can be found in many encyclopedias.[5] These often consist of wood poles lashed together and covered with a thatched roof. Stones often help stabilize the poles which are placed into postholes. In most cases, after 100 years, nothing but the piles of stones and postholes will remain of such a dwelling. Thus, when we look into the archeological record, we must look for just such an arrangement of material. The artifacts found inside such an arrangement will be different from those found outside. Inside such an arrangement, one should look for evidence of a hearth in the form of ashes or a stone wall surrounding ashes. Occasionally small walls of stone were built.

At Kostenki II, along the Don River in the Ukraine, the circular foundations of huts have been found. *Kost* is the Russian word for 'bone' and the nearby village was named after the large quantities of mammoth bones villagers found. It dates to 19,000 years ago.[6] The hut was 420 square feet and storage pits surrounded it. The hut was constructed of mammoth bones. Beneath this, in a layer 21-24 thousand years old, Kostenki I was found. Hearths were found regularly spaced along two lines. Bone was burned in the hearths as fuel and outside there were pits containing bone for fuel.

Anatomically modern man left archaeological evidence of huts at Pavlov in Moravia dating to 25,000 B.P.[7] These huts were round, oval, or pentagonal, and contained hearths and postholes within their boundaries. Several circular huts, 18 feet across, were found at Dolni Vestonice that radiocarbon dating says are between 25,000 and 29,000 B.P.[8] The lifestyles of these men were not that much different from the lifestyle led by modern technologically primitive peoples. Limestone blocks supported the poles that supported the roof. Inside the second hut, several hollow bones were found which are believed to be flutes or pipes. They also engaged in making ceramics.[9]

At Cueva Morin, 34,000 years ago, someone, Neanderthal or modern human, built what may have been a low stone wall[10] and a possible rectangular structure.[11] This was a Mousterian site, but there were two graves in which the bodies had totally decayed. The size of the graves indicates that the bodies had been 6 feet tall. This is decidedly taller than most Neanderthals.[12]

Neanderthal sites show much evidence for building walls, floors and habitations. At the Neanderthal site of Baume des Peyrards, large limestone blocks were aligned along the back wall of the shelter. The area between the back of the shelter and the line of stones outlined an area 11.5 meters long by 7 meters wide. Along the center of this area were hearths, former fireplaces. Inside the stone line the dark soil was rich in artefacts. Outside the stone alignment; the soil is yellow and lacking in artefacts.[13] The Neanderthal site at Kebara Cave is around 40-42,000 years old.[14] The evidence left at this site clearly shows that Neanderthal organized his space similarly to the way modern humans do. Hearths were concentrated in the center part of the cave, as modern humans would do. The scraps of bone and scrap stone flakes were removed from the central area of the cave, where the hearths were, and piled up along the north wall.[15] This shows cleanliness according to Bar-Yosef and Vandermeersch.[16]

Hayden, in a review of the cultural capacities of Neanderthals, cites several structures built by Neanderthals. At Pech de l'Aze, the Neanderthals apparently built a 30 centimeter high wall. Inside this wall were all of the Mousterian deposits.[17] At Grotte de Rigabe, a half meter high wall protected a hearth.[18] Grotte du Prince had a curved wall with an opening in front. It enclosed ashes and obviously protected a hearth. At Villerest, a Neanderthal structure that resembles later structures built by anatomically modern men was found. This feature had a perimeter of stone blocks and measured 2.5 by 1.5 meters across and was more rectangular than the later ones.

Bachler excavated a wall at the possible Neanderthal religious site at Drachensloch,[19] and a second wall at Wildskirchli a few miles away.[20] This can be see in Figure 27. This wall is *not* a random collection of stones which fell from the ceiling of the cave. Each stone is of a uniform size and the arrangement is quite regular and linear. A perishable wall made of skin or brush was constructed in such a fashion as to protect the Neanderthal cave occupants at Arcy-sur-Cur, 34,000 years ago. Hayden wrote:

"In the Classic Neandertal period (Wurm I-II), there are a number of indications of walls or physical barriers. A 'separation between ground littered with debris and virgin ground is so brutal and clear' in the Grotte du Renne (Schoppflin Gallery) at Arcy-sur-Cure, that excavators thought there must have been some sort of perishable wall separating the living area (only 10 m^2) from the damper, unused part of the cave. This part of the Grotte du Renne represents an intact Mousterian living surface in a sealed cave that was only opened in 1954 and recorded by Leroi-Gourhan."[21]

Figure 27 Neanderthal wall from Wildkirchli

This site also yielded evidence of circular huts 3 meters round.[22]

In Romania, at Ripiceni-Izvor, 40-44,000 years ago, multiple Neanderthal habitations have been found. In fact, at this site, five of the six different Neanderthal levels held evidence of structures. An arc of limestone 8 feet long by 3 feet wide surrounded deposits with charcoal and faunal fragments. Flint was also found inside the arc. The excavator, Paunescu, believes that this was the base of a lean-to. A second type of structure was 30 feet by 7 feet and contained over 500 flint artifacts and mammoth tusks and molars. The largest structure found was 24 feet by 18 feet, and contained approximately 70 limestone blocks, charcoal, bone, mammoth molars and tusks. One end of this structure appeared to have been the site of a hearth since charcoal was found there.[23]

Probably the oldest unchallenged open-air habitation site that has been found was at Molodova, Russia.[24] Dating from 60,000 years ago, a mammoth bone structure was found which was built by Neanderthals. The structure measured 30 feet by 24 feet and contained an extremely dense quantity of stone artifacts and bones.

It also contained fifteen hearths. The structure is very similar to that found at Kostenki constructed by anatomically modern men. If Neanderthals and anatomically modern men built similar structures in similar ways, it would seem to imply that they are equal in abilities in this area. It would also imply a similarity in the essence of their beings. However, in spite of the archaeological similarities of activities between Neanderthals and modern men, some believe that Neanderthal was merely imitating modern men. Concerning the Neanderthal's huts at Molodova Chris Stringer and Clive Gamble state,

> "But while they could emulate they could not fully understand. We suspect, for example, that the structures at Molodova and Arcy-sur-Cure more resembled 'nests' than the symbolic 'homes' of the Moderns at Kostenki or Dolni Vestonice."[25]

The fact is, our homes could equally be termed 'nests' if we chose to view man as nothing more complex than a chimpanzee. But to call a Neanderthal structure which appears as complex as a structure built by anatomically modern man a 'nest' seems to go beyond the evidence and into the world of bias.

Several hundred meters inside Bruniquel cave in southern France, a square stone structure measuring five meters by four meters was found.[26] It was constructed of stalagmite and stalactite pieces. In a hearth, a burnt bear bone was found which gave a radiocarbon age of greater than 47,600 years BP. Because of the antiquity of this find, it was almost certainly constructed by Neanderthals. No *Homo sapiens* have been found in France from that time period. This find shows that Neanderthals had the cognitive abilities to explore deep caves. It further demonstrates that they had some form of artificial lighting, since the cave at that depth was totally dark.

Eighty thousand years ago, men inhabited a rock shelter at Tor Faraj, Jordan. It is unclear whether Tor Faraj was occupied by anatomically modern men or Neanderthal. This shelter was organized exactly as modern hunter-gatherers organize their spaces. There was a line of three hearths, spaced three meters apart, that paralleled the back wall of the cave Henry et al. write,

> "Even at this early stage of the research, however, it is clear that the patterns observed in the cultural residues recovered from the living floors of the shelter strongly resemble those of modern foragers, implying that the occupants of the shelter were organizing their behaviors 70,000-80,000 years ago in much the same way as modern humans."[27]

As far back as one goes, Neanderthals built living structures and organized their space. Schick and Toth write,

> "And at the cave of Lazaret in France, estimated to be about 200,000 years old and also located on the French Riviera, an Acheulean assemblage was associated with a pattern of stone blocks measuring about thirty feet by twelve feet along one cave wall, with two concentrations of charcoal inside. Several areas with quantities of small marine shells suggested to the excavators that hominids were bringing dried seaweed (and, as hangers-on, the tiny shells) for bedding material. Once again the evidence is sparse and, often, controversial. It is likely that structures from this time period were ephemeral and would not have left behind much archaeological visibility."[28]

Lazaret even gives evidence for the sleeping habits of Neanderthal. Mellars relates,

> "Finally, de Lumley recorded some interesting if rather enigmatic data on the distribution of small marine molluscs in the deposits - principally those of *Littorina neritoides*. He argues that these must have been deliberately introduced into the site and suggests that they may have been attached to masses of seaweed possibly used as bedding for the site occupants. As he points out, the distribution of these molluscs seems to be tightly concentrated around the edges of the two main hearth areas and he suggests that these would be obvious places for sleeping, immediately adjacent to the hearths. He goes on to argue that the distribution of foot bones of various fur-bearing animals (mainly wolf and fox) show a broadly similar distribution, and suggests that these could derive from the remains of animal skins used as coverings for the bedding areas. Clearly, there are other possible explanations for these patterns (such as the introduction of

seaweed as a source of food) and it would be premature to accept de Lumley's interpretations without caution. Nevertheless, the distribution of these components shows an obvious pattern and must have specific implications for the character of the activities carried out in the cave."[29]

The Neanderthals went to the sea, collected seaweed for bedding and covered themselves with fur blankets. The behavioral evidence of sleeping habits left at Lazaret, 200,000 years ago, is exactly what would be found at a site occupied by modern men. Before one says that this is rather primitive bedding, the question needs to be asked, "Is this really any different from the straw or feather bedding our ancestors used merely a few hundred years ago?" Indeed, many people in America today sleep on pillows of eiderdown.

As far as I could determine this is the oldest evidence of Neanderthal habitations. This is not, however, the earliest evidence for habitations. Archaic *Homo sapiens* also left some evidence that he too built structures in which to live, and structured his use of living spaces, 300-400,000 years ago.[30] The cleanliness which Bar-Yosef and Vandermeersch attributed to the Kebara Neanderthal might also apply to the inhabitants of Lunel-Viel. They piled their bone refuse against the wall, as Hayden says, "like trophies."[31] More than that, post holes, 15 centimeters deep and 2.5 centimeters in diameter, were found in association with rocks which were used to stabilize the poles. At Olorgesailie, Kenya, huts were found which date to 400,000 years ago.[32]

Probably the most interesting find of habitations was that at Bilzingsleben, Germany, dating anywhere from 300,000 to 425,000 years ago. This fascinating site[33] was an entire village constructed by *Homo erectus*. This find clearly shows that he also organized his space much as modern humans do. Mania, Mania and Vlcek describe it as follows,

"The home base of early man from Bilzingsleben was situated on a shore terrace close to the outflow of a karst spring into a small lake. Previous excavations revealed a division of the camp site into different activity areas and outlines of three simple shelters with hearths and workshops set up in front of them. Five to 6 m from the dwelling structures, an artificially paved area with a diameter of 9 m was found. According to the archaeological evidence, special cultural activities may have been carried out there.

"Along with large pebble tools (choppers, chopping tools, and hammerstones), small specialized tools of flint appear. Basic standard forms are knives, scrapers, denticulates and notches, simple points which are pointed-oval, Tayac and Quinson points, borers, and core-like tools. Edge retouches predominate, but also unifacial and bifacial retouches occur. Large scrapers, knives, chisel-shaped tools, wedges, bodkins, and work supports were manufactured from the compact bone, preferably of the straight-tusked elephant. Mattock- and cudgel-shaped tools were made from cervid antlers. Specific, deliberate manufacturing activities are recognizable in the workshops. Apart from the dissection of the animal prey, these tools served for the working of predominantly organic materials which in turn were used for the manufacture of other tools and objects of daily use. Wood was also a frequently used raw material. Numerous calcified remains of wood artifacts were found at the site. Some bone tools display deliberately engraved sets of lines which we regard as expressions of abstract thinking, perhaps as graphic symbols."[34]

When apologists, like Hugh Ross, exclude *Homo erectus* from humanity and the spiritual realm, they apparently never mention places like Bilzingsleben. In *Homo erectus* we have a man who, at least in later times, lived in a village constructed very much as technologically primitive modern humans do today. In order to save a theological position--that evolution did not occur-- old-earth, anti-evolutionists must disinherit *H. erectus*. As noted in a previous chapter, there is one mention in the literature that the Bilzingsleben *Homo erectus* actually drew a picture of a four-footed animal, something which no dog, gorilla and chimpanzee is able to do!!![35]

Recently, an ancient case of modern-like use of space came to my attention. It is the living habitation of *Homo erectus* at Wonderwork Cave. Bednarik writes:

"The arid deposits seem to indicate a fairly stable sedimentation rate, and on the basis of the dated upper quarter of this sequence it is estimated that the unique succession of occupation layers spans about 800,000-900,000 years. Artefacts (mainly 'handaxes') occur down to the basal layer, and this site provides the longest continuous human occupation evidence in the world.

"Peter Beaumont's work has for many years produced rich evidence of early human capacities for instance through his work on Middle Stone Age ochre mining. According to him, the ongoing excavations in Wonderwork Cave have already yielded clear evidence for a 'home base' form of organization, spatial patterning of activities, the regular production of fire, the use of grass bedding and other indicators. Perhaps most importantly, each and every level of the site has produced an abundance of red ochre fragments, and a variety of unmodified items that must have been introduced into the cave intentionally: exotic quartz crystals and small 'pretty', coloured pebbles, two of which bear a natural pattern. These finds, along with those of other African sites, as well as sites in Asia and Australia, render the traditional claims that the advent of the Upper Palaeolithic in Europe marks a major change in cognitive evolution highly suspect. As Dr. H. J. Deacon (University of Stellenbosch) has recently stated, the southern African evidence permits the inference 'that the people in the earlier Late Pleistocene had cognitive abilities that are comparable to those shown by their Holocene and modern descendants' (Philosophical Transactions, Royal Society London, Sept. 1992)."[36]

While not the most impressive, the oldest evidence of habitations is from Olduvai Gorge and it is nearly 2 million years old.[37] When Mary Leakey was clearing a site, she noticed that the blocks of lava, some stacked up on top of each other, formed a rough circle about 13 feet in diameter.[38] Helson Mukuri, a member of the team, mentioned that this pattern of stones resembled the pattern of stones formed by the huts of certain African nomadic tribes. There were fewer fauna bones inside the circle than there were outside. (Sounds like another example of "cleanliness" as was the case at Kebara). This suggests that 2 million years ago mankind was manufacturing simple housing.

The shelter at Olduvai has been challenged by Lewis Binford who argued that this was no real evidence of human habitation, but merely the fortuitous arrangement of stones. Binford's argument carried the day until now most anthropologists do not accept this as a habitation. While Binford is correct that this is not a provable case of a habitation, one could make a similar statement about equivalent remains of huts left 100 years ago by modern man. Considering that other *H. erectus* were building habitations, one cannot rule it out either. Binford, after all, spent a career downplaying the abilities of fossil man; he would not have viewed *H. erectus* as a hunter.[39] Yet this year, as we shall see, *Homo erectus* was proven to be a hunter.[40] Binford's view of the capabilities of fossil man is probably too low. As Schick and Toth note,

> "It is worth noting that many of the hut structures of modern hunter-gatherers in Africa and Australia would probably leave little or no signs thousands of years in the future. We cannot verify early huts or shelters, but we cannot rule them out either."[41]

Based upon these criteria, future archaeologists would have to assume that technologically primitive humans today were incapable of organizing their space either.

Pavements

There are other constructed items found in the fossil record built by fossil man. Pavement and paving were constructed in *Homo erectus* times. Pavements have been viewed in two fashions by researchers. There are those who believe that they are made intentionally and those who try to account for them without human intervention. Some of the cases are quite difficult to account for based upon a natural explanation. Furthermore, the creation of pavements continues unabated as we go back from modern man to *Homo erectus*.

Cro-magnon man was anatomically fully modern. His summer living sites were often located in moist places (like caves and rock shelters). Living in mud is unpleasant. Often they would build a pavement upon which to pitch their tents and protect them from the dampness. In the cold European winters, the ground was frozen mud. Lighting a fire on the frozen soil would cause it to melt. To solve this problem, they would heat the stones and lay them down on the ground as a pavement. The heat would cause the stones to sink into the frozen ground. They would thus create a stable platform upon which to pitch their tents or burn their fires.[42]

In the case of anatomically modern Cro-magnon men, there is no question that they produced the pavement. But when it comes to Neanderthal, when similar structures are found at their sites, everyone seems to question the intentionality of the pavements. Hayden cites numerous examples of pavements made by Neanderthals or earlier men. At La Ferrassie, a rectangle of stones 16 by 10 feet was found. The stones had been placed carefully.[43] Seven slabs of hardened clay appear to have been placed near a hearth to grill meat.[44] A 6 square meter pavement was found at Grotte d'Aldene, a site from the Mindel/Riss interglacial. This means that it is around 200,000 years old. The stones were flaked and the smooth, unflaked sides were placed facing up. The fact that the pavement is made of limestone makes it unlikely that the pavement was formed by unused tools since limestone was rarely used in the manufacture of stone tools. At Baume Bonne, another Neanderthal site, a pavement containing 185 pebbles per square meter was excavated.[45] The pavement was 27 feet by 8 feet. Finally there is the 300,000 year old, *Homo erectus*-built pavement at Bilzingsleben, Germany.[46] Mankind has been behaving the same for hundreds of millennia. What ancient man did not have is our technology.

Do Clothes Make the Man?

"The LORD God made garments of skin for Adam and his wife and clothed them. " Genesis 3:21 (NIV)

Thus, man became clothed, according to the Bible. Since the advent of clothing was intimately connected with the Fall, clothing on any being should theologically be indicative of a post-Fall individual. So how far back in time does clothing appear? Since organic materials do not preserve well, the earliest piece of cloth we have is from 7,000 B.C.—only 9,000 years old.[47] If this is the case, then how do we find evidence for clothing prior to this time? There are several pieces of evidence that point to the making of clothing. These include the sewing needle, scrapers used to tan hides, a characteristic pattern of microscopic scratches on stone tools, the climate where fossil men lived, etc.

Today we wear clothing that is cut and sewn to fit our bodies. Such form-fitting clothing was not the norm in the warmer climates even in modern times. Robes, and loose fitting togas were the main form of clothing in the Mediterranean region during the Roman empire. The Greeks wore a peplos which consisted of two rectangles of woollen cloth held together by two pins on the shoulders. The cloth extended to the ankles and was belted by a girdle. On Crete women wore bell-shaped skirts and short jackets. The men wore simple loincloths. Babylonians wore skirts of leather or goatskins. Medes are considered to be the inventors of pants,[48] but as we shall see, there is evidence that pants were in existence at least as long ago as 23,000 years.[49] The ancient Egyptians (2,000-2,700 B.C.) wore a belted loincloth. On cold days an animal hide or linen cape was worn over the shoulders.[50]

While the linens mentioned above are made of cotton or wool, prior to the agricultural revolution such materials would not have been available. The domestication of cotton, a major fiber constituent of woven clothing, took place at the beginning of the agricultural revolution around 6-10,000 years ago. Since all wild species of *Gossypium* have no lint, it is unlikely that cotton was originally used for its fiber. The only other reason to domesticate the species would have been either for its oil or its fibrous stalk. The lint-filled cotton boll developed as an added benefit and, indeed, became the only reason cotton is harvested today.[51] Clothing prior to the development of cotton lint had to have been made with other fibers. Similarly, wool as we know it today would not have been harvested prior to the domestication of the sheep, which occurred 7,000 years ago.[52]

Prior to the advent of modern fibers like wool and cotton, mankind used animal hides, sewing them together to produce clothing. Bone objects, which are identical in size and shape to sewing needles, first appear in the archeological record 26,000 years ago from central Europe.[53] Buttons are found from the same time. These objects cannot be interpreted in any other fashion. In the arctic, such sewing needles were used to sew the hides together tightly and increase the warmth of the clothing in those cold environments. At Sungir, Russia, 130 miles northeast of Moscow, three humans were buried 24,000 years ago in an exquisite set of clothes, probably made from animal hides. Over 13,000 beads had been sewn onto the caps, shirts, pants and moccasins of the three individuals.[54] A needle would have been required for the sewing of these objects.

The importance of the invention of the needle cannot be underestimated. Brian Fagan writes:

> "And these needles enabled *Homo sapiens sapiens* to abandon ill-formed skins and to sew form-fitting arctic clothing, everything from underclothes to parkas, fur-lined hoods, and special footwear for hunting and trapping in subzero temperatures."[55]

Fagan has further suggested that it was the harsher climatic conditions which kept man out of Siberia before he possessed the ability to manufacture form-fitting sewn clothing.[56] Fagan believes that all clothing prior to this time was poorly fitted.[57] As we saw earlier, there is evidence of form-fitted clothing prior to the appearance of the needle in the fossil record; we will see below that there is evidence for the habitation of Siberia 300,000 years ago.[58] Habitation of such a cold region as Siberia would require tight-fitting waterproofed clothing.

We have one earlier evidence of woven material, but we don't have the woven material itself. Four fragments of pottery found at Pavlov I, dating to 27,000 years ago, showed clear impressions of a woven material having been impressed into the pottery prior to its firing. It is difficult in this case to know whether the woven material is a basket or textile but the twining of the fibers surprised the archaeologists because they did not realize that weaving had "reached such an advanced state" that early.[59]

As we go back into Neanderthal times, the direct evidence for clothing becomes harder to find but it still exists. As we saw in a previous chapter, a Neanderthal was buried at Hortus, France with the paw bones and tail bones of a leopard arranged around him as if there had been a leopard cape on his back.[60] The paws and tail bones of modern shamanistic capes also contain these bones. While this is evidence of Neanderthal religion 50,000 years ago, it is also direct evidence of Neanderthal clothing 50,000 years ago.

The other piece of evidence for clothing among the Neanderthals is deduced from the fact that they lived in glacial Europe when the climate was terribly cold. No hominid could survive the winters, naked in the snow. Although human beings do have a tremendous ability to withstand the cold given minimal clothing, they can't live naked outside during a northern European winter. Andre Leroi-Gourhan tells of the Tierra del Fuegians living at the southern tip of South America,

> "The Fuegians still lived, a few years ago, in little groups of a few families, just large enough to feed themselves without going beyond the resources of their hunting territory. The main part of their food was acquired by hunting and fishing, but in order to bear the periods of famine, they must have depended on the food resources provided by wild plants and small animals, even insects. They lived in rounded huts built with branches and brush and huddled around a small hearth, fed as much as possible by pieces of broken bones. They had great resistance to cold: they wore only a square piece of skin slung across their backs, just big enough to cover them when they squatted, their backs turned against the wind. Their domestic implements were reduced to the minimum: baskets woven in the simplest manner, harpoons with stone points, rocks used as hammers or pounders, thick end-scrapers, and knives. It is impossible to find in the recent world a clearer example of what the Neanderthal mode of life must have been."[61]

Other adaptations could have helped Neanderthal survive the bitter winters. Boyce Rensberger says,

> "Recently, there has been some evidence that skin colors are linked to differences in the ability to avoid injury from the cold. Army researchers found that during the Korean War blacks were more susceptible to frostbite than were whites. Even among Norwegian soldiers in World War II, brunettes had a slightly higher incidence of frostbite than did blonds."[62]

While these facts are quite true, it is also quite possible that form fitted clothing was manufactured without a needle. Many late Stone-Age Europeans would stitch skins together by making small cuts on the edge of the fur and pulling leather thongs through the holes with a crochet-like hook.[63] In this way warm clothing could be made without needles.

Going further back into the times of *Homo erectus* we find no direct evidence of clothing. However, two indirect lines of evidence remain. First, when a stone tool is used to scrape animal hides, a characteristic pattern of wear is left on the tool. This is the case no matter whether the tool is stone or bone. Microscopic examination of stone tools shows that some were used for working hides as long ago as 300,000 years.[64]

The second indirect evidences of clothing are the locales where *Homo erectus* lived. By 300-400 thousand years ago, *Homo erectus* had occupied Germany[65] and Siberia.[66]. Both of these localities would require clothing of some type. We must not forget the oldest evidence for the need of clothing, the Dmanisi mandible, a mandible of *Homo erectus* which was found in strata dated to 1.6 million years old in Georgia, on the Black Sea. All these regions get very cold during the winters and any human living in one of them would require clothing. Over the last 200 years, Germany has had winter temperatures as low as -20° F, Georgia as low as -20° F, and Siberia as low as -75° F.[67] No one could survive such temperatures without clothing. Thus, indirect though it be, the archeological evidence for clothing goes back an amazing 1.6 myr. If *Homo erectus* wore clothing, then how can Christian apologists eliminate him as a post-Fall individual? No animal makes clothing.

The most intriguing evidence of working hides comes from Swartkrans, a robust australopithecine site in South Africa. Robert Brain discovered some bone tools with a glassy polish. Johanson, Johanson and Blake relate,

> "Some of the fossil bones looked so worn at the tip that they must have been used for several days. Bob began to wonder if the hominids carried these digging sticks with them. Then he noticed that the wear scratches on some specimens were obscured by a glassy polish. A similar sort of polish occurs on modern bone tools used by hunter gatherers to burnish hides. Bob speculates that the hominids may have made hide bags to carry tools and tubers, and the glassy polish formed as the bones rubbed against the leather. A few tiny, awl-like pieces of bone—the sort of tools that could be used to puncture leather— were also uncovered at Swartkrans."[68]

If this is true, then australopithecines were working hides somewhere prior to 1 million years ago. While this certainly is not proof that australopithecines had clothing of any sort, Christian apologists should be prepared to handle future discoveries in this regard.

Man, the Carpenter

Mankind works in wood. Our houses are made of wood; our garden implements have wooden handles. Our tents, until about 20 years ago, were staked with wooden stakes. Until I began to write this section, I was under the impression that the number of wooden artifacts from the archeological record were few. I figured that I would whip out a short paragraph on man the carpenter and then go on. But mankind has been making wooden tools for a very long time. And in spite of the rarity of preservation, much evidence has remained.

Some Christians have tried to eliminate tool use and tool making as evidence of humanity. Frair and Davis write:

> "Many animals use simple tools. For example, certain wasps tamp down the earth of their burrows with stones held in their jaws, and certain birds and apes use pieces of wood in obtaining food. Some patterns of tool use, or even simple tool preparation, appear to be instinctive but others are learned."[69]

and

> "It is conceivable that nonhuman toolmakers existed in the past, as in the case of *Australopithecus(q.v., infra)* who does not seem to have been anatomically human, but who may have made tools either instinctively (as a wasp builds its nest of paper) or as an aspect of some kind of 'cultural' tradition."[70]

Hugh Ross writes:

> "Evidence of man's spiritual dimension would include divine worship, shown by religious relics, altars, and temples. From the Bible's perspective, decorating, burial of dead, or use of tools would not qualify as conclusive evidence of the spirit. Moreover, nonspirit creatures such as bower birds decorate their nests, elephants bury their dead, and chimpanzees use tools."[71]

We discussed in a previous chapter the evidence that elephants do not bury their dead, and we will show here that chimpanzee toolmaking and use is not the same as human toolmaking and use. One thing that must be made absolutely clear: *only* humans use a tool in order to make another tool. Schick and Toth write,

> "With a stone flake, a wooden branch is slowly honed to a sharp point by scraping. Such an implement could have been used as a formidable spear, digging stick, or skewer for carrying meat. Final shaping was done by grinding the point against a rough rock. The use of one tool to make another tool is one more human characteristic."[72]

This fact cannot be emphasized enough. Those Christian apologists who exclude beings who use tools to make other tools, are excluding those who behave exactly as we do. The only real reason for me to treat you as a human is that you behave like one. You treat me as a human for precisely the same reason. We do not ask about a person's "spirituality" before we accord him status as a human. Mankind uses tools to make other tools. No chimpanzee has ever been observed to use a tool to make another tool. This denigration of human activity which old-earth anti-evolutionists must engage in (in order to leave room for Adam) leaves little evidence of any observational difference between us and the animals. If spirituality, which is difficult to demonstrate observationally, is the only characteristic by which we differ from the animals, then what is to prevent one from assuming that animals are spiritual? What is to prevent me from assuming that an atheist is not human since he lacks spirituality? You say he has the capacity for spirituality? How do you know since he exhibits none? We accord humanity to an atheist because he acts like we do.

As to humans being like bowerbirds, Bednarik writes,

> "One could argue that these habits might have been the hominid equivalent of the behavior of certain birds, notably the bower birds of Australia, who collect brilliantly colored or reflective objects. But this is not a very convincing argument, and is in any case hardly relevant to the issue at hand. Firstly, the bower bird produces no stone tools or polished wooden planks and does not use hematite crayons or fire, nor is it likely to develop its present skills to those attributed to Middle and Upper Paleolithic humans."[73]

The first ancient example of woodworking comes from the Roman Empire. I include it not because anyone doubts that Romans were human but because this particular piece of woodworking is just simply interesting. It also illustrates a range of activities that humans engage in when working wood. Ivar Lissner wrote,

> "About twelve miles from the Tyrrhenian Sea, the Tiber meets a small but stubborn obstruction: an island which has been there from time immemorial. From here, over a wooden bridge, people could cross the Tiber in comfort. It was a very old bridge, older even than the Bronze Age which was succeeded by the Iron Age about 1000 B. C. Not a single nail was driven into the bridge and nothing but wood was used for its construction, for wood was still vested with the sanctity which Europe's great forests once possessed in the misty obscurity of prehistoric times. It still held the magic properties of the sacred trees once worshipped by the white inhabitants of primeval Europe."
>
> "There, at the left end of the bridge and in the heart of the fertile plain of Latium, rose the eternal city of Rome. It was there that Roman dancers used to gather each spring to prance about in wild war dances, while their weapons clanged, the woodwork groaned, and weird songs floated across the river to the opposite bank. Priests guarded the sanctity of this bridge and supervised its cults, and one of these priests was the bridge-maker, or the pontifex."
>
> "That was over a thousand years before Christ's birth; and yet today, 3000 years later, the man who builds the bridge between earth and heaven, which all true believers must cross, is still called pontifex."[74]

I do not agree with Lissner's theology, because it is Christ that built the bridge between man and God, but the history of the woodwork on the bridge, the etymology of the word "pontiff" and its effect on today's world is fascinating. To make a bridge, the ancient Romans had to pound posts into the earth for its supports. The bottoms of these posts would probably have been pointed so that they would enter the earth more easily. They also had to shape the wood and turn it into planks. The carpenter would have used an axe-head (metal or stone) attached to a wooden handle. Guards stood watch with wooden spears, and spears with points attached. In all of this, the ancient Romans performed the same types of woodwork that were performed by their ancestors hundreds of thousands of years ago.

Wooden implements and artifacts, even from the Roman times, are rare. From 1500 B. C. comes the discovery at Gwisho Springs, Zambia: a digging stick identical to one used by modern San peoples. The stick was preserved in the hot spring waters.[75] The oldest evidence of a felled tree and the oldest direct evidence of boats is from Star Carr, England.[76] The evidence of a boat comes from the preservation of a wooden paddle, not the actual boat. Both the felled tree and the paddle date to 7,500 B. C. From Australia come the oldest clubs and boomerangs dating to 8,000 B. C.[77] The oldest arrows are the 100 wooden arrows found at a site near Hamburg, Germany, dating from 10,500 years.[78]

Monte Verde, an archeological site in South America, yielded many wooden objects, "including mortars, several wooden hafts (or handles) containing stone scrapers, an assortment of digging sticks, and vast amounts of worked wood and other small objects."[79] Indians were making these wooden objects 12-14,000 years ago.[80] Dillehay, the excavator of Monte Verde, also discovered the wooden foundations of huts. Prior to this, wooden tools and objects are very scarce in the archeological record. In fact, there are more wooden artifacts in the Lower Paleolithic than there are in the Upper Paleolithic period.[81]

When one finds evidence that a stone tool was attached to a handle, it constitutes indirect evidence of woodwork. Hafting is the name used for attaching a stone point to a wooden shaft or handle. Arrowheads of Indians are hafted to the arrow shaft. Ancient axe heads were hafted or affixed to wooden handles. An axehead that is attached to a handle normally shows a notch on each side. An axe head from New Guinea dating to between 37,000 and 40,000 years ago shows the type of taper in the middle that hafted axe heads of more recent times show.[82] This would mean that New Guineans were accomplished carpenters 40,000 years ago. In Syria, at Umm el Tlel 36,000 years ago, men were gluing their tools onto wooden handles with bitumen.[83] Neanderthals did the same thing in Europe more than 40,000 years ago.[84] Prior to these discoveries, the earliest evidence for the gluing of stone tools onto wooden handles was from 8,200 B.C. This is a clear indication that what is considered the earliest example of anything in the archeological record is most likely *not* the earliest actual existence.

The issue of hafting is one that has implications for the intelligence of the Neanderthals. Lewis Binford and others have argued that the lack of hafting in Neanderthal tools meant that they did not have the intelligence to plan for the future. But the discovery of hafting changes things. Hayden counters;

> "Binford's arguments concerning foresight and the lack of curation cannot be sustained in light of the now copious evidence for hafting on Levallois and Mousterian points with thin bulbar areas, lateral notches, impact fractures and repaired tips; Mousterian scrapers with thinned bulbs, convergent and other scraper types with clear hafting wear, tanged tools of the Aterian and hafted leaf-shaped points in Greece. At some sites such as Konigsaue and Kerlich, the hafts, molded hafting resin (with imprints of the wood haft and chipped stone bits), as well as the complete hafted tools have been recovered. In and of itself, hafting is a strong indicator of curation, foresight and mental templates of tool designs employing different materials shaped to predetermined specifications.
>
> "Furthermore, lithic raw material analyses clearly demonstrate that raw materials originating 30-80 and more kilometers from sites were intensively used for tools and exhibited intensive degrees of resharpening, whereas this was not generally the case for locally obtained materials."[85]

The fact that Neanderthals mounted their tools[86] clearly shows that they could think through a sequence of steps to a future, distant goal--the obtaining of food. The manufacture of the stone tool required many manufacturing steps. Once it was made, the tool-maker needed to go get bitumen (or already have obtained it). The bitumen was then heated and used to glue the stone to the wood. Then, after all this, the tool was ready to be used in a hunt for food.

From Japan a wooden plank, that had been shaped by human activity, was recovered from strata dating from 50,000 to 70,000 years ago.[87] The board was 26.9 centimeters long, five centimeters wide and 3-7 millimeters wide. The board was artificial because the plane of the cutting cut across the tree rings and both sides of the board have an oval indentation. What it was used for no one knows. The board was made from a species of mulberry (*Cudrania tricuspidata*). This tree is extinct in Japan but lives in Korea and China. Could this board be a modern import? No. Radiocarbon dating of the wood itself yields a date of 54,000 B.P. Further evidence of the boards antiquity was that it was found in a layer with stone points that are from a culture that lived between 50-60,000 years ago. It is uncertain whether it was anatomically modern man or archaic *Homo sapiens* who made the object.

Evidence of a wooden spear made by anatomically modern man 80,000 years ago[88] has been found in a setting of violence.

"Careful study of the hip region of one of the men buried at the Mount Carmel cave of es-Skhul has shown that at death, or shortly after, this individual had received a dreadful wound from what must have been a wooden spear like those described from a previous age; the weapon had been driven in with such force that the head penetrated the head of the femur and emerged into the pelvic cavity."[89]

A seven-foot-long wooden spear was found at Lehringen, Germany, in 1950,[90] between the ribs of an extinct Straight-Tusked Elephant, is 110,000 to 130,000 years old.[91] Since anatomically modern man was not in Europe at this time, the spear had to have been made by Neanderthal. It also means that Neanderthal was a hunter, not a scavenger, and that he was making wooden spears and wooden objects, just like modern men do.

There is one other issue that Christian apologists must face. Originally, according to Scripture, man was a plant-eater. In Genesis 1:29 God gives man every seed-bearing plant for food. After the Flood, God gave Noah permission to eat meat. Genesis 9:2-3 (NIV) says, "Everything that lives and moves will be food for you. Just as I gave you the green plants, I now give you everything." If one postulates that Adam was created separate from all previous hominids, these verses require that the earliest men were vegetarians. However, the earliest anatomically correct men were *meat-eaters* as is evidenced by the butchery marks made on more than 5400 animal bones found among the 120,000-year-old human remains at Klasies River Mouth Cave.[92] In fact, the researchers stated that their behavior was quite modern. So the apologists who want to avoid evolution and harmonize the Scripture by having Adam be anatomically modern, are forced to implicitly reject what the Scripture says about early man being vegetarian. Moving further back in time, if we believe these verses, then the fact that Neanderthal was hunting elephants at Lehringen 120,000 years ago or so means that, theologically, Neanderthal was also a *descendant* of Noah, not an ancestor. We shall see that *Homo erectus* also made wooden spears and hunted flesh.

Further evidence of Neanderthal woodworking and butchery can be found in the fact that burned flint tools, along with bones and peas, were found in Neanderthal fireplaces.[93] The stone tools found with the hearths show microscopic wear traces that indicate hide working and woodworking. Microscopic examination of these tools from five Paleolithic sites clearly shows that they were using the stone tools for woodworking.[94] As long ago as 100-200 thousand years ago the tools of all forms of mankind have this microscopic wear caused by woodworking.[95] And some of the woodworking by Neanderthal was to make posts for his huts. At Combe Grenel, Neanderthals clearly dug a posthole for one of their habitations.[96] The posthole was 20 centimeters deep and was circular in cross-section with a diameter of 4 centimeters. The post's buried end had been sharpened exactly as we would sharpen a stake or post going into the ground.

As we go back to a time prior to 200,000 years ago, when the men on the planet are classified as either archaic *Homo sapiens* or *Homo erectus*, we see no lessening of the evidence for working wood. At Florisbad, Orange Free State, South Africa, an archaic *Homo sapiens* skull came from strata dated 259,000 years ago.[97] The skull was found in association with a wooden throwing stick used for hunting.[98] Beyond this, the throwing stick was decorated with art—parallel lines had been inscribed on the sides of this instrument. At Clacton-On-Sea, England a spear tip made from yew wood was found in 300,000 year old strata, and at Kalambo Falls, Zambia a wooden club was found dating between 200-400 thousand years old.[99] On the Clactonian spear, clear striations show that a stone tool had been used to sharpen it. Clive Gamble, in the well worn tradition of doubting any intelligence and hunting abilities among hominids, has suggested that these spears were not hunting weapons but a "snow probe".[100] As Mellars points out, the fact that the Lehringen spear was found lying in the bones of an elephant would present problems for that interpretation. In spite of this evidence, Lewis Binford, who has done much to downplay the cognitive and technical skills of the ancient people wrote as late as 1992,

"It is even possible to argue that systematic hunting was not even practiced by early hominids until after the appearance of modern humans, perhaps around 60 000 years ago, instead of in the very ancient past of more than 2 million years ago."[101]

and

"Systematic hunting may thus have been a behaviour whose appearance roughly coincided with that of our own species."[102]

But this view was demolished in an astounding find that was announced in 1997.

Hartmut Thieme, excavating at Schoningen, Germany, announced the incredible discovery of three, possibly four, 400,000-year-old wooden spears. This discovery destroys Gamble's and Binford's low view of the intellectual capabilities of *Homo erectus* and archaic *Homo sapiens*.[103] The spears are made and balanced exactly as a modern Olympic javelin is made.[104] Thieme writes:

> "All three spears, although of different lengths, were manufactured to the same pattern, with the maximum thickness and weight at the front; the tails are long, and taper towards the proximal end. In all of these respects they resemble modern javelins, and were made as projectile weapons rather than thrusting spears or lances."[105]

Theime further found the earliest evidence of what is known as a composite tool in which one of the parts was wooden. Many anthropologists had believed that ancient men were not smart enough to perform this type of tool manufacture. This is a tool that is made up of two components that have been conjoined. A stone axe-head attached to a wooden handle is a composite tool. An arrowhead attached to a shaft is a composite tool. Theime writes:

> "Lake-shore deposits from the Reinsdorf Interglacial level 1 contained numerous flint artefacts and three worked branches of the common silver fir, *Abies alba*. The wooden tools (length 170-320 mm, maximum width 36-42 mm) have a diagonal groove cut into one end. The grooves vary in length between 33 and 47 mm in the front, and 10 to 30 mm at the back; groove breadth varies between 4 and 9 mm at the distal end. It is postulated that the grooves were for holding flint tools or flakes; if this supposition is correct, these implements represent the oldest composite tools yet discovered."[106]

Robin Dennell, who wrote a commentary in Nature concerning this find, described the situation as follows,

> "The implications of the Schoningen spears are no less extraordinary than the degree of their preservation. First, these are unquestionably spears, and second, as such they must have been used for hunting large mammals. Why are these simple inferences so significant to Paleolithic archaeologists? The reason is simply that hunting has become profoundly unfashionable in discussions of the Lower, and even Middle, Palaeolithic over the past twenty years. Until the 1960s, stone tools associated with large mammal remains were routinely explained as indicating the butchery of animals that had been hunted. Important examples were the initial interpretations of the very ancient 'living floors', 1.8 million years old, at Olduvai in Tanzania, and the alleged elephant hunters at the much younger, Middle Pleistocene sites of Torralba and Ambrona, Spain. Then came a long reappraisal of how these stone tools and hominid and other mammal remains were found together. This was driven initially by Brain's re-examination of the australopithecine deposits at Swartkrans, South Africa, by Binford's criticisms of the assumptions and methodologies used at Olduvai and, later, by patient analysis of cut-and gnawing-marks, surface weathering and skeletal parts frequency at Koobi Fora and Olduvai, often under the inspiration of the late Glynn Isaac.
>
> "By the early 1980s, few of the claims for big-game hunting in the Lower and even Middle Palaeolithic could be substantiated, as evidence was too disturbed, fragmented and poorly preserved to show deliberate, purposeful hunting. Scavenging by both carnivores and hominids seemed a more reasonable inference, and some even suggested that big-game hunting did not occur until the appearance of fully modern humans in the Upper Pleistocene, about 40,000 years ago. To fit this picture, the Clacton and Lehringen spears were down-graded to digging-sticks or, imaginatively, snow-probes for locating buried carcasses."[107]

The Associated Press article quoted Dennell as saying, "They're not going to throw it at a squirrel in a dark night. These people were serious about hunting." Indeed they were. The implications of this are quite clear. *Homo erectus* was a meat-eater as were Neanderthal, archaic *Homo sapiens* and all anatomically modern humans. Given this, theologically, they must all be descendants of Noah rather than ancestors, assuming the

Scriptural description of the original diet of man is correct.

One possibly older spear has been mentioned in the literature, but has been rejected by some authorities. This comes from the 450,000- year-old site of Karlich-Seeufer, an open-air site in Germany.[108] This article also contains the earliest English language reference to the Schoningen spear I have been able to find.

Bilzingsleben, Germany, is one of the world's most interesting *Homo erectus* sites. As we saw earlier, the Bilzingsleben *Homo erectus* site has yielded strong evidence of huts and an area for working wood. Maybe some of the ancient spears were made at this 300-420,000 year old site.[109] Mania and Mania relate,

> " To the south-east, beside each dwelling, were two workshops, indicated by centrally placed anvils. More workshops, manifested by the distribution of stone, bone and wood remains bearing evidence of working, appeared in the large area in front of the dwelling structures. Around it extended a zone of manufacturing activity where blanks, half-finished pieces, finished tools and manufacturing debris of stone, bone, antler and wood were found."[110]

Somewhere between 240,000 and 750,000 years ago, a hominid made a wooden plank and polished one side. This wooden plank was found at Gesher Benot Ya'aqov, Israel,[111] and was planed from a fairly large branch. In English units, it is 10 inches long, 5 inches wide and 1.5 inches thick. The branch had to be at least 5 inches in diameter. The possibility that mankind was making wooden planks up to 750,000 years ago fits quite well with the evidence for ocean crossings by *Homo erectus* at about the same time. This will be considered below.

The oldest evidence of woodworking goes back an amazing 1.5 million years. Lawrence Keeley of the University of Illinois studies the microwear on stone flakes from Koobi Fora in Kenya. The flakes showed clear evidence that these stone tools were used to cut wood.[112] Stephen Mithen writes,

> "It is most likely that the working of wood to make artifacts stretches back to the common ancestor, 6 million years ago."[113]

Christians who believe that Adam was a late creation (e.g. 35-100 thousand years ago) must consider that the evidence we have just seen is inconsistent with such a view. Where in this long tradition of woodworking is there a sudden change in behavior that would be consistent with a sudden change in cognition that should have accompanied the acquisition of spirituality? Surely such a complex behavior as the building of a village, the making of spears, and the carpentering and polishing of planks would be beyond the reach of mere animals. In any event there is no sudden change in behavior which should be expected from the acquisition of spirituality.

To Weave a Basket

As noted earlier, the earliest basket found in the archaeological record is 9,000 B.C. There is, however, possible evidence of earlier basketry in the form of plant remains attached to a Neanderthal tool from Combe-Grenal from the time of the Wurm I glaciation, 73,000 years ago.[114] Anderson-Gerfaud writes:

> "However, we were able to identify at least one plant- harvesting tool from the Middle Palaeolithic—a convex scraper on a blade from a Wurm I level (Typical Mousterian) at Combe-Grenal, described earlier. This particular tool was significant in that it was clearly used with a curved, 'harvesting' motion, and edge damage on the edge opposite the one used suggests that it may have been used in a haft. We then examined the tool with the scanning electron microscope to search for any minute fragments of residue material which might clarify its use. A residue located near the working edge, in a slight depression in the tool surface was found by comparison with microscopic cellular fragments (e.g. siliceous phytoliths) we extracted and studied from living plants) to be from a grass, or possibly a sedge (*Cyperaceae*) or a rush (*Junicus*).
>
> "It is not clear whether this tool was used to procure edible seeds, although certainly seeds of grass and probably sedge are edible foodstuffs. It is more likely that this tool, like other Palaeolithic tools of its nature, was used to gather or process plant materials for various artisanal or maintenance purposes (e.g. construction, basketry, fuel, etc.). Indeed, obtaining the plant

stems seems to be the primary goal of harvesting plants with a stone tool, at least until cereals with domestic characteristics of ripening evenly and holding their grain at maturity are documented. This is probably because ripe seeds of wild grasses (and cereals, if we consider the Near Eastern Epi-Palaeolithic period) are efficiently gathered by hand-picking, or by stripping or rubbing of the plant over a basket, for example."[115]

While not conclusive of basketry, it does point to an entire area of Neanderthal behavior that does not preserve well in the archeological record.

The Master of the Seas

What was mankind doing with this woodworking skill? For one thing, he was building boats. The earliest preserved fishing boats and paddles from China were made between 4000 and 5000 B.C.[116] The oldest preserved water craft in the world was made in 6,400 B.C. It is from Pesse, Netherlands, and is a simple dugout canoe. From 7,400 B.C., at Star Carr, England a boat paddle was found.[117] There is evidence of over-ocean travel which may extend back to 1.8 million years.[118] The reason for the extremely late preservation of direct evidence of water travel is due to two factors. First, as we have seen, wood does not preserve well unless it is waterlogged. Secondly, due to the fact that sea levels were much lower during the past glacial periods, boats which could have traveled over the ocean tens of thousands of years ago would have been built and stored along the former coastlines. Unfortunately, those former coastlines are now up to 300 feet below sea level today. While they might have been preserved, having been waterlogged for all this time, such wooden vessels would be almost impossible to find today. But that does not remove the evidence for ocean travel.

The evidence consists of the human occupation of islands and continents which would require travel over the ocean. Obsidian trade is probably the earliest evidence of sailing on the Mediterranean. Obsidian from the island of Melos was found in Peloponnesia in deposits of the 8th millennium B.C.,[119] but this same obsidian has also been found in even earlier deposits, Upper Paleolithic levels, at Franchthi Cave in Peloponnesia.[120]

In the Pacific, a distinctive pottery, known as Lapita pottery, was spread throughout Polynesia between 3400 and 2000 years ago and is evidence of ocean travel. But the islands of New Britain and New Ireland, on the eastern end of New Guinea, were never joined to the Sahul landmass. (During the glacial ages, Australia and New Guinea were one land mass, which is called the Sahul.) Thus to reach these islands someone had to travel approximately 15 miles over the ocean. The earliest known occupation of New Ireland, the farther island from New Guinea, was 33,000 years ago at Matenkupkum Cave.[121] These two islands were occupiable by hopping from island to island, each of which was visible from the previous island. However, the next earliest occupation of a Micronesia island was that of Buka in the Solomon Islands and it was not visible from New Ireland. Buka Island is 105 miles east of New Ireland. To travel there required a trip across open ocean over the horizon, but due to the height of the mountains on each island, the mariner was never out of site of land. Theoretically, Buka, 1,654 feet tall, could only be seen from 49 miles away.[122] Haze probably restricts this in reality to maybe 40 miles or so. The mountains on the south end of New Ireland rise to 6,738 feet and could probably be seen from 80-100 miles away. This is sufficient for a "safe" trip from New Ireland to Buka, which was occupied 28,740 years ago. Further north, Okinawa was occupied by humans 32,000 years ago and it too was never connected to the Asian mainland.[123]

The boats used by Polynesians when James Cook encountered them were marvels of craftsmanship. The Polynesians manufactured multi-hulled, multi-plank boats, propelled by paddles and sails and they were extremely fast. Some Tahitian canoes were 65 feet long, longer than many power cruisers. One canoe that Captain Cook saw was longer than his own ship. Polynesian ships were all made with stone-age tools.[124] Metal was unknown among the Polynesians prior to its introduction by the West. Trees were cut, planks were shaped by stone adzes (a type of axe), and then the planks were sewn together with vegetable fibers to form the decks.[125] This sewn form of construction was used in China where three planks were sewn together to form the sanpan, which literally means "three planks."

The navigational skills of the early Polynesians must have been developed over the millennia. McCrone observes,

> "It is a great myth of modern man that we are inherently rational. We talk about deductive logic as if it were wired into our brains and as if it were the main difference between us and other

animals, but formal logic is a very recent creation of modern man. The artificiality of Western logic is highlighted by the navigation feats of the Truk Islanders in the South Pacific, who regularly sail hundreds of miles between little coral islands, finding their way by 'feel.' The trained Western navigator would find his way around by charts and measurements. If asked at any time where he was, he could point to a map and give a logical step-by-step account of the course he must follow to get to his destination. The Truk Islander, however, has a mental picture of where the island lies over the horizon and points his boat in the right direction until he gets there, keeping an eye on the waves and the winds and the general look of the sun and stars. Without any conscious calculation, he can keep a feel of where he should be heading even while continually tacking from side to side when sailing into the wind.

"The Truk style of thinking is seen as primitive because the islanders are unable to articulate the rules by which they are maintaining their course. When asked, they shrug and point to where they think the island lies. Early humans must have existed for hundreds of thousands of years with such natural methods of thought, doing what they felt was right from experience or custom. Today we would describe such thinking as intuitive or instinctive, but it is really just allowing our rich nets of knowledge to surface in the conscious plane and then taking heed of the images they present. Modern man has been so trained to trust only thoughts driven along by logical chains of words that this inner knowledge is treated nervously. We may sometimes act on hunches and gut feelings, but we are happier when we can talk of the logical reasons for doing something. Indeed, we will often find an intuitive answer and then look for logical reasons to back up our decision."[126]

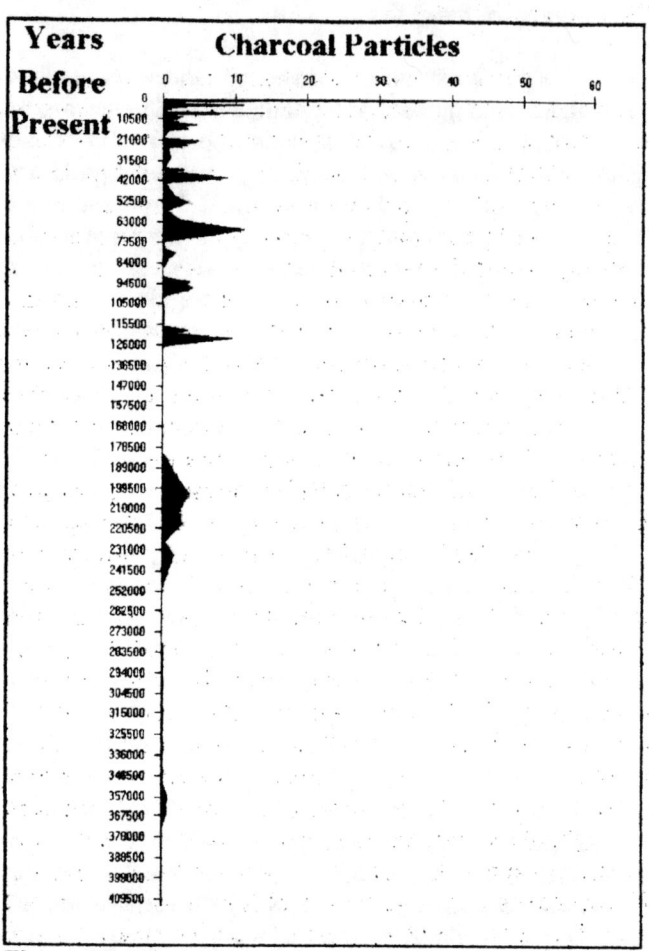

Figure 28 Charcoal particle density in the core from Lake George, Australia.

In the 1960's it was discovered that Australia had been inhabited for 10,000 years. Subsequent discoveries revealed an increasingly old colonization of the continent. Until last year, it was widely believed that Australia was first inhabited by humans 60,000 years ago. The timing of the colonization of Australia is quite important for over-ocean voyaging. It is quite impossible that Australia could have been seen during glacial periods with low sea levels from the highest peak on Timor, even on the clearest day. Timor is the location that is closest to the Australian coast and is believed to be the jumping off point for Australia's colonization.[127] During glacial periods, the trip to Australia required a trip of about 100 miles over open ocean.[128] Mankind would have had to have traveled out of sight of land with no foreknowledge that there was a land out there. Furthermore, once on the ocean's surface the voyagers could not see more than 3-4 miles in any direction. This means that they had to cross more than 25 consecutive horizons before reaching Australia. A trip like this shows that the first colonists of Australia possessed tremendous navigational confidence to go that far from land. The Timor Current flows southwest along the coast of Timor and would tend to shunt boats away from Australia.[129] Assuming this current existed at the time of colonization, the discovery of Australia would have had to have been an intentional voyage.

At Jinmium, Australia, a place on the northern coast close to the region archaeologists believe was first colonized, rock faces of a cliff are engraved with thousands upon thousands of small circles measuring about 1.2 inches in diameter. In 1996, R.L.K. Fullagar, the head of a team dating the site, reported on their results.[130] The dates were surprising and are going to force a reappraisal of the current views of the abilities of anatomically primitive peoples.[131] The oldest art on the wall was buried up to 5 feet deep,[132] and dating showed that it was 75,000 years old. But this was not the last surprise. There was evidence in the form of stone artifacts, starch grains and red ochre in sediments dated between 75,000 and 116,000 years ago. And in sediment dated even older, between 116,000 and 176,000 years ago, still more stone artifacts with sharpened edges, and starch grains were found.[133] This raises the possibility that Australia was inhabited as long ago as 176,000 years! There is more evidence of early habitation of Australia from a core taken from Lake George, New South Wales (Figure 28).[134] The core showed that 130,000 years ago there was a sudden increase in the number of prairie and forest fires on the continent.[135] Aborigines are well known for using fire to kill game. Thus, the increase in charcoal particles is likely to have been associated with the onset of these practices in Australia. This increase in soot was the most dramatic change in the entire Lake George core. Another core, this one from the ocean bottom off the Queensland Coast, is consistent with the Lake George core. This core shows an increase in charcoal particles associated with a decline of forest pollen around 140,000 years ago.[136] In Figure 28, one can clearly see the change in the pattern of charcoal particle density around that time. These cores, together with the artifacts found at Jinmium, paint a consistent picture that Australia was first colonized between 130-150,000 years ago. This is also in line with the fact that there was a major glacial episode centering around 140,000 years ago.[137]

What is the importance of this date? Quite simply, archaic *Homo sapiens* would have been the first inhabitant of Australia! Based upon present evidence, anatomically modern man was not on earth until 120,000 years ago (with some speculation that he had been in Africa for 200,000 years but no where else). He first appeared at Klasies River Mouth, South Africa and he did not spread to China until 67,000 years ago.[138] The anatomically modern human skeletons found in the Niah Cave, Borneo are less than 40,000 years old.[139] The oldest human remains currently known from Australia are the Lake Mungo skeletons dating to around 32,000 years ago.[140] The Lake Mungo remains are anatomically modern. Thus, while we lack evidence of the exact type of human who first inhabited Australia, the present evidence would suggest either archaic *Homo sapiens* or *Homo erectus*. It has recently been discovered that the last of the *Homo erectus'* lived a mere 27,000 to 53,000 years ago.[141]

This issue raises a point requiring a short diversion in our examination of the technology of the ancients. Christians have a problem if we exclude all beings except anatomically modern man from the human family. The evidence clearly shows a single colonization of Australia and a completely continuous culture since that initial occupation. If Australia was inhabited 140,000 years ago, then the only candidates for that colonization were the ancestors of modern aborigines--archaic forms of humanity. If archaic humans colonized Australia *and* they still have descendants there, then *if* the ancestors were not human but were mere animals, as some Christian apologists claim, then what is the theological status of Australian aborigines? Obviously the aborigines are fully human and they are descendants of Adam. However, the position advocated by many apologists would appear to exclude them. The more robust skeletons of the aborigines and slightly enlarged brow ridges link them to archaic humans who also shared these features. Speaking of the Sangiran skull, which was found in Indonesia, Shreeve writes,

> "Though some 700,000 years old, the face eerily resembled those of far younger, modern human fossils from Australia. The 'robust' Australian *sapiens* were as modern in brain size as any in the world, but they showed the same facial projection--big browridges, thick bones, sloping foreheads, and heavy molars--that Wolpoff saw in Sangiran 17. Many living Australian aborigines carry the same traits today."

> "Decades before, Weidenreich had suggested a connection between the *erectus* fossils of Java and modern Australian aborigines. But Weidenreich had only skullcaps to work from, like those of Solo Man from Ngandong. In the face of Sangiran, Wolpoff saw the missing anchor to Weidenreich's Australian lineage. Another researcher had suggested that the anatomy of Australian aborigines bore 'the mark of ancient Java.' For Wolpoff, Sangiran was stunning proof. In their arguments, Alan Thorne had been trying to convince Wolpoff that regional features would appear first at the remote edges of the hominid range, farthest away from the African birthplace of the earliest hominids. And here sat Sangiran 17, three quarters of a million years old and about as far from the African 'center' as one could get--but already full-fledged native

Australasian."[142]

The question of why these traits continue from ancient times to modern aborigines is a thorny one. The only acceptable answer that I am able to think of is that archaic humans, like archaic *Homo sapiens* and *Homo erectus* are fully human and children of Adam. In this way the ancestors of the aborigines were also human and spiritual. Any other solution raises the potential of excluding aborigines from humanity and thus the problems of racial discrimination.

Returning to the issue of boats, we find even earlier evidence of ocean crossings. The island of Flores, Indonesia has always been separated from mainland Asia by two straits. At the time of lowest sea levels there was a strait separating Bali from the combined islands of Lombok-Sumbawa and another strait separated Lombok-Sumbawa from Flores. Thus, to get to Flores would require two ocean voyages.[143] Sondaar and his team found stone tools associated with an extinct stegodon, a distant relative of the mammoths. At first I thought that this would require that Flores be attached to the Asian mainland because the very concept of mammoths swimming the ocean seemed ridiculous. Since many readers will probably think the same, I must mention what I have found out about mammoths. Mammoths, amazingly, were very good swimmers. They were found on the Channel Islands off California. Lister and Bahn relate,

> "Research indicates that there has not been a land bridge from the Channel Islands to the California mainland in recent geological history. The founding Columbian mammoths must therefore have arrived by swimming the 21 miles (35 km) or so from the mainland, plausible given the strong swimming powers of elephants."[144]

Since mammoths were also found on Malta in the Mediterranean,[145] and Malta was never attached to Europe, the mammoths must have swum there also. Bathymetric charts show that even when the sea level was 200 feet below the present levels, a 40 mile swim would have been required to reach Malta. Apparently they can swim quite well.

However, men do not swim as well as the probosceans. A swim in open ocean of several miles is not something that humans do with great regularity. While modern swimmers do swim marathon type distances, they only do so after years of training. Is it conceivable that ancient man would swim the distances to reach Flores? Throughout history men have studiously avoided being abandoned in the sea to sink or swim. Thus, given the fact that *Homo erectus* was engaged in woodwork as far back as 1.5 million years ago, the manufacture of boats may go back far into our distant past.

Very recently, there have been some speculation on possible ocean crossings occurring from Africa to Italy as long as 900,000 years ago.[146] This is based upon the discovery over the past few years of early fossil men inhabiting southern Europe. The oldest of these, is the Ceprano man found in Italy and dated to 900,000 years. The fossil material appears related to North African fossil men which then raises the question of how he got to Italy when there is currently no evidence of fossil men this early in Turkey, Greece and the Balkans. The article states,

> "The Italian scientists see more resemblance between Ceprano man and *erectus* fossils of about the same age that were found in the 1950s across the Mediterranean Sea in Algeria.
>
> "'That suggests that the origin of the people of Ceprano might be North Africa,' says Segre. If so, scientists must figure out how those early humans reached Italy without leaving traces in Turkey, Greece, or other points en route. Segre wonders if in the past western Europe and Africa were connected by land. Although other scientists doubt that idea, Segre cites the discovery of early stone tools in Sicily."[147]

and

> "'Large question marks' are what British paleontologist Alan Turner sees on potential passages from Africa through Gibraltar and Sicily. Even with a glacial sea-level drop such travel would have required a sea crossing. (Coastlines today and about 700,000 years ago are shown.) Evidence for the first routes into Europe remains scanty. Did early *Homo* find a way to cross deep water? Turner also suggests that large carnivores, like the saber-toothed cat and hyena above, probably

created obstacles to sustained hominid settlement north of the Mediterranean until about 500,000 years ago."[148]

There is the obvious possibility that future discoveries will find *Homo erectus* in Turkey and Greece at this time. But if not, the crossing of the ocean would not be too much out of line with the crossing of the ocean to Flores, Indonesia at a slightly later time. If the Indonesians could build a boat to cross the sea, then why not the north Africans?

Should the very new Ceprano man be ultimately disproven or moved to a later time, there is still a question concerning how the earliest Europeans got to Europe. The second oldest men found in Europe are from Gran Dolina, Spain, and date to more than 780,000 years ago.[149] If these are the first men in Europe, they are also the farthest from the land route into Europe, which goes through Turkey, Greece and the Balkans. The utter isolation of the Gran Dolina men accentuates the need to explain their existence in that locality. Even during glacial periods, there has not been a land bridge from Africa to Spain implying that these men might have crossed the Strait of Gibraltar. It is an intriguing possibility.

The Doctor was a Neanderthal

While many malpractice lawyers would give a hearty amen to the above statement, the earliest known surgery was performed by a Neanderthal. Birdsell wrote:

> "He was born with a withered right arm which was amputated above the elbow in life and healed successfully. This demonstrates a surprising skill in surgery in a Pleistocene society, and, more importantly, shows that a handicapped individual was able to survive. Presumably, it means that the society contained some altruism, enough to provide the essentials of food and shelter to its injured members."[150]

Man the Miner

Moles and rodents dig burrows but only for shelter. Man is the only being on earth who constructs subterranean mines for the purpose of removing inedible minerals for the manufacture of tools and other artifacts. Despite having worked in the mineral extraction industry for over 25 years, until I began this research I had no idea how far back in time mining extends. Like most people, I believed that the earliest mines were built sometime after the agricultural revolution began. I was familiar with the importance mines played in the success of the Roman Empire. Julius Caesar was nearly bankrupt when he was put in charge of the province of Spain, which had been seized from the Carthaginians during the second Punic War. Spain was the site of the world's most prolific gold and silver mines of that age. As was the custom, Caesar kept a portion of the profit of these mines, used the money to pay off his debts, and buy an army. The army then invaded Gaul and allowed Julius to became a member of the Triumvirate that ruled Rome. Because of this, Augustus, his nephew, was able to become the Imperator of Rome.[151] Because of the mines of Spain, the other world's empires named their kings after the Caesar family. These names took the form of Kaiser, Czar, Shah etc. Mines have played an important role in history.

There are two basic means of extracting minerals from the earth. There is surface mining (also called strip mining) in which the minerals are extracted without actually having rock over one's head. The second is subterranean mining in which a tunnel is cut into the rock and the miner has a ceiling of rock above his head. Both methods have been used by fossil men but the evidence for strip mining goes further back in time. The mines of Spain were subterranean.

The oldest identified iron mine of which I am aware is from Hungary and operated from 1100-800 B.C.[152] This is at the very beginning of what is known as the Iron Age. While iron is known from earlier times, the mass production of it didn't start until then. A piece of iron was found in the Great Pyramid and the earliest known iron objects are iron beads from Egypt, 6000 years ago.[153]

Australians, 20,000 years ago, were digging flint from the walls of Koonalda Cave in southern Australia.[154] The cave is a thousand feet long and the flint seam had to be reached by descending 180 feet deep into the darkness. The miners, carrying artificial lighting, then used their fingers to dig into the soft limestone containing flint, and indeed, they have left their finger markings on the walls of the cave. This site was used when the flint containing limestone outcrops were covered by high sea levels, cutting the aborigines off from their source of flint.

The mining of this cave illustrated a particular type of human reasoning. When the sea level rose, cutting them off from the flint outcrops, the aborigines had to form a mental hypothesis for the extension of the limestone into the subsurface. Knowledge of the cave would allow them to hypothesize that the limestone might be encountered if the cave went deep enough. Descending into the cave resulted in a payoff for their hypothesis. This is exactly the same type of reasoning that we in the oil industry use in looking for oil. Others also used this line of reasoning in their search for flint or other mineral species. Bednarik notes,

> "Underground mining involves quite a number of both technological and cognitive preconditions. To begin with, it requires a preparedness to enter an alien environment which most animal species avoid, or the behavioural flexibility to manage a perhaps genetically determined cortical response pattern to fear of caves. This already provides considerable insights into the level of conscious decision making required in this context. Next, most of the underground work presupposes the availability of artificial lighting, and there is some evidence of lamps and torches having been involved in these quests. It is also obvious from several of the sites that the mining activities must have been team work, involving at least two or three people, who no doubt had to co-ordinate various aspects of their efforts. We know that a variety of mining tools were involved, and we can assume that items such as pointed, perhaps fire-hardened wooden wedges were prepared outside the cave. At a few sites there is evidence of the use of scaffolding, which would imply even more planning. These observations together suggest that fairly complex planning patterns need to be postulated. Finally, some of the caves are of quite difficult access, and the sheer logistics of the mining operations conducted in them must have involved engineering skills of an order of magnitude few archaeologists would be currently prepared to credit any 'pre-Upper Palaeolithic' people with. Not only does the evidence for these abilities permit considerably more insight into the cognitive, intellectual, social and, presumably, linguistic skills of the people concerned than the futile and yet perennial arguments about language ability, the hyoid bone and Broca's area, there is still another factor to be considered.
>
> "I began this paper by explaining, in some detail the diagenetic conditions in which sedimentary silicas form, and why they occur primarily as tabular or quasi-tabular deposits. The geological reasons for this are known to us, and we can broadly explain the processes involved. But we have no reason to assume that the early miners were capable of rationalizing about these deposits in quite the same way. Yet the evidence seems to suggest, in some cases, that they were capable of predicting the occurrence and spatial extent of an as yet concealed geological feature. While it may be cognitively easy to follow an exposed seam, it is quite a different matter to undertake a calculated course of action that promises no immediate reward, and whose eventual reward is based entirely on the validity of an abstract prediction. Consider the procedure depicted in Figure 5: once the seam became inaccessible to methods of minimal labour input, a decision was made to remove the massive limestone overburden above the seam. This would have involved hours of back-breaking and most unpleasant work, without any guarantee of a reward- were it not for the expectation that the seam would in fact continue inwards. If it did not; the entire work effort would have been in vain. This implies that the miners were reasonably certain that the seam would continue horizontally. In other words they had the intellectual and cognitive capacity of observing and understanding a geological formation such as a tabular deposit: and they were capable of making an informed prediction with a sufficient degree of conviction to warrant the determined labour expenditure which we find documented."[155]

We will see that Neanderthal and archaic *Homo sapiens* were also capable of such reasoning.

The second oldest known coal strip-mining took place at Ostrava Petrkovice in Czechoslovakia, on the banks of the Oder River. The mining was carried out by anatomically modern men. This 32,000 year old site is located on a Carboniferous sandstone and is near the outcrop of the Ostrava coal seam.[156] Their hearths showed particles of coal and coke, which is consistent with the burning of coal. The excavators found several coal bracelets and coal discs that were in the process of being turned into bracelets when abandoned. The oldest coal strip mining was carried out by Neanderthal at Les Canalettes, France, 73,500 years ago.[157] The evidence for stripmining is indirect, just like that at Ostrava Petrkovice. But it is clear that since they burned coal in their hearths, they obtained it somewhere and strip mining is the most likely means. Only, in this case, there is direct evidence of the coal from the hearth. Apparently, France was so cold that there were few trees and Neanderthal resorted to the mining and burning of coal. When the weather warmed and trees returned, they gave up coal-mining.

Archaic *Homo sapiens* also engaged in subterranean mining. An amazing tale of geologic reasoning took place 33,000 years ago. Near the present-day village of Nazlet Khater, a gravel deposit containing cobbles of chert, up to 20 centimeters in diameter, suitable for making stone tools outcropped along a hill at the edge of the Nile. Apparently men collected cherts from this outcrop until such a time as the surface became picked over. They dug a 2 meter wide trench nine meters into the hillside in search for these cobbles.[158] Reasoning that there were more flints beneath the hill, these men moved up on top of the hill and mined the flint by sinking 5 foot deep shafts and then burrowing horizontally into the gravel. Radioactive dates at the base of one of the shafts yielded a date of 35,100 years ago.[159] Mining tools were left at the bottom of some of the shafts. These were axe-heads with concave sides used for hafting the tool to a handle. About half a kilometer northwest of the mining site a grave was discovered which contained an adult male. The man was an archaic *Homo sapiens*. Next to the skull was a 12-centimeter long axe head with concave sides, matching the tools found in the shafts.[160] The date of the skeleton was similar to the date of the mining activities. The owner of this skeleton was an early geologist.

Even earlier mining took place along the Nile. Near Qena, Egypt, 50,000 years ago,[161] ancient man dug a meter through the wind-deposited sands into a gravel layer to obtain chert.[162] Since no skeletons were found with the deposits, it is not known who did the digging.

Neanderthal mined for two different minerals: flint and red ochre. The flint was used for tools and the red ochre was used for burial rituals and, probably, body paint. At Lascabannes, in France, Neanderthals strip mined flint from a Jurassic outcrop.[163] They used this site to find and partially process the flints they obtained. Neanderthal tool manufacturing techniques, called Levallois, were among the most complex toolmaking techniques ever invented. As we pointed out earlier, no more than a dozen modern men have been able to master the skills needed to make Neanderthal tools. Since the technique was complicated, Neanderthals at Lascabannes would scratch the flint they obtained, to determine if it was of sufficient quality for tool-making. Flints not passing the test were found here. But Neanderthal did even more at this site. He partially prepared the flint for tool making by removing the outer part of the stone, called the cortex, in a particular fashion prior to taking the prepared cores elsewhere for tool finishing. All of this activity, the testing of the stone and the partial preparation of the core with the intention to carry the core elsewhere for the final processing, shows that Neanderthal people engaged in excellent future planning. Other sites, like Campsegret in the Bergerac region of France, Barbe Cormio and Lagrave, both in the Lot region of France, have also shown similar activities by Neanderthal.[164]

Veronika Gabon-Csank discovered a Neanderthal open-air mining site, similar in construction to Qena and Nazlet Khater 4, in which a deep shaft was dug in the search for flint. The site, Budapest-Farkasret, dates to 50,000 years ago and is clearly a Neanderthal site. She writes:

> "Dans le niveau le plus inferieur, parmi les outils en bois de cerf, nous avons trouve un racloir de type mousterien, et un chopping-tool."[165]

> "In the lowest level, among the deer bone tools, we found a racloir of Mousterian type and a chopping-tool." [Translation by Glenn Morton]

They dated the tools by comparison with the dates of similar tool types found in the same area.

> "C'est donc Budapest-Farkasret qui peut etre considere le premier site ou on connait, incontestablement, une mine une exploitation, a' ciel ouvert, et surtout des outils de mineurs a cette epoque."[166]

> "It is therefore Budapest Farkasret which can be considered the first site known which is incontestably an open air exploitation and especially the miner's tools of the same epoch." [Transl. Glenn Morton]

In this article, Gabon-Csank also mentions a mine in Switzerland near Bale that had Mousterian (Neanderthal) tools. I believe it is the same one described by E. Schmid.[167]

At Bara Bahau, France, at the end of a 116-meter cavern, evidence of Neanderthal mining was found.[168] The mining was carried out *before* the artwork that now occupies the cave was drawn. Bednarik believes that this was mining of Middle Paleolithic age. He further notes the geological reasoning used by the Middle Paleolithic people,

> "The cave entrance opens into a steep hill side overlooking the Vezere river, and among the rocky slopes next to it are numerous exposures of chert, with obvious quarrying traces. Therefore it would have been quite logical to search for further exposures of the strata within the cave."[169]

Neanderthals were using the same type of reasoning used by the men at Qena and Nazlet Khater 4, and the same as a modern day prospector.

In 1951 a red ochre mine was found at Lovas, Hungary, dating to 80,000 years ago.[170] One of the bone tools, an ulna, contained an intentionally drawn ladder pattern on it. Mezaros and Vertes believe that the ladder pattern is art. Another alternative might be the drawing of a ladder for the mine. We know that Paleolithic man used scaffolding to reach various places in caves. One gallery at Lascaux Cave has 20 sockets cut into the cave walls on both sides of the cave. These were used to support a wooden platform so that art could be drawn high on the cave ceiling.[171]

Mankind's search for pigments with which to paint themselves and other objects continues throughout history. A red mineral which easily powdered was found at Olduvai, 1.5 million years ago.[172] At Chowa, Zambia, the oldest manganese mine is found.[173] Manganese is a black pigment. The oldest subterranean mining, indeed the oldest mine in the world, is at Lion Cavern in Swaziland, where red ochre, a red pigment, was sought. Dart and Beaumont relate,

> "Lion Cavern is at the southern end of a steep scarp face of the Haematite Hill, Lion Peak, where ancient miners cut into the face of a cliff, more than 500 ft. high, a shelter-shaped working about 25 ft. wide, 30 ft. deep and 20 ft high. The floor of the shelter was covered by haematite soil and rubble. The first excavation, just inside the cavern's drip-line, bordered on the inner aspect of a large 5 ton haematite boulder fallen from above and almost blocking the entrance to the cavern."[174]

Radiocarbon dating 10 feet into the excavation yielded a 43,000 year old date.[175] Since this marks activity long after the onset of mining, it is believed that the mining operations began there 110,000 years ago.[176] Klein studied the fauna at the site and concluded that this date was correct.[177] Between 50 and 100 tons of red ochre were removed from this site.[178] If they went to all this trouble to obtain red ochre, it must have been very important to them. Red ochre could not aid in the acquisition of food. It was a material used in rituals of burial and body painting; there is no known utilitarian purpose for it. Some have suggested that it was used as a sunscreen, but dirt would work equally well and is obtained with much less trouble.

The collection of quartz crystals at Zhoukoudian, China 500,000 years ago is one evidence of mining for substances other than flint by *Homo erectus*. There is no flint found within two miles of the site.[179] Prior to this, there is no earlier evidence of mining except for the necessary "strip-mining" to find flints for tools. In this form, the search for flint for the manufacture of tools extends back to 2.6 million years ago at Gona, Ethiopia.[180] In this sense, mankind's search for and extraction of mineral wealth from the earth is a very long tradition.

Counting and Mathematics

A surprising little known fact is that numerals are not something that is innate to modern human beings. While, today, we are taught our numbers at an early age. But our numeral system was invented in a slow sequence of events which took place between 12,000 and 6,000 years ago. At the earliest stages of the agricultural revolution, the need for record keeping arose. Cities would build common storage areas for the farmer's grain. All the farmers of a region would put their produce into the same storage pit. When the population of the settlement was small, this caused no problem. When the population of the settlement grew to the point where a single person could not know everybody else, selfishness took over. The individual who put 20 baskets of grain into the storage pit, wanted to be able to remove 20 baskets of grain. To do this, required a record keeping system so no one would be cheated out of their grain. The solution to this problem was the invention of tokens. Between 8000 and 7500 BC farmers were given a clay token for each basket of grain placed in storage.[181]

The first appearance of tokens appears at the same time as a jump in the quantity of cereal pollen in the local soils. The initial system produced unmarked tokens each having a unique geometric form. Each shape was given for a particular product. A cone might be given for each jar of olive oil, another shape was given for each basket of wheat. This system worked well as long as the quantity of material deposited was not too large. As the economy grew, a man might have to keep track of many tokens. To avoid losing them, they were placed into a clay envelope which was then baked with the tokens inside. The tokens contained inside the envelope were impressed upon the outside of the envelope prior to firing. In this way, the contents of the envelope could be determined. An envelope with 5 disc impressions contained 5 discs. But once again, as the economy grew, the cities produced different products than the farms. A new set of complex, or decorated tokens was developed to keep track of these items. This took place around 3,500 B.C.

Eventually, the number of things which required record keeping increased to the point that the envelope system would not work. Holding hundreds of tokens became a bother. The solution to this was the use of a clay tablet containing the impression of a token for each item stored. Since there were no numerals yet, 50 baskets of wheat would be represented by 50 identical marks on the clay tablet. The entire concept of numerals was a foreign concept. There was only a concept of i,i,i not a concept of "three-ness". Three as an abstract number simply didn't exist. This is reflected in several ancient numeral systems. Roman numerals for 1, 2 and 3 consist of 1, 2 or 3 vertical lines. In Mandarin, the numbers 1, 2 and 3 consist of 1, 2 or 3 horizontal lines. But even this system of 50 impressions for 50 baskets of wheat became cumbersome. Numerals were invented which could be placed in front of the impressed token shape. Thus instead of writing "i,i,i,i,i,i" the scribe could write 6 i. Denise Schmandt-Besserat describes this transition,

> "Like the former tokens, the impressed signs continued to show the number of items counted by repeating the marking in one-to-one correspondence: one, two, or three small measures of grain were indicated by one, two, or three small wedges, and one, two or three bushels of grain were indicated by one, two or three circular markings. The same is true for the impressed markings indicating units of animal counts."[182]

Numerals are *not* innate to mankind, they are an *invention*. Schmandt-Besserat writes:

> "The invention of zero and place notation has been heralded as a major accomplishment of the civilized world, but the literature does not treat the advent of abstract numerals because of the common but erroneous assumption that abstract numbers are intuitive to humans. The token system is one piece of artifactural evidence proving that counting, like anything else, is not spontaneous. Instead, counting is cultural and has to be learned."[183]

Dean Falk notes of the above analysis,

> "It is instructive that it took up to 7,000 years for abstract clay tokens representing animals to occur after realistic images of a bull and deer first appeared at the entrance of Beldibi Cave in Turkey. It took another 5,000 years for humans to make the cognitive leaps from tokens to two-dimensional pictographs, numerals, and photmemes and another 1,500 years to establish a phonetic alphabet. Even though its domain was visual and, therefore, easily recordable, writing had a long evolution."[184]

Numerals are an invention of man. To expect *Homo erectus*, Neanderthal and early *Homo sapiens*, to have used numerals or writing prior to the economic need for these inventions, is to ask too much.
But agriculture was not the first need for counting. How did technologically primitive modern man enumerate things prior to the invention of numbers? Modern technologically primitive people used tally sticks to count time or objects. They would make a notch or mark on the stick, one mark for each item. Marshack explains,

> "A small notched message-stick of the Seneca, with a hole at one end for a knotted string, looks exactly like certain Upper Paleolithic notched bones and some Australian message-sticks.

It contains a marking of days, calling chiefs to a particular ceremony at a certain time. A small tally-stick of the Onondaga with twenty-seven notches was a 'condolence' record, listing twenty-seven chiefs who had died. A tribe of the Sioux at the end of the eighteenth and beginning of the nineteenth centuries had a 'slender pole about six feet in length, the surface of which was covered with small notches, and the old Indian who had it assured him [Clark] that it had been handed down from father to son for many generations and that those notches represented the history of his tribe for more than a thousand years back to the time when they lived near the ocean.'"[185]

Marshack gives another example of a calendar stick marking off 45 years of Pima Indian history. There was one unpainted notch for each year while painted notches and dots represented important events, like a meteor shower, a raid, a flood etc.[186] Upper Paleolithic man used very similar tally sticks to count the items important to their lives. These tally sticks are bones marked with parallel lines. Examples include, among many, many others, the Blanchard bone (172 marks)[187], Abri Lartet plaque (329 marks)[188], and the Ishango bone.(158 marks)[188].

Modern men were not the first beings to engage in this type of tally stick making. A tally stick was found at Cova Beneito which dates to 38,800 years ago and at the 53,900 year old site of Quneitra.[190] L. G. Strauss relates,

> "When all is said and done, however, there remains a residue of surprising objects-limited in number for no doubt both behavioral and preservational reasons. I have recently been struck, for example, by the discovery of a 'notationally' engraved bone and a perforated (fox?) canine apparently in a late Mousterian level at Cova Beneito (Alicante, Spain) and by the spiral-engraved flint nodule from a tuff-sealed Mousterian horizon at Quneitra on the Golan Heights. Bednarik takes us through a litany of cases that 'won't go away' and provides a reasonable discussion of their possible significance for human evolution."[191]

Bednarik, however, describes how such Neanderthal artifacts are received. He explains,

> "For instance, a set of parallel cut marks found in an Upper Palaeolithic context are inevitably seen as intentional, if not notational, while similar cut marks in a Lower Palaeolithic context are usually explained as incidental marks, attributable to some ultimately utilitarian process. A haematite pebble from an Upper Palaeolithic occupation stratum will always be accepted as evidence of pigment use, while its status will be questioned by many if it comes from a Lower Palaeolithic layer. Archaeologists apply these filters in a manner suggesting that they already know what the cognitive or intellectual capacities of hominids were. The truth is that no archaeologists possesses such insight, therefore the application of different standards to the evidence according to its age is clearly an unscientific practice."[192]

The 70,000 year old site of La Ferrassie yielded a stone with 35 parallel marks on it from the grave of a child. Marshack notes,

> "Even more significant, in a Neanderthal level at La Ferrassie in the Dordogne, a bit of stone in the grave of a child was intentionally engraved with linear sequences comparable to those in the Upper Paleolithic and to this one [Teviec—GRM] found in the Mesolithic."[193]

There are lots of other examples of Neanderthal-made tally sticks. Vallois and Vaufrey write:

> "Certain fragments [of bone] show orderly strokes which are made by a trial engraving. The other pieces carry a series of parallel marks, rather fine, of which the appearance is not like those made by butchery. Dr. Henri Martin has already pointed out at La Quina the transverse cutmarks are probably intentional and D. Peyrony at La Ferrassie [has pointed out that] a bone shows four series of transverse parallel marks."[194]

At the 90-100 kyr site of Prolom II, Crimea, contains horse teeth with up to five lines engraved on them.

Admittedly, at Prolom, it is uncertain whether these are objects of art or tally sticks, but they do raise the possibility. Other Neanderthal sites which contain bones with incised parallel lines include Dzhuruchula Cave in the Caucasus, Molodova and Pronyatin in the western Ukraine, La Ferrassie, Le Moustier and L'Ermitage in France.[195]

Going back to the beginning, the earliest possible tally sticks we have in the fossil record are those left on bone by *Homo erectus*. Bednarik writes:

> "In 1969, Bordes reported a rib fragment from the Acheulian of Pech de l'Aze as the earliest engraved object. It was thought to be in the order of 300,000 years old. More recently, several other Lower Palaeolithic engravings have been found: more than four on bone artefacts from Bilzingsleben, Germany, one on a small quartzite slab from the same site, and a set of engraved lines on an elephantine vertebra from Stranska Skala, Czech Republic. At both these sites, the finds come from strata containing skeletal fragments of Homo erectus."[196]

Bilzingsleben dates between 300 and 400 thousand years ago and the object with parallel lines found there, looks much like the parallel lined tally sticks of Upper Paleolithic man. These artifacts are evidence of counting capabilities in *Homo erectus*. His invention of the tally stick eventually led to the invention of numbers and then writing. *Homo erectus* was the first mathematician.

Food Processing

One of the implications of the Scriptural account is that Adam's immediate descendants processed plant matter for food. The evidence of food processing comes in the form of stone tools used to grind or process the food material. These objects were common features of early Neolithic sites over the past 10,000 years. There are four types of food-processors found in the archaeological record.[197] These are: rotary quern or rotary hand-mills, in which the rotation of the upper stone grinds the seeds into meal; mortars, in which the plant matter is pounded; grain-rubbers, where a stone is moved freely over a stone saucer; and saddle-querns, in which a to-and-fro motion was used. Rotary querns were used at mills along streams. The classical picture of an old mill, water wheel by its side, is nothing more than a rotary quern powered by water. Rotary querns are unknown prior to the agricultural revolution. It was probably a rotary grinder that Samson was forced to work (Judges 16:21). The saddle quern is referenced in Exodus 11:5.

But we must not think that plant processing is strictly a post agricultural technology. Even some archaeologists do not know their own literature. Steven Mithen says, "It is only quite recently, however, that archaeologists have been able to document a similar level of exploitation of plant foods by prehistoric hunter-gatherers."[198] He then points to Wadi Kubbaniya as an example of this relatively late use of plant exploitation. This is not true. Querns have been in use for at least 100,000 years. At Wadi Kubbaniya, in Lower Egypt, referenced by Mithen, infant feces were found containing undigested plant mush.[199] This means that someone had pounded the plant material for the purpose of weaning an infant. This fits quite well with the large number of grinding stones found at this site which dates to 19,000 years ago.[200]

At Cuddie Springs, Australia, plant pounding stones, presumably the pestles that accompanied mortars, were found that still had traces of plant tissue and starch grains on the polished surface.[201] The level from which it came dated 28,300 years ago.

While still controversial, the oldest evidence of plant processing and pounding stones comes from Jinmium, Australia, where a pounding stone was found in a shallow level dated between 75,000 and 116,000 years ago.[202] Another pounding stone was found with starch residue adhering to it in the oldest layer. This layer was deposited between 116,000 and 176,000 years ago.[203] The importance of this evidence of pounding stones located here, at this time, is that there were no anatomically modern men in Australia at this time. This means that archaic *Homo sapiens* engaged in plant processing. I have been unable to find evidence of similar behavior by Neanderthal, but then Neanderthal lived in Europe during the coldest periods of the ice ages where plants would have been scarce. In fact, the use of coal by Neanderthal 73,500 years ago at Les Canalettes may have been due to the lack of trees as I mentioned previously.

One other form of food processing must be discussed. This is the use of fire to cook various forms of food. The process of cooking tenderizes, sterilizes, and detoxifies the food and of all creatures on earth, only mankind uses fire to prepare his food. Since cooking relies on the prior control of fire, the oldest use of fire comes from Swartkrans, South Africa (1-1.5 million years ago)[204] and Chesowanja, Kenya (1.4 million years ago).[205] The two oldest cases of food preparation by means of fire that I have been able to discover are from Menez-Dregen in Brittany, France. The site is between 120 and 500 thousand years old. An even earlier site is the possible cooking which was carried out at Swartkrans, South Africa between 1 and 1.5 million years ago. Paul Bahn notes:

> "French archaeologists believe they have uncovered one of the world's oldest fireplaces, some 465 000 years old, in Brittany. These are by no means the oldest known traces of fire, which date back 1 to 1.5 million years in Africa (at Swartkrans Cave in South Africa, experiments suggested that the burnt animal bones had been cooked on a wood fire) but there is no proof that the fire in early sites of this kind was truly controlled.
> "The new French evidence comes from the site of Menez- Dregan, a collapsed cave on the shore near Plouhinec (South Finistere), which has been excavated for the past six years by a team led by Jean-Laurent Monnier."[206]

Bahn further writes that at Menez-Dregan:

> "Analyses prove that the stones have been heated to more than 500 degrees Celsius, indicating that the fire must have been carefully maintained.[207]

and further

> "...but at the Breton site, the bones of rhinos and big wild cattle by the fireplaces suggest that cooking was already in practice at this time." [208]

The oldest French chef was therefore half a million years old. But most interestingly, is the fact that at the Swartkrans site, in the level with the evidence for fire and cooking, the only fossils are of *Australopithecus robustus*. It is mere assumption that some species of our genus was at the site to handle the fire. For all we know, it might have been *Australopithecus* cooking the food.

One other aspect of food which relates directly to the inventiveness of Neanderthal is the acquisition and processing of marine shellfish and other aquatic foods. While we anatomically modern humans might want to believe that only we, with our intelligence and inventiveness were capable of realizing the potential of the sea, it is not true. The earliest examples of the exploitation of sea for food comes from Italy, 140,000 years ago.[209] Since modern man would not reach Europe for at least another 100,000 years, it is quite reasonable to attribute this activity to Neanderthals. Thus Neanderthals were able to invent a whole new industry which continues to this day.

After a good meal, French, Italian or otherwise, men throughout the world like to pick their teeth. My grandfather used to enjoy sitting out on the back porch, bourbon in hand, picking his teeth after dinner. How long has man engaged in this simple human pleasure? Apparently as long as he has been cooking. The long term use of toothpicks damages the teeth in a very characteristic pattern. According to Juan Luis Arsuaga Ferreras, such tooth damage was found on fossils from the 120-500,000 year old archeological site of Sima de los Huesos, Spain. He writes:

> "Several adult specimens, including skull 5, have grooves worn in the roots immediately under the crowns of rear teeth. This wear was apparently produced by the frequent use of toothpicks. Similar grooves have been observed in modern humans and other Homo fossils, the oldest in Ethiopian fossils dated between 1.5 and 2 million years ago."[210]

Thus, mankind has been cooking and picking his teeth for well over 1 million years.

Throwing

The use of a wide variety of tools is a hallmark of mankind. The simplest tool mankind uses is the stone thrown as a missile. While not in and of itself a technological innovation, the art of throwing stones and hitting targets accurately is a uniquely human activity.[211] Australian aborigines have been known to hit a target the size of a coin at 100 paces. Furthermore, they use stones effectively in their hunting. They throw stones into flocks of flying birds, knocking nuts out of trees, and killing wallabies. Calvin describes the physical requirements for accurately throwing at a target.

> "Throwing has a crucial timing step that is not present in hammering or clubbing or kicking: the projectile must be released at just the right instant. Too early, and the launch angle will be too high, the projectile overshooting the target. Too late, and the projectile hits the ground in front of the target. We can talk of the 'launch window' as that span of release times wherein the projectile will hit the target somewhere between its top and bottom. for a throw at a rabbit-sized target from a four meter distance (about the length of an automobile), this launch window is 11 ms wide. So at the end of a throw which started several hundred ms earlier, one must time the relaxation of the grip to stay within that 11 ms. The typical spinal motorneuron has an intrinsic timing jitter of at least that much when the cell is generating action potentials at 200 ms intervals.
>
> "But most people can, with practice, hit a rabbit-sized target at such a comfortable distance on perhaps half of the tries, so let us assume that the timing mechanisms are sufficient. Move the target out to twice the distance, eight meters, and practice enough to achieve hits on half the tries. The reason that this is so hard is that the launch window shrinks by a factor of eight to only 1.4 ms. The solid angle subtended by the target fell by a factor of four; furthermore, throwing twice as far with a reasonably flat trajectory means throwing twice as fast, so the time scale is halved. An electronic instrument would have no trouble in keeping its timing jitter to within a 1.4 ms window over several hundred milliseconds. But neurons would probably have to be redesigned from scratch."
>
> "Nature seldom redesigns, but it does one thing very well: it duplicates cells endlessly. And it seems that the way around cellular jitter is simply to assign a great many neurons to the same task, timing the launch, and then to average their recommendations. Need to halve the jitter? Just use four times as many cells as originally sufficed. To double the throwing distance and maintain hit rate, one needs an eight-fold reduction in jitter: that merely requires 64 times as many timing cells as originally sufficed. Triple the distance? Only 729 times as many cells."[212]

While *Australopithecus afarensis* had a hand which would have been capable of throwing, the earliest evidence of throwing comes later with *Homo habilis*. At Olduvai and Koobi Fora evidence was found for the development of a throwing ability. Isaac notes,

> "The archaeological evidence is more tenuous. Within the assemblages of early stone artefacts excavates at Olduvai Gorge in Tanzania are many unmodified stones that appear to have been carried in by the toolmakers—*Homo habilis*— around 1.9 million years ago. Many are of a suitable size and shape for throwing. There are few such stones at archaeological sites of comparable age at Koobi Fora in Kenya, but the hominids would have had easy access to cobbles and pebbles in the nearby streams. Acheulean sites of a million years in age or younger, such as Olorgesailie in Kenya, have many stones suitable for throwing.
>
> "Excavations of Mousterian sites have revealed that by 40 000 to 30 000 years ago, people occasionally stockpiled objects, probably for ammunition. Among the finds were more than 100 stone spheroids at La Quina in France and piles of dried clay balls-some of them small enough to have served as slingshot--at Achenheim in Germany."[213]

Gowlett adds to the possibilities for missiles as weapons,

> "At Olorgesailie in Kenya, a middle Pleistocene site dating to around 800 000 years ago, finds of manuports (unmodified stones) are of a type of lava available about a kilometre away, but unsuitable for toolmaking. By contrast, lavas suitable for handaxe-making were brought from volcanic mountains several kilometres distant."[214]

The only reason for someone to carry stones from a kilometer away and leave them unmodified is for use as ammo for throwing. It is obvious that the Olorgesailie people knew enough to use the better stone from further away to make their stone tools and the poor quality stone for throwing. While sparse, the evidence for stones thrown as weapons, a uniquely human trait, is as ancient as many of the other human characteristics we have been examining.

Calvin[215] and O'Brien[216] further suggest that the Acheulean hand axe was really a projectile thrown at game. If so, this would extend this unique human characteristic back 1.5 million years ago. Once again a human trait is found to exist in the earliest members of the genus *Homo*.

Bone Tools

In the tradition of ignoring evidence and downplaying the abilities of archaic humans like Neanderthal and *Homo erectus* which became so prevalent in anthropology over the past 20 years, Steven Mithen writes of the precursors to anatomically modern humans,

> "Why did early Humans ignore bone, antler and ivory as raw materials? Although there is evidence that Early Humans used pieces of unworked bone, such as for hammers when making handaxes, there are no carved artifacts made from bone, antler or ivory."[217]

Then he answers his question with,

> "The first was the absence of artifacts made from bone, antler or ivory. This can only be explained by recognizing that Early Humans could not think of using such materials for tools: these materials were once parts of animals and animals were thought about in the domain of natural history intelligence. The conceptual leap required to think about parts of animals using cognitive processes which had evolved in the domain of inert, physical objects appears to have been too great for Early Humans."[218]

If this is the case, then why did Bonifay find a bone awl at Regourdou Cave made by Neanderthal? Hayden writes:

> "Whatever the ultimate explanation may be for the increase in the use of bone in the Upper Paleolithic, there are occasional bone tools that clearly occur in Preneandertal and Neandertal contexts, as well as substantial use-wear traces of bone working in Mousterian industries and an example of an antler digging pick and awl from Regourdou. Thus, Mousterian groups were capable of making bone tools, but decided not to do so in most situations."[219]

Neanderthals made ivory beads, lance points, awls, and picks from bone at Grotte du Renne around 35,000 years ago.[220,221] An 80,000 year old bone mining pick from Lovas, Hungary had a ladder drawn on it.[222] A 250,000 year old digging stick made from a mammoth rib was found at Abri Vaufrey, France.[223] There were the ivory spear points from Torralba and Ambrona, Spain, dating to 400,000 years ago, reported by Howell and Freeman.[224] *Homo erectus* manufactured ivory lance points at Bilzingsleben 300,000 years ago.[225] Even the oldest tool based culture, the Oldowan from around 1.5 million years ago, has reports of bone tools.[226] While archaic humans didn't work a lot in bone, they were perfectly capable of it quite a long time ago. Similarly, the fully modern Tasmanian Aborigines did not make bone tools when Europeans found them.[227]

Animal Domestication

Everyone is aware of the major domestication events which took place at the beginning of the agricultural revolution. What is less well known is that some of these domestication efforts may have begun much earlier than anyone has realized. Recently there was a study of the mitochondrial DNA of wolves and dogs which produced a surprising result. It showed that dogs had split from the wolves somewhere between 135,000 to 400,000 years ago. This was surprising because the earliest domesticated dog, found in the fossil record, is from 14,000 years ago at Oberkassel, Germany.[228]

Vila et al, write:

> "The coyote and wolf have a sequence divergence of 0.07 +/- 0.002 and diverged about one

million years ago, as estimated from the fossil record. consequently, because the sequence divergence between the most different genotypes in clade 1 (the most diverse group of dog sequences) is no more than 0.010, this implies that dogs could have originated as much as 135,000 years ago. Although such estimates may be inflated by unobserved multiple substitutions at hypervariable sites, the sequence divergence within clade 1 clearly implies an origin more ancient than the 14,000 years before the present suggested by the archaeological record. Nevertheless, bones of wolves have been found in association with those of hominids those of hominids from as early as the middle Pleistocene, up to 400,000 years ago."[229]

To explain this discrepancy the researchers suggest that wolves took up residence with early man. Vila et al write:

""To explain the discrepancy in dates, we hypothesize that early domestic dogs may not have been morphologically distinct from their wild relatives. Conceivably, the change around 10,000 to 15,000 years ago from nomadic hunter-gatherer societies to more sedentary agricultural population centers may have imposed new selective regimes on dogs that resulted in marked phenotypic divergence from wild wolves." [230]

Wolves are a northern hemisphere, circumpolar species. Animals called wolves who live in the southern hemisphere belong to different genera. Because of the northern habitat of wolves, mankind could not have interacted with the wolf prior to leaving Africa. There are more than 30 subspecies of wolf so finding the ancestral wolf is problematic.[231] Two researchers have suggested that the Chinese wolf, *Canis lupus chanco*, is the dog's ancestor because it shares one trait with dogs that no other wolves have, The ascending ramus (the upright part of the mandible) on the lower jaw, is swept back like dogs.[232] It is interesting that this is one of the earliest canids found in association with man.[233]

Wolves do not normally live deep in caves yet they are found at Zhoukoudian Cave in China with *Homo erectus*. Olsen and Olsen relate,

"At the site of Chouk'outien, 42 kilometers southwest of Peking, some vertebrate remains were found in close association with Homo erectus pekinensis. An approximate date of 500,000 years ago can be set for the period of maximum Homo utilization of the cave's lower area. The canid found along with Peking Man has been described as Canis sp.. It was later renamed as Canis lupus variabilis, for, as was noted, it differed from the common wolf of the area only in that it was a bit smaller with a more slender muzzle and a weak sagittal crest."[234]

Site	Age (years)
Modern human sites:	
Kesslerloch Scweiszersbild	15,000
Timonovka Russia	23,000
Neanderthal sites:	
Wildkirchli, Switzerland	80,000
Lunel-Viel, France	200,000
Homo erectus sites:	
Zhoukoudian	500,000
Lan-tien	700,000
Yuanmou, Yunnan	700,000+

Table 4 Early sites with man and canid

The speculation is that the wolf was raised by Homo erectus which often tames it. This has been done by Buffon as well as people in this century. Zeuner writes:

"Buffon, as long ago as 1797, related that in Persia wolves were trained for shows, being taught to dance and exhibit a number of tricks. He himself reared several, and found them very docile and even 'courteous' during the first year. They never attempted to seize poultry or other animals when properly fed, until, they were about eighteen months old, when they began to do mischief. According to Blanford, the young of the Indian wolf are born with drooping ears, have all the habits of domesticated dogs and are readily tamed, and Mr. Cris Crisler has recently tamed them in Alaska. It is therefore easy to imaging conditions in which the domestication of wild dogs might have been carried out in more or less permanent living sites, such as camps or villages."[235]

There have been several suggestions as to what began man's relationship with the dog. The canid could serve as a watch dog during the night, giving warning when dangerous animals were approaching. Wolves could have aided in bringing down game, which otherwise would have escaped (although this is to be doubted). Finally, wolves could have served as a source of warmth on cold nights. The primitive Australian Aborigines sleep with their dogs on cold nights to keep warm. A really cold night is called a "five dog night"[236] The band "Three Dog Night" supposedly got their name from this saying. It was probably the warmth factor that was most important in beginning the domestication process. Aborigines probably treat the dingo as ancient man did. They will drive the dingos away at the beginning of a hunt considering them to be more of a nuisance than anything else.[237]

Whatever it was that early man saw in the wolf, it was Homo erectus whose bones are first found with canids. Table 4 shows other man-canid sites and their dates.[238] Once again, here is a human activity found among early man.

Conclusion

What we have seen is clear evidence that the ancient hominids, which some Christian apologists wish to exclude from the human family, were quite capable and inventive. They were behaving in a human fashion. If they behaved in a human fashion—performed similar work as modern men do— then how can we say that they were not human?

We will next look at the intelligence of these ancient men and the origin of the soul. Since the soul and spirit of man are the hallmarks of our divine origin, this issue is theologically important.

References

1. Jane Van Lawick-Goodall, In the Shadow of Man, (New York: Houghton Mifflin, 1971), p. 29-30
2. Jean Guilaine, "The First Farmers of the Old World," in Jean Guilaine, editor, Prehistory: The World of Early Man, (New York: Facts on File, 1986), p. 80-81.
3. Andrew Sherratt, "The Beginnings of Agriculture in the Near East and Europe," in Andrew Sherratt, editor, The Cambridge Encyclopedia of Archaeology, (New York: Cambridge University Press, 1980), p. 105-106.
4. Jean Guilaine, "The First Farmers of the Old World," in Jean Guilaine, editor, Prehistory: The World of Early Man, (New York: Facts on File, 1986), p. 80-81.
5. An example is from Lewis R. Binford, "Subsistence--a Key to the Past", in Steve Jones, et al, editors, The Cambridge Encyclopedia of Human Evolution, (Cambridge: Cambridge University Press, 1992), p. 368
6. Chris Stringer and Clive Gamble, In Search of the Neanderthals, (New York: Thames and Hudson, 1993), p.204
7. J.M. Coles and E. S. Higgs, The Archaeology of Early Man, (New York: Frederick A. Praeger, 1969), p. 298
8. J.M. Coles and E. S. Higgs, The Archaeology of Early Man, (New York: Frederick A. Praeger, 1969), p. 298
9. Chris Stringer and Clive Gamble, In Search of the Neanderthals, (New York: Thames and Hudson, 1993), p. 206
10. Richard G. Klein, The Human Career, (Chicago: The University of Chicago Press, 1989), p. 315
11. Paul Mellars, The Neanderthal Legacy, (Princeton: University Press, 1996), p. 313
12. The tallest Neanderthal is Amud at 5' 10". Jim Foley, Personal Communication, Oct. 3, 1997.
13. Brian Hayden "The Cultural Capacities of Neandertals ", Journal of Human Evolution, 24(1993):113-146, p. 134-135; Leslie Freeman, "The Development of Human Culture," in Andrew Sherratt, editor, Cambridge Encyclopedia of Archaeology, (New York: Cambridge University Press, 1980), p. 84-85
14. Yuri Smirnov "Intentional Human Burial: Middle Paleolithic (Last Glaciation) Beginnings," Journal of World Prehistory, 3:2(1989), pp 199-233, p. 219
15. Bar-Yosef, et al., "The Excavations in Kebara Cave, Mt. Carmel," Current Anthropology, 33(Dec. 1992):5:497-550, p. 531
16. Ofer Bar-Yosef and Bernard Vandermeersch, "Modern Humans in the Levant," Scientific American, 268(April,

1993):4:94-100, p. 99
17. Brian Hayden, "The Cultural Capacities of Neandertals: A Review and Re-evaluation," Journal of Human Evolution, 24(1993):113-146, p. 135
18. Brian Hayden, "The Cultural Capacities of Neandertals: A Review and Re-evaluation," Journal of Human Evolution, 24(1993):113-146, p. 135
19. Emil Bachler, Das Drachenloch, (St. Gallen, Druck der Buchdruckerei Zollikofer & Cie., 1921), plate 15, p. 80.
20. Emil Bachler, Das Alpine Palaolithikum der Schweiz, II, (Basel: Verlag Birkhauser & CIE., 1940), plate 112.
21. Brian Hayden "The Cultural Capacities of Neandertals," Journal of Human Evolution, 24(1993):113-146, p. 133.
22. Andre Leroi-Gourhan, The Hunters of Prehistory, (New York: Atheneum, 1989), p. 133
23. Stephen Mertens, "The Middle Paleolithic in Roumania," Current Anthopology, 37:3, June 1996, pp.515-521, pp. 519
24. Chris Stringer and Clive Gamble, In Search of the Neanderthals, (New York: Thames and Hudson, 1993), p.157
25. Chris Stringer and Clive Gamble, In Search of the Neanderthals, (New York: Thames and Hudson, 1993), p. 207
26. Robert G. Bednarik, "Neanderthal News," The Artefact, 18(1996): 104.
27. Donald O. Henry et al, "Middle Paleolithic Behavioral Organization: 1993 Excavation of Tor Faraj, Southern Jordan," Journal of Field Archaeology, 23(1996):31-53, p, 51
28. Kathy D. Schick and Nicholas Toth, Making Silent Stones Speak, (New York: Simon and Schuster, 1993), p. 281
29. Paul C. Mellars, The Neanderthal Legacy, (Princeton: University Press, 1996), p. 285
30. Paul Mellars, The Neanderthal Legacy, (Princeton: University Press, 1996), p. 295. The Mindel glaciation occurred 300-400 thousand years ago.
31. Brian Hayden "The Cultural Capacities of Neandertals ", Journal of Human Evolution, 24(1993):113-146, p. 133
32. Christopher Wills, The Runaway Brain, (New York: Harper Collins, 1993), p.108-109
33. Emanuel Vlcek, "A New Discovery of Homo erectus in Central Europe,"Journal of Human Evolution, (1978) 7:239-251, p. 250
34. D. Mania and U. Mania and E. Vlcek, "Latest Finds of Skull Remains of Homo erectus from Bilzingsleben (Thruingia)", Naturwissenschaften, 81(1994), p. 123-127, p. 124
35. von Rudolf Feustel, in H. Muller-Beck and G. Albrecht eds. 1987 Die Anfange der Kunst vor 30,000 Jahren. pp 60-63; see also Paul G. Bahn and Jean Vertut, Images in the Ice, (Leichester: Windward, 1988), p. 208 note 63.
36. Robert G. Bednarik, "Wonders of Wonderwork Cave," The Artefact, 16(1993), p. 61
37. M.D. Leakey, "Primitive Artefacts from Kanapoi Valley," Nature, Nov. 5, 1966, p. 579
38. Donald Johanson and James Shreeve, Lucy's Child, (New York: William Morrow and Co., Inc., 1989), p. 152-153
39. Ian Tattersall, The Fossil Trail (New York: Oxford University Press, 1995), p.208
40. Thieme, H. 1997. Lower Paleolithic hunting spears from Germany. Nature 385(Feb. 27):807.
41. Kathy D. Schick and Nicholas Toth, Making Silent Stones Speak, (New York: Simon and Schuster, 1993), p.214
42. Brian M. Fagan, The Journey From Eden, (London: Thames and Hudson, 1990), p. 151.
43. Brian Hayden "The Cultural Capacities of Neandertals," Journal of Human Evolution, 24(1993):113-146, p. 133
44. Brian Hayden "The Cultural Capacities of Neandertals," Journal of Human Evolution, 24(1993):113-146, p. 133
45. Brian Hayden "The Cultural Capacities of Neandertals," Journal of Human Evolution, 24(1993):113-146, p. 132-133
46. D. Mania and U. Mania and E. Vlcek, "Latest Finds of Skull Remains of Homo erectus from Bilzingsleben (Thruingia)", Naturwissenschaften, 81(1994), p. 123-127, p. 124
47. Science News, Vol. 144, p. 54; Science News, 144:418
48. Ivar Lissner, The Living Past, translated by J. Maxwell Brownjohn, (New York: G. P. Putnam's Sons, 1957), p. 102
49. Sungir dates to 23,000 years B. P. see Richard Klein, "Later Pleistocene hunters," in Andrew Sherratt, editor,

The Cambridge Encyclopedia of Archaeology, (New York: Cambridge University Press, 1980), p. 90
50. "Costume", The Software Toolworks Multimedia Encyclopedia, 1992, (Grolier Electronic Publishing, Inc. 1992).
51. Carl O. Sauer, Agricultural Origins and Dispersals, (Cambridge: MIT Press, 1969), p. 78
52. "Agriculture, History of", Encyclopedia Britannica 1982, 1, p. 325-326.
53. Donald Johanson and Blake Edgar, From Lucy to Language, (New York: Simon and Schuster, 1997), p. 99
54. Olga Soffer, "Sungir: A Stone age Burial Site", in Goran Burenhult, The First Humans, (San Francisco: HarperSanFrancisco, 1993), p. 138-139.
55. Brian Fagan, The Journey From Eden, (London: Thames and Hudson, 1990), p. 181
56. Brian Fagan, The Journey From Eden, (New York: Thames and Hudson, 1990), p. 232; Richard Klein, "Later Pleistocene Hunters," in Andrew Sherratt, editor, The Cambridge Encyclopedia of Archaeology, (New York: Cambridge University Press, 1980), p. 94.
57. Brian Fagan, The Journey From Eden, (London: Thames and Hudson, 1990), p. 181
58. Michael R. Waters, Steven L. Forman, and James M. Pierson,"Diring Yuriakh: A Lower Paleolithic Site in Central Siberia,", Science, 275(Feb. 28, 1997):1281-1283, p. 1283
59. B. Bowers, "Stone Age Fabric Leaves Swatch Marks," Science News, 147:276, May 6, 1995, p. 276.
60. James R. Shreeve, The Neandertal Enigma, (New York: William Morrow and Co., 1995), p. 52
61. Andre Leroi Gourhan, The Hunters of Prehistory, transl. Claire Jacobson, (New York: Atheneum, 1989), p. 94-95
62. Boyce Rensberger, "Racial Odyssey," in Elvio Angeloni, Editor, Annual Editions Physical Anthropology 94/95, (Sluicedock, Guilford, Conn.: The Dushkin Publishing Group, Inc., 1994), p.40-45, p. 42
63. "Clothing and Footwear Industry," Encyclopedia Britannica, 4, (Chicago: Encyclopedia Britannica, 1982), p. 750
64. Kathy D. Schick and Nicholas Toth, Making Silent Stones Speak, (New York: Simon and Schuster, 1993), p.162
65. H. P. Schwarcz et al, "The Bilzingsleben Archaeological Site: New Dating Evidence," Archaeometry, 30:1(1988):5-17.
66. Michael R. Waters, Steven L. Forman, and James M. Pierson,"Diring Yuriakh: A Lower Paleolithic Site in Central Siberia,", Science, 275(Feb. 28, 1997):1281-1283
67. Howard J. Critchfield, General Climatology, (Englewood Cliffs: Prentice-Hall, Inc., 1966), p. 210
68. Donald C. Johanson, Lenora Johanson, and Blake Edgar, Ancestors, (New York: Villard Books, 1994), p. 163-165
69. Wayne Frair and P. William Davis, The Case for Creation (Chicago: Moody Press, 1967), p. 66
70. Wayne Frair and P. William Davis, The Case for Creation (Chicago: Moody Press, 1967), p. 66
71. Hugh Ross, The Fingerprint of God, (Orange: Promise Publishing, 1991), p. 159-160.
72. Kathy D. Schick and Nicholas Toth, Making Silent Stones Speak, (New York: Simon and Schuster, 1993), p.159
73. Robert G. Bednarik, "On Lower Paleolithic Cognitive Development," 23rd Chacmool Conference Calgary 1990, pp 427-435, p. 433
74. Ivar Lissner, The Living Past, translated by J. Maxwell Brownjohn, (New York: G. P. Putnam's Sons, 1957), p. 371.
75. Kathy D. Schick and Nicholas Toth, Making Silent Stones Speak, (New York: Simon and Schuster, 1993), p. 157
76. Victor Barnouw, An Introduction to Anthropology: Physical Anthropology and Archaeology, 1, (Homewood, Ill: The Dorsey Press, 1982), p.185
77. Clive Gamble, Timewalkers, (Cambridge: Harvard University Press, 1994), p. 216-217
78. Richard Klein, "Later Pleistocene Hunters," in Andrew Sherratt, editor, The Cambridge Encyclopedia of Archaeology, (New York: Cambridge University Press, 1980), p. 90; Victor Barnouw, An Introduction to Anthropology: Physical Anthropology and Archaeology, 1, (Homewood, Ill: The Dorsey Press, 1982), p. 162
79. E. James Dixon, Quest for the Origins of the First Americans,(Albuquerque: University of New Mexico Press, 1993), p. 96
80. L. Luca Cavalli-Sforza, Paoli Menozzi and Alberto Piazzi, The History and Geography of Human Genes, (Princeton: Princeton University Press, 1994), p. 306
81. Robert G. Bednarik, "Palaeoart and Archaeological Myths," Cambridge Archaeological Journal, 2(1) (1992):27-57, p. 38 259,000 years ago Rainer Grun et al, "Direct Dating of Florisbad Hominid," Nature, 382, August 8, 1996,

p. 500-501.
82. Christopher Wills, The Runaway Brain, (New York: Harper Collins, 1993), p.104
83. Simon Holdaway, "Tool Hafting with a Mastic" Nature 380, March 28, 1996, p. 288-289
84. Eric Boeda, et al, "Bitumen as a Hafting Material on Middle Palaeolithic Artefacts," Nature, 380, March 28, 1996, p. 337
85. Brian Hayden "The Cultural Capacities of Neandertals," Journal of Human Evolution, 24(1993):113-146; p. 115-116
86. Paul Mellars, The Neanderthal Legacy, (Princeton: University Press, 1996), p. 114-115
87. Paul G. Bahn, "Excavation of a Palaeolithic Plank from Japan," Nature, 329(1987), p. 110.
88. Colin Groves, "Human Origins," in Goran Burenhult, The First Humans, (San Francisco: HarperSanFrancisco, 1993), p. 49
89. Grahame Clark and Stuart Piggott, Prehistoric Societies, (New York: Alfred A. Knopf, 1965), p. 60
90. Hallam L. Movius, Jr., "A Wooden Spear of Third Interglacial Age from Lower Saxony," Southwestern Journal of Anthropology, 6(1950):139-142.
91. Paul Mellars, The Neanderthal Legacy, (Princeton: University Press, 1996), p. 227
92. B.Bower, "Early humans Make their Marks as Hunters," Science News, April 12, 1997, p.222
93. Donald C. Johanson, Lenora Johanson, and Blake Edgar, Ancestors, (New York: Villard Books, 1994), p. 275
94. Paul Mellars, The Neanderthal Legacy, (Princeton: University Press, 1996); p. 119-120
95. Kathy D. Schick and Nicholas Toth, Making Silent Stones Speak, (New York: Simon and Schuster, 1993), p.289-290
96. Paul Mellars, The Neanderthal Legacy, (Princeton: University Press, 1996), p. 308
97. Rainer Grun et al, "Direct Dating of Florisbad Hominid," Nature, 382, August 8, 1996, p. 500-501
98. Thomas P. Volman, "Early Prehistory of Southern Africa," in Richard G. Klein, Prehistory and Paleoenvironments, (Boston: Balkema, 1984), p. 215; J. M. Lindly and G. A. Clark, "Symbolism and Modern Human Origins," Current Anthropology, 31(1990):233-261, p. 237
99. Kathy D. Schick and Nicholas Toth, Making Silent Stones Speak, (New York: Simon and Schuster, 1993), p. 271
100. Paul Mellars, The Neanderthal Legacy, (Princeton: University Press, 1996), p. 227
101. Lewis R. Binford, "Early Human Behaviour and Ecology," in Steve Jones et al, editors, The Cambridge Encyclopedia of Human Evolution, (Cambridge: Cambridge University Press, 1992), p. 366
102. Lewis R. Binford, "Early Human Behaviour and Ecology," in Steve Jones et al, editors, The Cambridge Encyclopedia of Human Evolution, (Cambridge: Cambridge University Press, 1992), p. 366
103. Hartmut Thieme, "Lower Palaeolithic Hunting Spears form Germany," Nature, 385(Feb. 27,1997), p. 808
104. B. Bower,"German Mine Yields Ancient Hunting Spears," Science News, 151(March 1, 1997), p. 134.
105. Hartmut Thieme, "Lower Palaeolithic Hunting Spears form Germany," Nature, 385(Feb. 27,1997), p. 809
106. Hartmut Thieme, "Lower Palaeolithic Hunting Spears form Germany," Nature, 385(Feb. 27,1997), p. 808
107. Robin Dennell, "The World's Oldest Spears," Nature 385(Feb. 27, 1997), p. 767
108. Sabine Gaudzinski et al, "Palaeoecology and Archaeology of the Karlich-Seeufer Open-Air Site (Middle Pleistocene) in the Central Rhineland, Germany," Quaternary Research, 46(1996):319-334, p. 332
109. H. P. Schwarcz et al, "The Bilzingsleben Archaeological Site: New Dating Evidence," Archaeometry, 30:1(1988):5-17.
110. Dietrich Mania and Ursula Mania, "Deliberate Engravings on Bone Artefacts of Homo Erectus," Rock Art Research 5:2(1988): 91-107, p. 91-92
111. S. Belitszky et al, "A Middle Pleistocene Wooden Plank with man-made Polish," Journal of Human Evolution, 20(1991):349-353.
112. Kathy D. Schick and Nicholas Toth, Making Silent Stones Speak, (New York: Simon and Schuster, 1993), p.160
113. Steven Mithen, The Prehistory of the Mind, (New York: Thames & Hudson, 1996), p. 27
114. Lawrence Guy Straus, "Southwestern Europe at the Last Glacial Maximum", Current Anthropology, 32:2, (April 1991), pp 189-199, p. 190
115. Patricia Anderson-Gerfaud, "Aspects of Behaviour in the Middle Palaeolithic: Functional Analysis of Stone Tools from Southwest France," in Paul Mellars, The Emergence of Modern Humans, (Ithica: Cornell Univ. Press, 1990), pp. 389-418, p. 400
116. L. Luca Cavalli-Sforza, Paoli Menozzi and Alberto Piazzi, The History and Geography of Human Genes,

(Princeton: Princeton University Press, 1994), p. 202

117. Grahame Clark and Stuart Piggott, Prehistoric Societies, (New York: Alfred A. Knopf, 1965), p. 106

118. Roy Larick and Russell L. Ciochon, "The African Emergence and Early Asian Dispersals of the Genus *Homo*," American Scientist, 84(Nov/Dec, 1996), p.548-550

119. Jean Vaquer, "Reliving Prehistory,"in Jean Guilaine, editor, Prehistory: The World of Early Man, (New York: Facts on File, 1986), p. 179.

120 Colin Renfrew, Archaeology & Language (New York: Cambridge University Press, 1987), p. 168

121. Clive Gamble, Timewalkers, (Cambridge: Harvard University Press, 1994), p. 228

122. The formula for the distance to the horizon is approximately 1.224 x square root of the height in feet. see "Horizon" Encyclopedia Britannica, V, (Chicago: Encyclopedia Britannica, 1982), p. 132-133

123. Clive Gamble, Timewalkers, (Cambridge: Harvard University Press, 1994), p. 228

124. John R. Whiting,"Boat and Boating" The Software Toolworks Encyclopedia, 1992 Ed. version 1.5. Text Copyright Grolier Inc. 1992

125. Howard I. Chapelle, "Boats and Boating", Microsoft Encarta, 1994.

126. John McCrone, The Ape That Spoke, (New York: William Morrow and Company, 1991), p. 134-135

127. Christopher Wills, The Runaway Brain: The Evolution of Human Uniqueness, (London: HarperCollins, 1994), p. 145. Willis says that the strait between Timor and Australia would have only been 70-90 kilometers wide (43-55 miles). But measurements from the 200 foot contour to 200 foot contour on maps shows that the gap would have been twice that. Bartstra agrees. see Gert-Jan Bartstra et al, "On the Dispersion of *Homo sapiens* in Eastern Indonesia: The Palaeolithic of South Sulawesi," Current Anthropology, 32:3(1991), p. 317-321, p. 317

128. Gert-Jan Bartstra et al, "On the Dispersion of *Homo sapiens* in Eastern Indonesia: The Palaeolithic of South Sulawesi," Current Anthropology, 32:3(1991), p. 317-321, p. 317

129. "Timor Current," Encyclopedia Britannica, IX, (Chicago: Encyclopedia Britannica, 1982), p. 1018.

130. R.K.Fullagar, D.M. Price & L.M. Head, "Early Human Occupation of Northern Australia: Archaeology and Thermoluminescence Dating of Jinmium Rock-Shelter, Northern Territory," Antiquity 70(1996):751-773, p. 754.

131. Brian M. Fagan, The Journey From Eden, (London: Thames and Hudson, 1990), p.127; Chris Stringer and Clive Gamble, In Search of the Neanderthals, (New York: Thames and Hudson, 1993), p.199

132. "Engravings could be World's Oldest Artwork" Dallas Morning News, Saturday, September 21, 1996, p. 14A

133. B. Bower, "Human Origins Recede in Australia," Science News Sept 28, 1996, p. 196

134. G. Singh, A. P. Kershaw and Robin Clark, "Quaternary Vegetation and Fire History in Australia," in Fire and the Australian Biota, (eds. A. M. Gill et al.) (Canberra: Australian Acad. of Science, 1981), pp 23-54; see also G. Singh, N. D. Opdyke, and J. M. Bowler, "Late Cainozoic Stratigraphy, Palaeomagnetic Chronology and Vegetational History from Lake George, N.S.W.", Journal of the Geological Society of Australia, 28(4), 1981, pp 435-452

135. A. Peter Kershaw, "Palynology, Biostratigraphy and Human Impact," The Artefact, 1993, 16:12-18, p. 17; Paul G. Bahn, "Further Back Down Under," Nature, Oct 17, 1996, p. 577-578, p. 578

136. Paul G. Bahn, "Further Back Down Under," Nature, Oct 17, 1996, p. 577-578, p. 578

137. A. McIntyre, W. F. Ruddman and R. Jantzen, "Southward Penetrations of the North Atlantic Polar Front: Faunal and Floral Evidence of Large-Scale Surface Water Mass Movements Over the Last 225,000 Years," Deep-Sea Research; 1972: 19:61-77, p. 71.
This is the Riss Glaciation.

138. L. Luca Cavalli-Sforza, Paoli Menozzi and Alberto Piazzi, The History and Geography of Human Genes, (Princeton: Princeton University Press, 1994), p. 203

139. Gunter Brauer, "The Evolution of Modern Humans: a Comparison of the African and non-African Evidence," in Paul C. Mellars and Chris B. Stringer ed. The Human Revolution. (Princeton: Princeton University Press, 1989), pp. 123-153, p. 145

140. L. Luca Cavalli-Sforza, Paoli Menozzi and Alberto Piazzi, The History and Geography of Human Genes, (Princeton: Princeton University Press, 1994), p. 344

141. C. C. Swisher III, W. J. Rink, S. C. Antón, H. P. Schwarcz, G. H. Curtis, A. Suprijo, Widiasmoro, "Latest Homo erectus of Java: Potential Contemporaneity with Homo sapiens in Southeast Asia" Science Volume 274, Number 5294, Issue of 13 December 1996, pp. 1870-1874

142. James R. Shreeve, The Neandertal Enigma, (New York: William Morrow and Co., 1995), p. 102-103

143. P.Y. Sondaar, et al., "Middle Pleistocene faunal turnover and Colonization of Flores(Indonesia) by *Homo erectus*," Comptes Rendus de l'Academie des Sciences. Paris 319:1255-1262, p. 1261

144. Adrian Lister and Paul G. Bahn, Mammoths, (London: Boxtree, 1994), p. 34
145. Adrian Lister and Paul G. Bahn, Mammoths, (London: Boxtree, 1994), p. 35
146. Rick Gore, "The First Europeans," National Geographic, July, 1997, p. 102
147. Rick Gore, "The First Europeans," National Geographic, July, 1997, p. 101
148. Rick Gore, "The First Europeans," National Geographic, July, 1997, p. 102
149. Rick Gore, "The First Europeans," National Geographic, July, 1997, p. 104.
150. J.B. Birdsell, Human Evolution, (Chicago: Rand McNally & Co., 1972), p. 285
151. Noble Metals, (Alexandria: Time-Life Books,) p. 23-28
152. R. Shepard, Prehistoric Mining and Allied Industries, (New York: Academic Press, 1980), p. 210.
153. "Iron", Microsoft Encarta Multimedia Encyclopedia, 1994.
154. Colin Renfrew, Past Worlds, (New York: Random House, 1995), p. 68
155. Robert G. Bednarik, "Early Subterranean Chert Mining," The Artefact, 15:(1992), pp 11-24, p. 20-21
156. R. Shepard, Prehistoric Mining and Allied Industries, (New York: Academic Press, 1980), p.231-232
157. I. Théry, J. Gril, J. L. Vernet, L. Meignen, J. Maury, "Coal Used for Fuel at Two Prehistoric Sites in Souther France: Les Canalettes (Mousterian) and Les Usclades (Mesolithic)," Journal of Archaeological Science, v 23, n 4, July 1996, p509-512
158. Robert G. Bednarik, "Early Subterranean Chert Mining," The Artefact, 15:(1992), pp 11-24.
159. P. M. Vermeersch, et al., "33,000-yr Old Chert Mining Site and Related Homo in the Egyptian Nile Valley," Nature, 309(May 24, 1984): 342-344
160. Robert G. Bednarik, "Early Subterranean Chert Mining," The Artefact, 15:(1992), pp 11-24.
161. P.M. Vermeersch and E. Paulissen, "The Oldest Quarries Known: Stone Age Miners in Egypt," Episodes, 12:1(March 1989), p. 35-36, p. 35.
162. P. M. Vermeersch, E. Paulissen, G. Gijselings and J. Janssen, "Middle Palaeolithic Chert Exploitation Pits Near Qena (Upper Egypt)," Paleorient (1986), 12(1):61-65.
163. Paul C. Mellars, The Neanderthal Legacy, (Princeton: University Press, 1996), p. 265
164. Paul Mellars, The Neanderthal Legacy, (Princeton: University Press, 1996), p. 160
165. Veronika Gabori-Csank, "Une mine de silex Paleolithique a Budapest, Hongrie," in H. L. Dibble and A. Montet-White eds. Upper Pleistocene Prehistory of Western Eurasia, pp 141-143. Monograph 54 University Museum. University of Pennsylvanian, Philadelphia., p. 142.
166. Veronika Gabori-Csank, "Une mine de silex Paleolithique a Budapest, Hongrie," in H. L. Dibble and A. Montet-White eds. Upper Pleistocene Prehistory of Western Eurasia, pp 141-143. Monograph 54 University Museum. University of Pennsylvanian, Philadelphia., p. 143
167. E. Schmid, "Ein Silex—abbau aus dem Mousterien im Berner Jura, Ur-Schweiz, xxxii(1968):4:53-65
168. Robert G. Bednarik, "Early Subterranean Chert Mining," The Artefact, 15:(1992), pp 11-24, p. 14
169. Robert G. Bednarik, "Early Subterranean Chert Mining," The Artefact, 15:(1992), pp 11-24, p. 14
170. R. Shepard, Prehistoric Mining and Allied Industries, (New York: Academic Press, 1980), p. 210.
171. Adrian Lister and Paul Bahn, Mammoths, (London: Boxtree, 1994), p. 95
172. D. Bruce Dickson, The Dawn of Belief, (Tuscon: The University of Arizona Press, 1990), p. 42-43
173. R. A. Dart and P. Beaumont, "Amazing Antiquity of Mining in Southern Africa," Nature, 216,(1967):407-408, p. 407.
174. R. A. Dart and P. Beaumont, "Amazing Antiquity of Mining in Southern Africa," Nature, 216,(1967):407-408, p. 408
175. R. A. Dart and P. B. Beaumont, "On a Further Radiocarbon Date for Ancient Mining in Southern Africa," South African Journal of Science, 67:1 (January, 1971), pp 10-11, p. 10
176. Robert G. Bednarik, "Early Subterranean Chert Mining," The Artefact, 15:(1992), pp 11-24, p. 15; P.M. Vermeersch and E. Paulissen, "The Oldest Quarries Known: Stone Age Miners in Egypt," Episodes, 12:1(March 1989), p. 35-36, p. 36
177. K. Klein, 1978. "Preliminary Analysis of the Mammalian Fauna from Redcliff Stone Age Cave Site, Rhodesia," Occasional Papers, National Museum of Southern Rhodesia A4(2): 74-80
See also Thomas P. Volman, "Early Prehistory of Southern Africa," in Richard G. Klein, Prehistory and Paleoenvironments, (Boston: Balkema, 1984), p. 215
178. R.G. Bednarik, "Reply" Current Anthropology, 36:4(1995), pp. 605-634, p. 626
179. Victor Barnouw, An Introduction to Anthropology: Physical Anthropology and Archaeology, Vol. 1, (Homewood, Illinois: The Dorsey Press, 1982) p. 141

180. S. Semaw, et al, "2.5-Million-year-old stone Tools from Gona Ethiopia," <u>Nature</u>, 385(January 23, 1997), p. 333-336, p. 335; Bernard Wood, "The Oldest Whodunnit in the World," <u>Nature</u>, 385(Jan.23, 1997), p. 292
181. Wayne M. Senner, "Theories and Myths on the Origins of Writing: A Historical Overview," in Wayne M. Senner, editor, <u>The Origins of Writing</u>, (Lincoln: University of Nebraska Press, 1989), p. 23
182. Denise Schmandt-Besserat, "Two Precursors of Writing: Plain and Complex Tokens," in Wayne M. Senner, editor, <u>The Origins of Writing</u>, (Lincoln: University of Nebraska Press, 1989), p. 38
183. Denise Schmandt-Besserat, "Two Precursors of Writing: Plain and Complex Tokens," in Wayne M. Senner, editor, <u>The Origins of Writing</u>, (Lincoln: University of Nebraska Press, 1989), p. 38
184. Dean Falk, <u>Braindance</u>, (New York: Henry Holt and Co., 1992), p. 187
185. Alexander Marshack, <u>The Roots of Civilization</u>, (New York: McGraw-Hill Book Co., 1972), p. 139
186. Alexander Marshack, <u>The Roots of Civilization</u>,(New York: McGraw-Hill Book Co., 1972), p. 139-140
187. Alexander Marshack, <u>The Roots of Civilization</u>,(New York: McGraw-Hill Book Co., 1972), p. 49
188. Alexander Marshack, <u>The Roots of Civilization</u>,(New York: McGraw-Hill Book Co., 1972), p. 50-51
189. Alexander Marshack, <u>The Roots of Civilization</u>,(New York: McGraw-Hill Book Co., 1972), p. 23
190. L.G. Straus et al, "A Review of the Middle to Upper Paleolithic Transition in Iberia", <u>Prehistoire Europeenne</u>, 3(January 1993), p. 11; Anthony Marks reviewing, "Quneitra" <u>Journal of Field Archaeology</u>," 19, 1992, p. 85-88, p. 86
191. L.G. Straus, "Comments" <u>Current Anthropology</u>, 36:4(1995), pp. 605-634, p. 623
192. Robert G. Bednarik, "Art Origins", <u>Anthropos</u>, 89(1994):169-180, p. 170
193. Alexander Marshack, <u>The Roots of Civilization</u>,(New York: McGraw-Hill Book Co., 1972), p. 349
194. Translation by Glenn Morton, from H. Vallois and R. Vaufrey, <u>L'Anthropologie</u> 58:5-6, pp. 433-443, p. 441.
195. Vadim N. Stepanchuk, "Prolom II, A Middle Palaeolithic Cave Site in the Eastern Crimea with Non-Utilitarian Bone Artefacts," <u>Proceedings of the Prehistoric Society</u> 59, 1993, pp 17-37, p. 35-36
196. Robert G. Bednarik, "Art Origins", <u>Anthropos</u>, 89(1994):169-180, p.170
197. E. Cecil Curwen, "Querns," <u>Antiquity</u>, 11(1937), 133-151.
198. Steven Mithen, <u>The Prehistory of the Mind</u>, (New York: Thames and Hudson, 1996), p.217-218
199. Gordon C. Hillman, "Late Palaeolithic Plant Foods from Wadi Kubbaniya in Upper Egypt: Dietary Diversity, Infant Weaning, and Seasonality in a Riverine Environment," in David R. Harris and Gordon C. Hillman, <u>Foraging and Farming</u>, (Boston: Unwin Hyman, 1989), p. 230
200. Gordon C. Hillman, "Late Palaeolithic Plant Foods from Wadi Kubbaniya in Upper Egypt: Dietary Diversity, Infant Weaning, and Seasonality in a Riverine Environment," in David R. Harris and Gordon C. Hillman, <u>Foraging and Farming</u>, (Boston: Unwin Hyman, 1989), p. 227
201. J.H. Furby et al, "The Cuddie Springs Bone Bed Revisited, 1991",in M. A. Smith et al, ed. <u>Sahul in Review Pleistocene Archaeology in Australia New Guinea and Island Melanesia</u>, 204-212, p. 208.
202. R.K.Fullagar, D.M. Price & L.M. Head, "Early Human Occupation of Northern Australia: Archaeology and Thermoluminescence Dating of Jinmium Rock-Shelter, Northern Territory," <u>Antiquity</u> 70(1996):751-773, p. 770
203. R.K.Fullagar, D.M. Price & L.M. Head, "Early Human Occupation of Northern Australia: Archaeology and Thermoluminescence Dating of Jinmium Rock-Shelter, Northern Territory,"<u>Antiquity</u> 70(1996):751-773, p. 764-765.
204. C. K. Brain and A. Sillen, "Evidence from the Swartkrans cave for the earliest use of fire," <u>Nature</u>, 336, Dec. 1, 1988, p. 464-465
205. J. A. J. Gowlett, J. W. K. Harris, D. Walton and B. A. Wood, "Early Archaeological Sites, Hominid Remains and Traces of Fire from Chesowanja, Kenya," <u>Nature</u>, 294, Nov. 12, 1981, p. 128
206.Paul G. Bahn, "Light My Fire," <u>The Artefact</u>, 18(1995):90
207. Paul G. Bahn, "Light My Fire," <u>The Artefact</u>, 18(1995):90
208.Paul G. Bahn, "Light My Fire," <u>The Artefact</u>, 18(1995):90
209. Lewis R. Binford, "Early Human Behaviour and Ecology," in Steve Jones et al, editors, <u>The Cambridge Encyclopedia of Human Evolution</u>, (Cambridge: Cambridge University Press, 1992), p. 367
210. Juan Luis Arsuaga Ferreras, "Faces from the Past," <u>Archaeology</u>, May/June 1997, p. 31-33, p.32
211. Barbara Isaac, "Throwing," in Steve Jones et al, editors, <u>The Cambridge Encyclopedia of Human Evolution</u>, (Cambridge: Cambridge University Press, 1992), p. 358
212. William H. Calvin, "The Unitary Hypothesis: A Common Neural Circuitry for Novel Manipulations, Language, Plan-ahead, and Throwing?" in K. R. Gibson and T. Ingold, eds., <u>Tools, Language and Cognition in Human Evolution</u>, (Cambridge: Cambridge University Press, 1993), pp 230-250, p. 246-247

213. Barbara Isaac, "Throwing," in Steve Jones et al, editors, The Cambridge Encyclopedia of Human Evolution, (Cambridge: Cambridge University Press, 1992), p. 358

214. J. A. J. Gowlett, "Early Human Mental Abilities," in Steve Jones et al, editors, The Cambridge Encyclopedia of Human Evolution, (Cambridge: Cambridge University Press, 1992), p. 344

215. William H. Calvin, "The Unitary Hypothesis: A Common Neural Circuitry for Novel Manipulations, Language, Plan-ahead, and Throwing?" in K. R. Gibson and T. Ingold, eds., Tools, Language and Cognition in Human Evolution, (Cambridge: Cambridge University Press, 1993), pp 230-250

216. Eileen M. O'Brian, "What was the Acheulean Hand Ax?" Natural History, July, 1984, pp 18-23.

217. Steven Mithen, The Prehistory of the Mind, (New York: Thames and Hudson, 1996), p. 121

218. Steven Mithen, The Prehistory of the Mind, (New York: Thames and Hudson, 1996), p. 121

219. Brian Hayden "The Cultural Capacities of Neandertals," Journal of Human Evolution, 24(1993):113-146, p. 119-120

220. James R. Shreeve, The Neandertal Enigma, (New York: William Morrow and Co., 1995), p. 336

221. James R. Shreeve, The Neandertal Enigma, (New York: William Morrow and Co., 1995), p. 336

222. R. Shepard, Prehistoric Mining and Allied Industries, (New York: Academic Press, 1980), p. 210.

223. Alexander Marshack, "Early Hominid Symbol and Evolution of the Human Capacity," in Paul Mellars, The Emergence of Modern Humans, (Ithica: Cornell Univ. Press, 1990), pp 457-498, p. 474

224. F. Clark Howell and L. G. Freeman, "Ivory Points from the Earlier Acheulean of the Spanish Meseta," in Martin Almagro Basch, editor, Homerage, (Madrid: Ministerio de Cultura, 1983), p. 41-61.

225. Robert G. Bednarik, "Concept-mediated Marking in the Lower Palaeolithic," Current Anthropology, 36:4(1995), pp. 605-634, p. 611

226. Mary Lecron Foster, "Symbolic Origins and Transitions in the Paleolithic," in Paul Mellars, The Emergence of Modern Humans, (Ithica: Cornell Univ. Press, 1990), pp 517-539, p. 522; Donald C. Johanson, Lenora Johanson, and Blake Edgar, Ancestors, (New York: Villard Books, 1994), p. 163-165

227. Joyce Flood, Archaeology of the Dreamtime, (New Haven: Yale University Press, 1990), p.176

228. Juliet Clutton-Brock, "Origins of the Dog: Domestication and Early History," in James Serpell, ed. The Domestic Dog, (Cambridge: Cambridge University Press, 1995), p. 10

229. Juliet Clutton-Brock, "Origins of the Dog: Domestication and Early History," in James Serpell, ed. The Domestic Dog, (Cambridge: Cambridge University Press, 1995), p. 10

230. Carles Vila et al, "Multiple and Ancient Origins of the Domestic Dog," Science, 276(June 13, 1997):1687-1689, p. 1689

231. Stanley J. Olsen and John W. Olsen, "The Chinese Wolf, Ancestor of New World Dogs," Science, August 3, 1977, p.533-535, p. 533

232. Stanley J. Olsen and John W. Olsen, "The Chinese Wolf, Ancestor of New World Dogs," Science, August 3, 1977, p. 533-535, p. 534

233. Stanley J. Olsen, Origins of the Domestic Dog, (Tuscon: The University of Arizona Press, 1985), p. 44-45.

234. Stanley J. Olsen and John W. Olsen, "The Chinese Wolf, Ancestor of New World Dogs," Science, August 3, 1977, p. 533-535, p. 534

235. Frederick E. Zeuner, A History of Domesticated Animals, (New York: Harper & Row, Publishers, 1963), p. 83-84

236. Josephine Flood, "The Archeology of the Dreamtime, (New Haven: Yale University Press, 1989), p. 205; Stanley J. Olsen, Origins of the Domestic Dog, (Tuscon: The University of Arizona Press, 1985), p. 88

237. Stanley J. Olsen, Origins of the Domestic Dog, (Tuscon: The University of Arizona Press, 1985), p. 88

238. References for the table are:
Frederick E. Zeuner, A History of Domesticated Animals, (New York: Harper & Row, Publishers, 1963), p. 80; Dennis A. Etler, "The Fossil Evidence for Human Evolution in Asia," Ann. Rev. Anthropol. 1996, 25:275-301, p. 281; Stanley J. Olsen and John W. Olsen, "The Chinese Wolf, Ancestor of New World Dogs," Science, August 3, 1977, p. 533-535, p. 535

INTELLIGENCE AND SOUL

There are two main schools of anti-evolutionary thought in Christendom: young-earth and old-earth. The young-earth creationists deny evolution and many other conclusions of modern geology, biology and astronomy. The great paradox of the young-earth position is that they generally accept the humanity of all forms of fossil men, in spite of the great morphological (or shape) differences between them and us. Young-earth creationists believe that *Homo erectus* and Neanderthal were either preflood men or degenerate survivors of the postflood world. Most old-earth creationists appear to accept all the conclusions of modern science *except* the humanity of fossil man. When it comes to fossil men, the old-earth creationists ignore anthropological data as much as young-earth creationists ignore geological data. The reason for this reversal of fortunes on fossil man is the need for old-earth creationists to have a place for Adam. Generally speaking, old-earth anti-evolutionists place Adam as the first anatomically modern human.

Christians who believe in evolution have an entirely different problem. Scripture clearly says that mankind was a special creation of God. God breathed and man became a living soul. However, if evolution occurred and man arrived here by means of a gapless process, then what we call the soul must have gradually arisen. Atheistic evolution requires that consciousness, spirit or soul, must arise by a slow process. The consciousness of a fly is certainly less than the consciousness of a chimpanzee, which in turn appears to be less than that of man. If Christians take that approach, they have a problem. Assuming the soul gradually arose, then God did not impart it to man in a special act. If we are not special creations of God, then many theological problems follow.

To examine the issue of the soul, we must first establish the possible ranges of intelligence for fossil man. A large part of what makes us human is our intelligence. If intelligence is a large part of our soul, then we need to examine this issue.

Several facts are indisputable. The range of human brain size for people of normal intelligence extends from 900 cubic centimeters to over 2,000 centimeters.[1] The average is about 1,300 cubic centimeters. Neanderthal had a brain size which, on average, is larger than that of modern man. All previous fossil hominids had brain sizes which were smaller than modern man's. The average *Homo erectus* brain size is 930 cc.[2] A large gorilla may have a brain size of 730 cc while *Australopithecus* cranial capacities range from 441-650 cc.[3] Chimpanzee brain size averages 400 cc. What does brain size mean for intelligence? Nothing. Entirely too much emphasis is placed on brain size. After a study of the issue of brain size and intelligence, I must totally agree with Lubenow. He writes,

> "The relationship between brain size and body size must be factored in, and the crucial element is not brain size but brain organization. A large gorilla brain is no closer to the human condition than is a small gorilla brain. The human brain varies in size from about 700 cc to about 2200 cc with no differences in ability or intelligence. That variation, more than a factor of three, is an incredible difference in size variation but indicates no difference in quality. Those brain-size charts are meaningless."[4]

Modern men with brains smaller than 1000 cc are known to have absolutely normal intelligence. Such brain sizes are quite common among the Veddahs, a tribe in India.[5] Anatole France is an oft cited example of a brilliant man with a small brain. His is reported to have been in the 1,000 cc range.

The really interesting fact is that many people with quite abnormally small brains are quite normal in social and intellectual abilities. <u>The Guiness Book of Records 1996</u> cites the case of Daniel Lyon of Ireland who had a 650 cc brain.[6] In spite of the fact that such a brain is the same size as that of a gorilla, the largest *Australopithecus*, and of *Homo habilis*, his intelligence was quite normal. However, with that size of brain, his skull would have been morphologically different from most of his fellow humans. In 1980, an article was published in <u>Science</u>, entitled "Is Your Brain Really Necessary." This article highlighted the work of Dr. John Lorber, who, from the mid-1960's, had been studying hydroencephalic children. This horrible disability arises when the pressure inside a young child's ventricles presses the brain into the walls of the skull, thus compressing and distorting the brain. Many children with this problem are severely retarded. The article says,

> "'There's a young student at this university,' says Lorber, 'who has an IQ of 126, has gained a first-class honors degree in mathematics, and is socially completely normal. And yet the boy has virtually no brain.' The student's physician at the university noticed that the youth had a slightly

larger than normal head, and so referred him to Lorber, simply out of interest. 'When we did a brain scan on him,' Lorber recalls, 'we saw that instead of the normal 4.5-centimeter thickness of brain tissue between the ventricles and the cortical surface, there was just a thin layer of mantle measuring a millimeter or so. His cranium is filled mainly with cerebrospinal fluid."[7]

Assuming a normal head size and a millimeter of brain around the skull gives this student a brain-size of around 108 cubic centimeters, the brain size of rhesus monkey! Obviously intelligence and social normality are not due solely to brain size. This case is not a fluke.

Lorber subdivides hydroencephalics into four groups: those with only slightly enlarged ventricles, those with ventricles filling 50-70 percent of the skull, those with ventricles filling 70-90 percent of the skull and those whose ventricles fill an amazing 95 percent of the cranium. Only 10 percent of the hydroencephalics are in this last group, but half of them are normally intelligent![8] Many of these people have quite above average intelligence, having an IQ up to 131.[9] Lorber cites the case of Grant, who was born on June 13, 1963. His head was rapidly enlarging at the age of 10 weeks and was referred to Lorber for treatment. Lorber writes,

> "After injection of air, a radiograph showed no evidence of cerebral cortex--the air collecting immediately below the inner table of the skull and moving freely in all directions. No brain was encountered during needling the skull on this occasion. But for the absence of symptoms and the fact that the CSF was fully normal at this stage, this could have been a case of gross subdural effusion."[10]

At 17 weeks, Grant was observed to laugh easily and he could lift his head off the bed both when placed on his stomach and on his back. Grant was able to grasp objects placed in front of him, in fact he grabbed at them. A second x-ray showed no evidence of the brain except the falx and the tentorium. At 21 months, Grant was alert, playful and could speak. He was able to feed himself using a spoon and fork. His parents, by this age, had toilet trained him. In spite of his handicap, intellectually he was normal for his age.

In 1981, Lorber wrote an article looking back on 20 years of research. He said,

> "Well over 500 CT scans were performed on patients, some of whom were over 20 years of age. These included some who already occupied responsible positions in life, including senior nurses, university graduates and members of executive councils. Many were never suspected of having hydrocephalus, although looking back on their past history this diagnosis could have been made much earlier. They obviously had slow progressive hydrocephalus which did not detectably interfere with their life style. By the time they had a CT scan, some had such enormously dilated ventricles there was hardly any brain left above the level of the tentorium. They retained the midbrain cerebellum and pons but what was virtually missing was the part of the brain we attribute to superior intelligence; the centres for the fine control of movements and the appreciation of visual and auditory stimuli.
>
> "The systematic CT scan study showed there were many older children and adults who had grossly dilated ventricles with very thin residual brain and yet did not suffer from physical defects and had normal intelligence. Some were outstandingly intelligent with IQs well above the 'bright normal' range. I can only presume hydrocephalus with only moderately raised intracranial pressure can slowly progress over many years to reach eventually extreme degrees without ever causing symptoms. It is possible that specific functions of the brain, such as the motor cortex, may be relocated elsewhere from early infancy onwards or that we do not need such a large quantity of brain and only need to use a very small part of it under normal circumstances."[11]

Other people with other types of brain damage also can have normal intelligence. The two halves of the brain are called hemispheres. These hemispheres are connected by a cable of neurons called the *corpus callosum*. Occasionally people will develop severe epilepsy in which the epilepsy starts in one brain hemisphere and spreads to the other hemisphere. The resulting electrical storm can threaten to destroy the healthy hemisphere. In severe cases, the only solution is to remove the diseased half of the brain. This procedure is done only as a last resort to save the life of the individual. It is called a hemispherectomy.

Effectively this procedure results in a human being with a brain size of around 600-650 cubic centimeters. This is smaller than the brain size of a gorilla, and is within the range of the australopithecines. Yet the effect of this drastic reduction in brain size does not result in a corresponding decline in intelligence. Most patients with hemispherectomy end up with IQs averaging one standard deviation below normal.[12] It is not what we would consider an advantage, but it is certainly great enough to be *human.*

Another type of procedure, called a hemidecortication severs the frontal cortex from the rest of the brain. Even this procedure when performed early in life does not totally destroy intelligence. Patients with this procedure have post-operative IQs averaging 70.[13] However, one of these patients has been reported to have an IQ of 103.

Smith performed a long-term study of 64 infants who had hemispherectomies, 36 on the left side and 28 on the right. He wrote:

> "At a 25-year follow-up; each had obtained a college degree and had enjoyed a successful career as an executive, following a right hemispherectomy in one case and a left hemispherectomy in the other. Thus, as Smith noted, the findings demonstrate that at birth each of the two cerebral hemispheres contains the neuroanatomical and substrate necessary for the development of normal or even superior adult language and verbal and nonverbal cognitive functions."[14]

All of them had graduated from college. To conclude, Lubenow is correct. Brain size has *nothing, absolutely nothing* to do with the possible intelligence of the individual.

An issue which is related to intelligence is the issue of innovation. Humans are believed to be very innovative. Wilcox makes a large issue out of the concept that Neanderthals were not inventive. He says that in less than half the time that Neanderthals walked the earth, anatomically modern humans were walking on the moon. He implies that because Neanderthals didn't walk on the moon they couldn't be children of Adam.[15] Put this way, it is a very strange requirement for spirituality.

One can illustrate the logical flaw in this argument by replacing the subject and object:

In less than 1/7 the time that the Romans were the superpower, Americans have walked on the Moon.

In less than 1/25 of the time the Chinese have walked the earth, Americans walked on the Moon.

By Wilcox's logic, only Americans would be spiritual. Put any group you want into this sentence and you can exclude them from humanity. Obviously, the Romans and the Chinese are human. But if we are to use Wilcox's logic, they are not human. The argument from innovation would exclude from humanity all those *Homo sapiens* who have lived in the stone age for the past 20,000 years, while Western Europe and Asia moved forward into the copper, bronze and iron ages.

Beyond this, the argument simply is mathematically false. The oldest skeleton claimed to be Neanderthal is that from Ehringsdorf, Germany, dating 230,000 years old.[16] The youngest Neanderthal skeleton is 30,000 years old.[17] Thus, Neanderthal walked the earth for 200,000 years. The earliest claim for the existence of anatomically modern humans is 130,000 years ago at Klasies River Mouth Cave, South Africa.[18] Modern men have lived on earth 130,000 years. Simple division shows that modern men have been on earth for *more than half the time (65%) that Neanderthal has been on the earth.* This is a useless and false argument.

Planning Depth

The characteristics which the Bible ascribes to man are, knowledge of future consequences and an ability to plan for those circumstances. Scripture says,

> "And the LORD God commanded the man, "You are free to eat from any tree in the garden; but you must not eat from the tree of the knowledge of good and evil, for when you eat of it you will surely die." Genesis 2:16-17 (NIV)

By this, the Bible indicates that man was clearly capable of understanding consequences. After the fall, other consequences were apparent. God indicated to Adam and Eve that a Messiah was coming. (Genesis 3:15). Eve, by her reaction when Cain was born, may have thought she had given birth to the

Messiah. She said; "I have begotten a man, the Lord"[19] By the way this is stated in the Hebrew, it sounds very much as if Eve thought Cain was God. This may have had something to do with the way she raised him, possibly contributing to his murderous ways.

This knowledge of the future and how to plan for it is quite different from what animals have.

> "Of course, humans are the supreme planners. A chimpanzee can hold on-line the location of previously stored food and go for it when permitted to do so. However, Savage-Rumbaugh tells us that unlike humans, chimpanzees have difficulty attending to more than one task at a time, do not plan much ahead, and seem to have no concept of death (perhaps the ultimate in planning ahead). Because it has expanded so much during hominid evolution, the frontal lobe is the one part of the brain in which chimpanzees and human sulcal patterns are easily distinguishable."[20]

The lack of multitasking ability in chimps has important implications.[21] Because of this limitation, chimpanzees are unable to use a tool to make another tool. They are unable to hold the two goals and their relationship in their minds. This inability of chimpanzees is applicable to all creatures except man. This will be important when we see how Neanderthal multitasked to create stone tools to make wood tools with which to kill prey.

Planning for the future takes many forms that can be deduced from the fossil record. Planning applies to the plans for the manufacture of a tool; to the time over which the activities persist, and planning applies to the training of the next generation. Chimpanzees make tools, but the sequential steps required to make chimpanzee tools are few. The much celebrated manufacture of termite fishing sticks uses only five sequential actions. These are:

1. Pick twig
2. Remove leaves
3. Stick twig in termite nest
4. Remove stick
5. Lick termites off stick

The use of hammerstones and anvils to crack coula and panda nuts in the Ivory Coast[22] is another form of tool use by chimpanzees. But this requires only four steps:

1. Find hammerstone
2. Find anvil
3. Smash nut
4. Eat nut.

Another tool, a leaf sponge, requires a similar number of steps. Such leaves are used for two purposes. Leaf sponges are used to retrieve water from the hollow at the branch of a tree and to wipe the skin of the chimp. The steps are:

1. Find leaf
2. Chew leaf
3. Dip leaf in water
4. Suck water from leaf

1. Find leaf
2. Chew leaf
3. Wipe skin

Recent studies have clearly shown that apes and chimpanzees are only able to hold a maximum of six simultaneous ideas in their minds.[23] With this as a limit, even the ancient forms of man engaged in much more planning than mere monkeys or chimpanzees. Gowlett relates:

> "The most basic step in making a stone tool—striking a flake from a cobble—is relatively simple. To strike a sequence of flakes, in such a way that each one helps in the removal of others, demands much more ability, not so much in terms of manual dexterity as in control by the brain. Each step must be evaluated for its own merits, and its consequences for future steps.

No ape in captivity has yet been induced to strike more than a single flake from a cobble. Stoneworking cannot be operated by an inflexible and mechanical routine, because there are too many possibilities at every strike. Control of the process can be maintained only by constant re-evaluation 'in the mind's eye'--a phrase that suggests the importance of visual imagination."[24]

The sequential steps in manufacturing a stone tools require a long sequence of operations. The rock must be chosen. It can not be just any rock. Sandstone, shale and limestone cannot be flaked unless they are very fine grained and permeated with silica. To make a Neanderthal tool, a Levalloisian flake, one must take a flattish rock of appropriate thickness, and then trim it on the edges. There might be as many as 10 blows, directed more or less vertically, required to trim the edges of the core. Then the top face of the stone is trimmed with up to 11 blows, directed more or less horizontally, removing 11 flakes. These flakes are directed at such an angle as to taper the top side of the rock. After this tapering, one large flake is removed from the upper, prepared surface by means of a blow directed horizontally, splitting the flat rock in two along its flattest section. The upper, large flake is suitable for the manufacture of numerous Mousterian flake tools. The lower part of the rock with the trimmed edge, is now ready for an entirely different sort of tool making process. Continuous flaking can take place around the perimeter of the core. In this way, the Neanderthal, who apparently invented this technique, intricately planned to make maximum use of the available flint supplies. Little flint went to waste.[25] The complexity of the Neanderthal toolmaking techniques is such that there were five or six different multistep strategies to achieve the end.[26]

The planning did not end there. These tools were most often used to work wood.[27] They made spears, like those found at Lehringen, Germany, dated 125,000 years ago. Such wooden spears were used to kill game for food. As we saw, there is also evidence that Neanderthal would glue[28] or tie his stone tools onto a stick[29] and then the weapon was suitable for killing game. Many of the stone tools hafted during the Mousterian were made and hafted so similarly that it appears that the Neanderthals were planning to be able to replace parts in their tools.[30] This is exactly what a 20th century man, named Eli Whitney, the inventor of the cotton gin, did for the manufacture of flintlock rifles. Few people are aware of his contribution to mass production techniques. He manufactured standardized interchangeable parts so that when a part broke, it could be replaced easily. Neanderthal man anticipated Whitney by at least 80 millennia! Thus we can see that the planning depth and the number of steps through which Neanderthals were able to conceive, were much more numerous than those of the chimpanzee. The steps could be outlined as follows:

1. Find a suitably shaped flattish stone
2. Make sure it is of the correct mineralogy (Flint, etc.)
3. Trim the edges (approx. 10 actions)
4. Trim the top (approximately 10 actions)
5. Shear off the top of the tool
6. Manufacture a chopper (many steps)
7. Find suitable sapling
8. Chop the sapling down with the chopper
9. Make flake from previously made core to shape the tree
10. Shape the sapling appropriately into a shaft
11. Make the spear head to a standard size
12 Make or find the glue or obtain thongs for tying spear head to shaft
13 Glue or tie spear head to shaft
14. Find the prey
15. Throw spear at prey.
16. Make more flakes from core to butcher prey
17. Butcher prey
18. Eat the prey

Paul Mellars says that the making of a Levalloisian tool involves six separate steps which must be planned out from the beginning of the manufacturing process.[31] If this is so, then the above list should contain at least two more steps in the chain of reasoning.

As one can see, there was an entirely different quality to the tool-making of Neanderthal than there was to the tool-making of chimpanzees. At various stages along the total process of obtaining food, the initial core preparation techniques produced further input to the goal of obtaining food. This type of planning requires a human type of memory, with a multitasking mind. As we saw above, chimpanzees cannot multitask; Neanderthal, like anatomically modern man, could.

Recently, the amazing discovery of three wooden spears at Schoningen, Germany, provide evidence that *Homo erectus* engaged in exactly the same line of reasoning outlined above. In early 1997, Hartmut Thieme of the Hannover, Germany, Institute for the Preservation of Historical Monuments, announced the discovery of 400,000 year old wooden spears. Robin Dennell describes the implications,

> "The Schoningen spears now provide unambiguous evidence that large animals were killed in this manner by 400,000 years ago.
>
> "The spears have other exciting implications. First, the time and skill needed to make them: each is made from the trunk of a 30-year-old spruce tree; in each, the end with the tip comes from the base of the trunk, where the wood is hardest; and each has the same proportions, with the center of gravity a third of the way from the sharp end, as in a modern javelin. These represent considerable investment of time and skill--in selecting an appropriate tree, in roughing out the design and in the final stages of shaping. In other words, these hominids were not living within a spontaneous 'five-minute culture', acting opportunistically in response to immediate situations. Rather, we see considerable depth of planning, sophistication of design, and patience in carving the wood, all of which have been attributed only to modern humans." [32]

The importance of these spears to our quest is that *Homo erectus went through a 15 step process for acquiring food.* The only difference between the Neanderthal and *Homo erectus* is the hafting of the stone tools on the shaft of the spear. *Homo erectus* engaged in a very human sequence of acquiring food.

1. Find a suitably shaped stone
2. Make sure it is of the correct mineralogy (Flint etc)
3. Detach a large flake
4. Use hard hammerstone to remove flakes alternately from opposite edges of the large flake.
5. Use hard hammerstone to shape the tip to make it narrow
6. Use antler or bone to remove small chips from edge
7. Find suitable sapling
8. Chop the sapling down with the hand-axe
9. Make flake from previously made core to shape the tree
10. Shape the sapling appropriately into a shaft
11. Find the prey
12. Throw spear at prey.
13. Make more flakes from core to butcher prey
14. Butcher prey
15. Eat the prey

This too is far beyond the chimpanzee level of tool-making. The complexity of this process is human. Fred Smith writes of Neanderthal in the Middle East:

> "For a number of reasons, it has recently become the vogue to suggest that Neandertals were not really quite 'human' in behavior. Specifically, it has been suggested that Neandertals were incapable of any type of 'planning depth' in terms of lithic technology or subsistence, did not organize their use of habitation space in a systematic manner, and probably did not bury their dead. The lithic, faunal, and feature data from Kebara show that on the first two of these points Neandertal capabilities do not differ in any major way from those of the early 'modern' Skhul-Qafzeh Levantine hominids or from available descriptions of Upper Paleolithic people. More recent work in the Levant, partially based on the Kebara data, may indicate some differences in resource utilization between Tabun B and Tabun C Mousterian, but there is still no evidence of *qualitative* behavioral difference between Neandertals and early modern people in the Levant." [33]

Temporal Planning Range

Chimpanzees have almost no temporal range to their planning. The Boesch's studied the distribution of the distances chimpanzees carried hammerstones to crack nuts in the Ivory Coast;[34] 83% of them came from within 160 feet, about the depth of most modern suburban homes lots. Chimpanzees can only plan a few minutes ahead. The maximum distance they bring a hammerstone is about a quarter of a mile. It would take them about 20 minutes to walk this distance so that this represents the maximum temporal planning depth a chimpanzee is capable of. One study of chimpanzee tool use showed that 85-94% of all tools were acquired from within 2 meters of the food.[35]

What is the maximum planning depth for a Neanderthal? Neanderthals were capable of planning days in advance, at the very least. One method of measuring this temporal planning depth is by looking at the distance stone tools were carried. This can be determined when a chemically unique flint stone is found as a tool in a distant region. By measuring the distance from source to final resting place, and assuming a walking speed, one can come up with a minimum time over which it was planned for this errant stone to be turned into a tool. This time is the minimum time, not the maximum time of temporal planning depth. During the Middle Paleolithic, 130,000 to 40,000 years ago, during the time of the Neanderthals, flint stones and other materials were transported distances of up to 186 miles![36] What is fascinating, given our discussion of the planning abilities of Neanderthal, is that the flints which were carried that far are the best quality flints.[37] This means that Neanderthal was able to calculate a cost/benefit ratio. Is the carrying of this heavy stone, nearly 200 miles, worth the effort. If it was poor quality, then the answer was no. If it was high quality, then the rock was carried. Given that a human can walk about 48 miles per day, this represents about a four-day travel, assuming a forced march at a speed of 4 miles per hours for 12 hours per day. This, at the very least, represents 576 times the temporal planning depth chimpanzees are capable of. Neanderthal was certainly no chimpanzee! In reality, considering that the Lewis and Clark expedition averaged about 3 miles per day, the planning depth of the Neanderthal was probably much greater than what we calculated.

One type of evidence shows that Neanderthal had at least a 6-month planning depth capability. In the Perigord region of France, there are numerous caves along both sides of the valleys. Neanderthal had a strong preference for inhabiting caves on the north or northwest side of the valley.[38] This would seem strange until one realizes that the southern exposure allows sunlight to come into the cave mouth, warming it. Today there are measurable temperature differences of up to 25 degrees centigrade in caves on the north versus the southern side of the valley. There is also the factor of wind. During the winter the cold wind comes mainly from the north, blowing directly into the caves on the southern valley slopes. Six months planning is 13,000 times more temporal depth than a chimpanzee. It doesn't end there.

Neanderthal generally sited his habitations near the major migration routes of animals.[39] This implies a planning depth of at least a year, since the animals would migrate with the seasons. True to a multitasking human mind, which the Neanderthals possessed, they also located their sites near outcrops of good quality flint.[40] The process of site selection solved *two* problems at once and thus displayed an ability to handle multicomponent problems. Mankind is the only animal that consciously plans that far in advance on problems of that complexity. Neanderthal is named *Homo sapiens* for good reason.

What kind of planning depth can we ascribe to *Homo erectus*? Even at the earliest appearance of *Homo erectus* he was regularly transporting objects up to 6 miles (10 kilometers). The longest documented distance of chimpanzee transport is .8 kilometers, less than half a mile. Schick and Toth observe,

> "In fact we can now document in some cases the distances over which hominids transported rocks from their sources to the Oldowan sites where they left them. These distances range from only a few tens of meters (with the site in the immediate vicinity of the rock source) to several kilometers (with the site in the immediate vicinity of the rock source) to several kilometers (with the rock source some distance from the site). Again, these longer distances appear to indicate systematic carrying of stone by hominids beyond anything seen so far among nonhuman primates and suggest intelligent behaviors involving foresight and planning evolved in tandem with the earliest stone technologies.
>
> "This is a much more complicated pattern than many would have suspected from this remote period of time. It bespeaks to us an elevated degree of planning among these early hominids than is presently seen among modern nonhuman primates."[41]

Two hundred thousand years ago some man or men carried stone tools containing a rare Jurassic fossil for a distance of 120 miles (193 km) from the nearest surface source known for this fossil.[42] At the 400,000 year old[43] site of Olorgesailie stone tools were carried 30 miles (48 km).[44] These imply a *Homo erectus* minimum temporal planning depth of at least one day,[45] 72 times greater than that of a chimpanzee.

One type of planning which humans regularly do is planning to pass their knowledge on to the future generations. There have only been two recorded instances of this among chimpanzees. Mithen notes,

> "Moreover, active teaching appears to have been observed. The Boeschs report two instances in which mothers saw their infants having difficulty in cracking nuts and provided demonstrations of how to solve the problem. In one case the correct positioning of a nut on an anvil prior to striking was demonstrated, while in the other the proper way to grip a hammerstone was shown to an infant who immediately seemed to adopt the grip with some success."
>
> "What is remarkable, however, is that such active teaching, or even passive encouragement, should be so rare. The two instances that the Boeschs describe constitute less than 0.2 per cent of almost 1,000 maternal interventions in nutcracking seen during 4,137 minutes of observation. Why don't they do more of it?"[46]

Only man teaches his young. Of course, this is very difficult to document in the archeological record. Only two examples have come to light.[47] At Tuc d'Audoubert Cave near Ariege in France, fifty small heel prints were found in clay near two sculpted bison. This site has been interpreted as the gathering of adolescents for the performance of a puberty rite. The second is an engraving of human figures on the walls of the Grotta dell'Addaura on Sicily. Both of these instances are applicable to anatomically modern man and do not shed light on the status of Neanderthal, archaic *Homo sapiens*, or *Homo erectus*.

The Soul Man

> "And the Lord God formed man of the dust of the ground, and breathed into his nostrils the breath of life; and man became a living soul." Genesis 2:7 (King James)

The soul has been variously defined. In the Old Testament, the soul is something which cannot exist without the body.[48] The New Testament broadens the term to "mean the whole person, alive or after death; it can designate the immaterial part of a person with its many feelings and emotions; and it is an important focus of spiritual redemption and growth."[49] The soul can be defined as that which makes us who we are; our self-consciousness. If this is the case, then Christians who believe in evolution are faced with the difficulty of explaining the soul. When and how did self-consciousness arise? Was the emergence of self-awareness sudden or gradual? The Scripture would imply that it was sudden.

If our soul is somehow equated with our self-consciousness, then it is our soul which gives us the ability to empathize with others. We see someone who is hurt and understand what they are experiencing and feeling. Our care for them comes from our desire that if such an injury were to happen to us we would want some help. In other words, our self-consciousness allows for our feelings to be understood and acted upon. Thus, if the soul is what allows the acting upon our feelings and emotions, we must look for evidence of feelings in the fossil record to find the oldest evidence of the human soul. This is difficult since feelings, per se, do not fossilize. But we can look for actions consistent with feelings: Actions associated with care, compassion, and love. Humans have a compassion for their fellow man in that they will care for their companions who are sick. They will bring water and food to the sick. No other animal engages in this type of behavior. While elephants will stand by a sick and dying elephant, they do not bring him food and water. They have nothing with which to carry the water in anyway. The finding of human compassion in the fossil record would define the onset of humanity.

It has long been known that Neanderthal's show much evidence of treating their companions with compassion and care. Klein writes:

> "However, the same skeletal pathologies and injuries that show that the Neanderthals lived risky lives and aged early also reveal a strikingly 'human' feature of their social life. The La Chapelle-aux-Saints and Shanidar 1 individuals, for example, must have been severely incapacitated and would have died even earlier without substantial help and care from their

comrades. This implicit group concern for the old and sick may have permitted Neanderthals to live longer than any of their predecessors, and it is the most recognizably human, nonmaterial aspect of their behavior that can be directly inferred from the archeological record." [50]

Some Christians have accepted such evidence and accept the humanity of Neanderthal but not of Homo erectus. John Wiester wrote:

"There is additional evidence of those qualities associated with humanity at the Shanidar cave. The analysis of undeveloped bone structure indicates that another man, known as Shanidar I, was a severe cripple from birth. His right arm was entirely useless and may have been amputated just above the elbow. Extensive bone scar tissue indicates that he was blind in his left eye. He was apparently cared for by his people until his death at age forty, a very old age by Neanderthal standards. This is the first sign of compassion and tenderness in the archeological record." [51]

There is evidence for the care for a dying comrade preserved in the fossil record of *Homo erectus*. This would seem to go beyond what can be expected of a mere ape. The case is deduced from a fossil known as KNM-ER 1808. It exemplifies the care of a human, even if 1808 looked a lot different from us. The same year, 1992, Walker, Zimmerman and Leakey published an article in Nature, which gave the briefest set of facts about a 1.7 million-year-old case of a dying woman.[52] While they did not make it explicit in that article, the dying woman had been cared for.

In 1973, Kamoya Kimeu, a hominid fossil hunter of mythic renown discovered the fragmented bones of a Homo erectus. The bones were so fragmented that many parts were quite small and could only be retrieved by sifting the soil. The *Homo erectus* skeleton fragments were mixed in with the bones of hippos, crocodiles, and turtles, among others. The fragments of this individual were easy to pick out from the 40,000 bones of other species because the H*omo erectus'* bones were terribly diseased and deformed. They had a fabric known as woven bone and it covered every part of the skeleton except the cranium. Even when the fragments were glued back together, it was difficult to recognize what part of the body each bone came from. After a period of careful study, the fossil turned out to be the first nearly complete skeleton ever found of a Homo erectus. Unfortunately, the diseased bones allowed very little to be learned of the normal anatomy of H. erectus. This fossil was given the museum number KNM-ER 1808. The KNM-ER stands for Kenya National Museum-East Rudolf. Geologic dating revealed that the fossil was 1.7 million years old, making this one of the oldest erectus fossils around. The bones had belonged to an adult female erectus.

The diseased bone material covered almost all of her skeleton except for the head. It consisted of two parts. There was a normal core where the osteocytic lacunae are parallel. The osteocytic lacunae are tiny spaces in bone where the bone's cell once lived. Surrounding this normal core was a half inch of 'woven' bone, thickest on the limb bones and almost nonexistent on the skull. Woven bone has bloated and highly irregular osteocytic lacunae and was deposited near the end of 1808's life. This fabric develops for one of three reasons: 1) when the creature grows very rapidly, 2) when fractures heal and 3) when a disease is operative. Since there is a core of normal bone which represents an adult-sized skeleton, rapid growth as a cause can be ruled out. Since the woven bone was all over the skeleton except the skull, fractures didn't seem very likely as a cause. This left open disease, but what disease?

Alan Walker consulted with doctors at John Hopkins, looking for a diagnosis. The consensus seemed to settle onto a diagnosis of hypervitaminosis A. This type of disease is found among modern health fadists who take too much vitamin A. But since 1808 could not go to the local pharmacy and buy vitamin A supplements, how did she get mega-doses of this vitamin? Walker suggests that she obtained it in the same way that some arctic explorers got it -- by eating carnivores. More specifically, she got it by eating carnivore livers. When carnivores eat their prey, they obtain fairly large doses of vitamin A. Vitamin A is then stored in its liver, where it is never broken down or detoxified. Carnivores, like dogs, leopard seals, polar bears, or killer whales, eat other animals, including their livers. Because a carnivore eats so many livers, its liver becomes a veritable warehouse of vitamin A."[53]

Sir Douglas Mawson, the sole survivor of an ill-fated Antarctic expedition, provides an excellent example of hypervitaminosis A in an arctic environment. The disease causes excruciating pain and exacts a horrible death upon its victims. On Nov. 11, 1912, Mawson and two companions, Ninnis and Mertz, left their base camp to explore a large area on three sleds. Their path would take them across 2,000 miles of Antarctica. Because of the length of the journey they had stashed some food on the path and only carried small quantities with them. On their return, they traveled too slowly and ran out of food. They abandoned one sled with all unnecessary equipment. They then sorted their gear, putting the scientific equipment onto the lead sled, and placing the food in the trailing sled. This was done because of their fear of crevasses. They had the scientific sled take the lead, reasoning that if it fell into a crevasse there would be no big loss. But if the food fell into a crevasse, they would be in dire straits. Unfortunately, fate had a different idea. On Dec. 13, 1912, Ninnis, driving the food sled, fell into a crevasse. Ninnis and the dog team died. The lead sled, with the scientific equipment, had made the crossing but apparently had weakened the ice bridge to the extent that it could no longer support the weight of the food-carrying sled. Mawson and Mertz were 320 miles from base camp with only enough food for ten days. As they continued on their trek, they began to kill and eat the sled dogs. The dog meat was tough and chewy. The livers were soft and better tasting. They ate liver which turned out to be a fatal mistake.

They began to suffer from dizziness, stomach cramps, nausea, and balance problems. Their hair fell out and their skin cracked and peeled off in strips. Their joints throbbed with pain. Delirium and dementia set in. Walker and Shipman write:

> "Any sort of movement produced terrible pain, for what they were experiencing was exactly what happened to 1808. The excess vitamin A they had eaten--Mawson's biographer reckons they ate sixty toxic doses-- caused the periosteum, the tough, fibrous tissue that encases each bone, to rip free from the bone with each pull of a muscle. (The muscles are anchored on bones through the periosteum.) Between the periosteum and bone, torn apart blood vessels spilled their contents, forcing further separation of the tissues. In the case of 1808, the blood formed huge clots, which ossified--turned to bone--before she died." [54]

As time went by, the two men began to suffer from another vitamin related disease, scurvy. Scurvy causes your gums to bleed and your teeth to fall out. Since Mertz was the first one to loose his teeth, Mawson did the compassionate thing--he gave Mertze the soft liver and he chewed the tough dog muscle. A few days later, they lost another third of their meager rations to another crevasse accident. Finally, reduced to one sled and two dogs, they discarded more equipment, killed the final two dogs and continued. Mertz got worse and worse and finally was unable to walk. Mawson pulled the sled with Mertz in it. Around January 5, 1913, Mawson and Mertz dug a snow cave. Mawson wanted to attempt to nurse Mertz back to health. But Mertz was no longer interested in eating. Mertz died. Mawson buried him 100 miles from base camp.

Mawson continued, finally reaching an ice cave containing a cache of food only 5 miles from the base camp. He was able to "feast" on biscuits, pemican, pineapple and powdered milk. Due to bad weather, Mawson had to stay in this cave for seven more days. The ship he was supposed to take back to Australia had been scheduled to leave January 15th and it was now early February. Finally the weather cleared and Mawson continued. When he reached the base camp, he finally found some good fortune, the ship had not left. His good friend, Frank Bickerton, greeted him with "My God! Which one are you?"[55]

This tale of tragedy and survival illustrates the symptoms of this dread disease. Walker and Shipman note that KNM-ER 1808 would probably have been immobilized with pain, as was Mertz. Yet the advanced state of ossification of the torn periosteum testifies that she lived for several weeks or even months in this condition. She must have been in excruciating pain. Walker and Shipman write,

> "The implication stared me in the face: someone else took care of her. Alone, unable to move, delirious, in pain, 1808 wouldn't have lasted two days in the African bush, much less the length of time her skeleton told us she had lived. Someone else brought her water and probably food; unless 1808 lay terribly close to a water source, that meant her helper had some kind of receptacle to carry water in. And someone else protected her from hyenas, lions and jackals on the prowl for a tasty morsel that could not run away. Someone else, I couldn't help thinking, sat with her through the long, dark African nights for no good reason except human concern. So, useless as 1808 was for telling us much about normal *Homo erectus* morphology, she told us something quite unexpected. Her bones are poignant testimony to the beginnings of sociality, of strong ties among individuals that came to exceed the bonding and friendship we see among baboons or chimps or other non human primates"[56]

It is highly unlikely that 1808 could have spent her dying days next to a water hole, as one Christian friend suggested. Water holes attract predators and scavengers at all hours of the day and night. A body lying motionless is most certainly going to attract vultures. Predators also come to water holes to drink and would find a sick 1808 an easy victim. After describing African water holes and hominid inability to defend themselves at night, Lew Binford (1983, p. 68) writes:

> "The place I would never choose to establish a camp in the African savannah is next to a water source! Nevertheless, archaeologists tell us that our hominid ancestors habitually located home bases in exactly these places. At this point, it becomes relevant to ask whether the three criteria used by the East African researchers really permit the reliable recognition of home-base occupation sites."[57]

The obvious necessity for water to be carried to 1808 gives an interesting insight into cognitive and technological skills of people at that time. Richard Leakey writes:

> "The use of skin as a means of carrying small objects, as in the case of gourds or skins for water transport, does require a degree of abstract thought which in today's terms is basically human. The ability to make stone tools to a set and regular pattern is also a consequence of abstraction: the maker of the tool has to 'see' the finished tool in the raw material before making the tool. The collection of raw material for making stone tools has to be guided continually by reference to an abstract ideal. There are no grounds for doubting that early hominids were making such mental decisions, and the capacity for abstraction obviously existed. Is it, therefore, improbable that a primate with such a brain could have used it in the manufacture or fabrication of simple carrying utensils? The archaeological record is ambiguous at present and we can only speculate."[58]

One other objection must be discussed. Is the reaction of the *Homo erectus* care-giver merely like that of an ape? The answer is unequivocally no. Jane Goodall gives an account of how chimpanzees treated an injured comrade. There was a polio outbreak in the chimp tribe Goodall was watching and it afflicted many of the chimps. One named McGregor was paralyzed by the polio. When this newly paralyzed chimp dragged himself back to camp, Goodall reports the tragic reception his troop members gave him. Her account, quoted at length is very poignant,

> "McGregor's condition was patently far worse. Not only was he forced to move about in an abnormal manner, but there was the smell of urine and the bleeding rump and the swarm of flies buzzing around him. The first morning of his return to camp, as he sat in the long grass below the feeding area, the adult males, one after the other, approached with their hair on end, and after staring began to display around him. Goliath actually attacked the stricken old male, who, powerless to flee or defend himself in any way, could only cower down, his face split by a hideous grin of terror, while Goliath pounded on his back. When another adult male bore down on McGregor, hair bristling, huge branch flailing the ground, Hugo and I went to stand in front of the cripple. To our relief, the displaying male turned aside.
>
> "After two or three days the others got used to McGregor's strange appearance and grotesque movements, but they kept well away from him. There was one afternoon that without doubt was from my point of view the most painful of the whole ten days. A group of eight chimps had gathered and were grooming each other in a tree about sixty yards from where McGregor lay in his nest. The sick male stared toward them, occasionally giving slight grunts. Mutual grooming normally takes up a good deal of a chimpanzees time, and the old male had been drastically starved of this important social contact since his illness.
>
> "Finally he dragged himself from his nest, lowered himself to the ground, and in short stages began the long journey to join the others. When at last he reached the tree he rested briefly in the shade; then, making the final effort he pulled himself up until he was close to two grooming males. With a loud grunt of pleasure he reached a hand toward them in greeting—but even before he made contact they both had swung quickly away and without a backward glance started grooming on the far side of the tree. For a full two minutes old Gregor sat motionless,

staring after them. And then he laboriously lowered himself to the ground. As I watched him sitting there alone, my vision blurred, and when I looked up at the groomers in the tree I came nearer to hating a chimpanzee than I have ever been before or since.

"For several years Hugo and I had suspected that the aggressive adult male Humphrey was McGregor's younger brother. The two traveled about together frequently and often the older male had hurried to Humphrey's assistance when he was being threatened or attacked by other chimps. It was during the last days of Mr. McGregor's life that we became convinced these two males were siblings:no bond other than that of a family could have accounted for Humphrey's behavior then--and afterward.

"In the whole period Humphrey seldom moved farther than a few hundred yards away from the old male--although even he never actually groomed McGregor. Sometimes Humphrey went away across the valley to feed, but within an hour or so he was back,resting or grooming himself near his paralyzed friend. On the first day of his return to camp McGregor climbed quite high in a tree and made a nest. Suddenly Goliath began to display around him, swaying the branches more and more vigorously, slashing the old male on the head and the back. Gregor's screams grew louder, and he clung to the rocking branches tightly. At last, as if in desperation, he let himself drop down through the tree from branch to branch, until he landed on the ground. Then he started to drag himself slowly away. And Humphrey, who had always been extremely nervous of Goliath, actually leaped up into the tree, displaying wildly at the much higher ranking male, and for a brief moment attacking him. I could hardly believe it.

"One day Mr. McGregor managed to pull himself right up to the feeding area, up thirty yards of very steep slope, to join a large number of chimpanzees who were eating there. We were able to give him a whole box to himself so that for a while, at least, he was part of the group again. When the others moved away up the valley, Gregor tried to follow. But whether he dragged himself on his belly, or hitched himself backward, or laboriously somersaulted, he could move only very slowly, and the rest of the group were soon out of sight."[59]

Thus, the compassion of Homo erectus was quite human. There was no compassion on the part of the chimpanzees. Christians need to understand that from a spiritual perspective, archeological evidence for the soul extends far back in time, at least to the earliest *Homo erectus*.

The Development of the Soul

Self-consciousness is much very difficult to document from the fossil record. Stephen Mithen published an interesting book, <u>The Prehistory of the Mind</u>, endeavoring to examine this issue. He claims at the start of his book,

"I will thus provide the hard evidence to reject the creationist claim that the mind is a product of supernatural intervention."[60]

Anyone making such a claim as Mithen does deserves to have his evidence examined. His equivalent term for Christian soul is "cognitive fluidity". Mithen begins his attempt by examining the structure of the human mind.

According to many modern evolutionary psychologists, the mind is composed of various mental modules. Each module is specifically used for the solution of a particular type of problem and begins operating at different times during the development of the human. While most of these modules are associated with particular circuit systems in the brain, we generally are unable to identify the circuit. According to Mithen, the human soul is developed in three phases in which various modules come into play during development.

In phase 1 there is only one active module, and it is called 'general intelligence'. General intelligence is what solves general problems of survival. It involves trial-and-error learning and controls the decision-making processes. A creature with general intelligence learns at a slow rate with many errors. According to Mithen, two-year-olds have general intelligence, but they do not have self-consciousness. In adult minds there is little evidence left of this general intelligence module.

The second phase is the phase in which specialized modules are built. These are the social module, the natural history module, the language module and the technical module. The module of "social intelligence" allows interaction with other human beings socially. It is what allows us to know from body language and facial expression that our wives or husbands are mad at us. Like all of the mental modules it is not content free. It brings a consistent structure to the domain it controls. An example is the smile. All human cultures use the facial expression of a smile to alleviate tension, display friendship, and to laugh. A picture of a laughing human conveys almost universal knowledge of the internal mental state of the laughter. Similarly, crying carries a universal message. Social intelligence is also used in our political maneuvering, such as the forming of alliances with other humans.

Another important mental module is the "natural history intelligence." This module brings an understanding of the physical world: plants, animals, earth and physics. As the child grows, the classification of animals and plants in their environment becomes very important. Babies intuitively understand that a living object is quite different from an inanimate object.[61] Children understand that if a horse is made to wear stripped clothing, the horse has not become a zebra. A dog is still a dog even if it only has three legs and can't bark. All known peoples have common notions of vertebrates, plants, patterns of naming these objects (like wolf, red wolf, Mexican wolf, etc), and an overall taxonomy which matches closely the modern biological system. The physics portion of this module operates rather early in a child's life, allowing a baby to know intuitively concepts learned in high school.[62] Experiments with babies show that they already have concepts of solidity, inertia and gravity. The child could not have learned this from experience and so these concepts appear to be hardwired into the brain. The natural history module is what collects and recalls information on the medicinal uses of plants.

Around the age of two, another module comes into operation. This is the "language module." As we saw earlier language is innate and is associated with the circuitry of Broca's and Wernicke's areas. We also saw that the language module is content-rich. Grammar is innate, and even if children are raised in an environment with no grammar, they will provide the grammar. Without this module humans would be unable to communicate.

The final intelligence is "technical intelligence." It is this module that allows the manufacture of artefacts, the manipulation of objects, and the control of innovation and invention. It allows us to see that if we attach a spear loosely to another stick in a particular fashion, we can use the second stick, a spear-thrower, to throw the spear with much greater force. It is what allows us to see that if we bang two rocks together in a particular pattern, we can create a stone tool that can be used to cut objects.

The final phase of the development of the soul comes when all the barriers between the various modules are broken down and the individual achieves "cognitive fluidity." In this stage, any problem can be solved by sharing information. In order to understand cognitive fluidity, it is best to look at a case where it doesn't exist. Chimpanzees are exquisite in their social interactions. They are cunning, ambitious, deceptive, and cognizant of their social standing. They are full of intrigues when it comes to overthrowing the alpha chimp. Yet they are unable to entertain the notion of using a tool to further their social position. According to Mithen, they have no cognitive ability to put the two pieces of information together. Even chimpanzee food sharing is not used to further social status.[63] Cognitive fluidity is what makes us conscious of all our aspects of ourselves and our world. It is what gives us the "subjective feelings of 'knowing', 'remembering' and 'perceiving'."[64] It is the global integrator; the I.

Mithen says that social intelligence was the first to evolve.[65] It evolved after the lemurs and monkeys split sometime after 55 million years ago. Natural history intelligence and technical intelligence evolved between 1.4 and 1.8 million years ago when *Homo erectus* appeared.[66] It was technical intelligence that gave *H. erectus* the ability to make the Acheulean hand-axe. According to Mithen natural intelligence allowed *H. erectus* to effectively scavenge. However, as we have seen, *Homo erectus* was a hunter. Language evolved, according to Mithen, between 500 and 200 thousand years ago.[67] Partial cognitive fluidity evolved with the advent of anatomically modern men 130,000 years ago and full cognitive fluidity arose around 60,000 years ago.[68]

Mithen's claim that the natural history module arose between 1.8 and 1.4 Myr ago, long after the split from gorilla and chimpanzee, is falsified by the discovery over the past several years that baboons, chimpanzees and gorillas have quite a lot of natural history knowledge about the medicinal value of plants in their environment.[69]

To support his case, Mithen makes several assertions which are not supported by the data we have just seen. These assertions form the basis of a means to determine the cognitive fluidity or humanity of a fossil people. According to Mithen, a people lacking the majority of these features would not have the same type of consciousness as modern humans experience. A *Homo erectus* making a stone tool would have no consciousness of making the tool. Erectus would be on auto pilot because, Mithen says, he lacked cognitive fluidity. According to him, cognitive fluidity allows the integration of various areas of knowledge. And he says early humans did not possess this. He defines "early humans" as Neanderthals, archaic *Homo sapiens*, and *Homo erectus*. His test for determining cognitive fluidity is as follows:

1. Early humans did not make bone, antler or ivory tools.
2. Early humans did not make tools designed for specific purposes.
3. Early humans did not make multicomponent tools.
4. Early humans stone tools show limited degrees of variation across time and space.
5. Early human settlements imply small groups.
6. Distribution of artifacts suggest limited social interaction.
7. Absence of personal decorations.
8. No evidence for ritualized burial among early humans.

In this way Mithen claims to prove that the human mind evolved.

There are two things very wrong with Mithen's list. The first is that his definition for the presence of mind would exclude the perfectly human Tasmanians of the 18th and 19th century. The second error is that early man did engage in the activities he claims they didn't. We will use his list to examine both aspects of this problem.

The Tasmanians were technologically the most primitive society ever found. Prior to the European conquest of Australia, their society had been isolated from all other human contact for 8-12 thousand years.[70] This isolation had many curious effects on the Tasmanians technology, which show that Mithen's list is untenable.

1. Mithen says, "Early humans did not make bone, antler or ivory tools." A glance at the bone tool section of the previous chapter will easily dispel this canard. Furthermore, the lack of bone tools does not indicate a lack of cognitive skills. The Tasmanians did not make bone tools when they were found. Their only tools were stone tools. Archaeology has confirmed that Tasmanians ceased making bone tools about 3,500 years ago and this caused them to cease sewing animal skins together to make clothes. Eight thousand five hundred years ago, when sea levels rose isolating the Tasmanians, they produced one bone tool for every 3 or 4 stone tools. By 4000 years ago, they were producing only one bone tool for every fifteen stone tools. By 3,500 years ago, they were no longer making bone tools. When Europeans found them, they merely tied the animal skins around them with animal skin. Tasmanians fail Mithen's test. Their small numbers and isolation caused them to lose this technology.

2. Mithen says, "Early humans did not make tools designed for specific purposes." The Tasmanian tools were less complex than those of the Neanderthals.[71] The tools were made from prepared cores.

3. Mithen says, "Early humans did not make multicomponent tools." This is the supposed lack of early man attaching their stone tools to handles. This "hafting" of a stone tool to a handle is supposed to imply advanced thinking concepts. Yet, according to Flood, "These Tasmanian stone tools were all hand-held; the concept of hafting was apparently unknown in Tasmania, although hafting was employed in mainland Australia in the Pleistocene."[72] Apparently the isolation caused the Tasmanians to lose this piece of technology also. Tasmanian tools were also very simple. An infamous article once compared the tool technology of the Tasmanians with chimpanzees.[73] The most complex Tasmanian tool was a baited hide. Such a tool would not survive in the archeological record, and so could not be found among early human artefacts.

4. Mithen says, "Early Human stone tools show limited degrees of variation across time and space." Due to only 8-12 thousand years of isolation, the constancy of the Tasmanian stone tools might not be in the same league as the Achulean hand axe. But if this requirement is supposed to be an inventiveness criteria, the Tasmanians were not inventive in the sense we use the term since they were actually losing technology. Their isolation caused what one man called, the "slow strangulation of the mind."[74]. However, the Tasmanian tools did get smaller over the past 4,000 years.[75]

5. Mithen says, "Early human settlements imply small groups." So? The entire population of the island was believed to be no more than 2,000 when the white man first settled the island in 1803.[76] The Tasmanians apparently lived in groups no larger than forty individuals.[77] This is a small group, probably no larger than many Neanderthal societies, yet no one today questions the humanity of the Tasmanian. The Tasmanians fail another of Mithen's tests.

6. Mithen says, "Distribution of artifacts suggest limited social interaction." By the time George Robinson lived with the Tasmanians between 1830 and 1836, they had no "large ceremonial gatherings."[78] The social interaction was limited by the small size of the populations. Tasmanians lose again.

7. Mithen says, "Absence of personal decorations." This is the only criterion which Tasmanians pass. During the 18th century, they wore shell necklaces.[79]

8. Mithen says, "No evidence for ritualized burial among early humans." There was none of this among

the Tasmanians either. The Tasmanians burned the bodies and smashed the bones of their loved ones.[80] Tasmanians also displayed little evidence of religion during their last days in Tasmania. This is hardly the behavior that Mithen expects of cognizant humans. Another of Mithen's tests failed.

According to Mithen's criteria, one would be forced to conclude that the Tasmanians are not fully cognizant, or cognitively fluid. This, of course, would be a ridiculous conclusion. Their technology was poor, but their cognitive fluidity, or their humanity, was just fine. The problem lies in Mithen's criteria. When we turn this criteria upon fossil man, we find similar flaws in Mithen's logic.

Mithen claims that these criteria prove that the cognitive skills of early man were not up to the modern standard. The problem is that none of his assertions about the abilities of fossil man are true, as we have seen. We will examine each of his points.

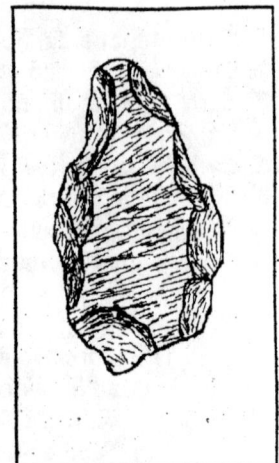

Figure 29 Oldest hand ax 1.4 myr.

1. *Early humans did make bone, antler or ivory tools.* Early humans did make bone tools. Neanderthals made a bone awl and ivory spear points and *Homo erectus* made ivory spear points at Torralba and Bilzingsleben.[81]

2. *Early humans did make tools designed for specific purposes.* Neanderthal and *Homo erectus* made javelins for the express purpose of hunting big game.[82]

3. *Early humans did make multicomponent tools.* These are tools which are made of two different materials, in this case flint and wood. At Schoningen, Germany, in 1996, wooden handles were found which are believed to be part of multicomponent tools.[83] These tools are from a time in which *Homo erectus* lived. Neanderthal also hafted his spear points.[84]

4. *Early human's stone tools do not show limited degrees of variation across time and space.* The concept of a static tool repertoire can only be claimed of the Acheulean hand-axe, which was made from 1.5 to .3 million years ago. Even this tool changed over time. The earliest Acheulean hand-axe was quite crude compared to the final ones. Compare Figures 29 and 30. And if, Eileen O'Brian is correct, that the hand-axe was really a projectile (thrown like a discus) which landed point first most of the time, then the variation in shape would be limited by the laws of physics.[85] Neanderthal tools were entirely different from those of the earlier *Homo erectus*, being made by entirely different (Levalloisian) techniques that made dual use of the Levalloisian core to make flakes and tools from different parts of the core. Blade tools, which used to be considered characteristic of the Upper Paleolithic less than 40,000 years of age, have now been found from strata dating older than 250,000 years.[86] So far, modern men have not been found in strata of this age. Thus, at 250,000 years ago we have only "early humans" who are producing three different types of tools, Acheulean, Levalloisian and flake tools.

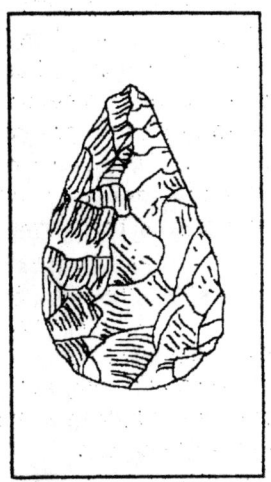

Figure 30 Late Hand-axe circa 300 kyr

5. *Early human settlements imply small groups.* This is irrelevant. Small groups of modern humans today still possess cognitive fluidity.

6. *Distribution of artifacts do not suggest limited social interaction.* We saw that even from the earliest times material was being moved across rather large distances. Certainly the movement of certain materials 300 kilometers during the Neanderthal period says something quite significant.[87] It is quite likely that the transportation of materials over such distances implies trade with other groups. This is a social interaction. *Homo erectus* transported material as much as 48 km.

7. *There is no absence of evidence for personal decorations.* From the earliest times red ochre was found with *Homo erectus*.[88] It was often used for body paint. Neanderthals made personal ornaments, such as necklaces.[89]

8. *There is evidence for ritualized burial among early humans.* This is probably the most ridiculous assertion made by Mithen. There is the Shanidar "flower" burial mentioned in a previous chapter.[90] There is the young man at Le Moustier, who was buried covered with red ochre and food offerings.[91] Recently a young Neanderthal child was discovered with a small triangular piece of flint placed over her heart.[92] Neanderthal intentionally buried his dead. Furthermore, as we have seen, the burial pit at Sima de los Huesos provides evidence for the ritual disposal of corpses more than 300,000 years ago.[93]

From this we can see that all of Mithen's reasons for rejecting the creation of man's mind are wrong. In conclusion, we can state that the soul of man, that which makes us human, has much evidence back to nearly 2 million years ago. Was Adam created at that time? No. The evidence for the widespread occupation of the world by *Homo erectus* 2 million years ago is immense. The earliest African *Homo erectus* is 1.9 million years old.[94] Stone tools clearly made by *Homo* were found in Pakistan,[95] and France around 2 million years ago.[96] In Java, *Homo erectus* dates to 1.8 million years.[97] *Homo erectus* was also discovered at Dmanisi, Georgia dated to 1.6 million years ago.[98] Because *Homo erectus* was so widespread when he first appeared, one thing is almost certainly true. *Homo erectus* was on earth long before 2 million years ago. He needed this time to spread out as far as he did by 2 million years ago.

In a previous book, this author has suggested that Adam was created 5.5 million years ago. The time of this creation was suggested because it allows for two things: the incorporation of a plausible locale for the Flood of Noah, and because it fits with the appearance of hominids. While I cannot prove yet that *Homo* was in existence that long ago, there are two possibilities. First, Ernst Mayr may be correct when he suggested that *Australopithecus* was really *Homo*. Second, it just might be that small populations of humanity lived in isolation long before they burst upon the archeological scene. In any event, one thing is absolutely true. The current apological approaches are absolutely falsified by the anthropological data. The view presented in this book and my prior book[99] is consistent with the data of anthropology. The evidence of language, the evidence of culture, the evidence of art, all show that the soul of man is at least 2 million years old.

References

1. John Relethford, <u>Fundamentals of Biological Anthropology</u> (Toronto: Mayfield Publishing, 1994), p. 192
2. Ralph L. Holloway, "The Cast of Fossil Hominid Brains," <u>Scientific American,</u> July 1974, p. 113.
3. Ralph L. Holloway, "The Cast of Fossil Hominid Brains," <u>Scientific American,</u> July 1974, p. 111 Jim Foley, Personal communication 10-3-1997.
4. Marvin L. Lubenow, <u>Bones of Contention,</u> (Grand Rapids: Baker Books, 1992), p. 83
5. Jacquetta Hawkes, <u>Prehistory,</u> (New York: Mentor Books, 1963), p. 94
6. <u>Guiness Book of Records 1996,</u> (New York: Facts on File, 1995),p. 14
7. Roger Lewin, "Is Your Brain Really Necessary," <u>Science,</u> Dec. 12,1980, p. 1232.
8. Aaron Smith, "Early and Long-Term Recovery from Brain Damage in Children and Adults: Evolution of Concepts of Localization, Placticity, and Recovery,", in C.R. Almli and S. Finger, editors, <u>Early Brain Damage: Research Orientations and Clinical Observations,</u> 1, 299-323, p. 310
9. John Lorber, "The Medical Treatment of Hydrocephalus Using Isosorbide," <u>Mod. Probl. Paediat.,</u> 18(1977):178-180, p. 178
10. John Lorber, "Hydranecephaly with Normal Development," <u>Develop. Med. Child Neurol.</u> 1965, 7, pp 628-633, p. 629-630
11. John Lorber, "Is your Brain really Necessary?", <u>Nursing Mirror,</u> April 30, 1981, p. 20
12. Bryan Kolb, <u>Brain Plasticity and Behavior,</u> (Mahwah, New Jersey: Lawrence Erlbaum Assoc. 1995), p. 87-88
13. Faraneh Vargha-Khadem and Charles E. Polkey, "A Review of Cognitive Outcome After Hemidecortication in Humans," in F. D. Rose and D. A. Johnson, <u>Recovery from Brain Damage,</u> (New York: Plenum Press, 1992), pp. 137-151, p.139-140
14. Aaron Smith, "Early and Long-Term Recovery from Brain Damage in Children and Adults: Evolution of Concepts of Localization, Placticity, and Recovery,", in C.R. Almli and S. Finger, editors, <u>Early Brain Damage: Research Orientations and Clinical Observations,</u> 1, 299-323, p. 308
15. David Wilcox, cited by Percival Davis and Dean H. Kenyon, <u>Of Pandas and People</u>, 2nd edition (Dallas: Haughton Publishing Co., 1993), p. 110-111; He says the same thing in David Wilcox, "Adam, Where Are You? Changing Paradigms in Paleoanthropolgy," <u>Perspectives in Science and Christian Faith,</u> June, 1996, pp. 88-96
16. Chris Stringer and Clive Gamble, <u>In Search of the Neanderthals,</u> (New York: Thames and Hudson, 1993), p. 66
17. Robert G. Bednarik, "Neanderthal News," <u>The Artefact,</u> 19(1996):104.
18. Donald C. Johanson, Lenora Johanson, and Blake Edgar, <u>Ancestors,</u> (New York: Villard Books, 1994), p. 239
19. New American Standard Bible,(La Habra, California: The Lockman Foundation, 1971), note Genesis 4:1.
20. Dean Falk, <u>Braindance,</u> (New York: Henry Holt, 1992), p. 64-65

21. Dean Falk, Braindance, (New York: Henry Holt, 1992), p. 57
22. C. Boesch and H. Boesch, "Mental Map in Wild Chimpanzees: An Analysis of Hammer Transports for Nut Cracking", Primates 25(2):160-170, p. 162
23. Audrey E. Cramer and C. R. Gallistel, "Vervet Monkeys as Travelling Salesmen," Nature, 387(May 29, 1997), p. 464
24. J. A. J. Gowlett, "Early Human Mental Abilities," in Steve Jones et al, editors, The Cambridge Encyclopedia of Human Evolution, (Cambridge: Cambridge University Press, 1992), p. 341
25. Paul Mellars, The Neanderthal Legacy, (Princeton: Princeton University Press, 1996), p. 61-63
26. Paul Mellars, The Neanderthal Legacy, (Princeton: University Press, 1996), p. 88
27. Brian Hayden "The Cultural Capacities of Neandertals," Journal of Human Evolution, 24(1993):113-146, p. 119-120
28. Eric Boeda, et al, "Bitumen as a Hafting Material on Middle Palaeolithic Artefacts," Nature, 380, March 28, 1996, p. 336-337; Simon Holdaway, "Tool Hafting with a Mastic" Nature 380, March 28, 1996, p. 288
29. Rose L. Solecki, "More on Hafted Projectile Points in the Mousterian," Journal of Field Archaeology, 19(1992):207-212, p. 211
30. Brian Hayden "The Cultural Capacities of Neandertals," Journal of Human Evolution, 1993, 24:113-146, p. 122
31. Paul Mellars, The Neanderthal Legacy, (Princeton: University Press, 1996), p. 88
32. Robin Dennell, "The World's Oldest Spears," Nature 385(Feb. 27, 1997), p. 767-768
33. Fred H. Smith, "Comments: The Excavations in Kebara Cave, Mt. Carmel," Current Anthropology 33:5(1992): 497-550, p. 540
34. C. Boesch and H. Boesch, "Mental Map in Wild Chimpanzees: An Analysis of Hammer Transports for Nut Cracking", Primates 25(2):160-170, p. 162
35. Iain Davidson and William Noble, "The Archaeology of Perception: Traces of Depiction and Language" Current Anthropology, 30:2, April 1989, p. 134-135.
36. Chris Stringer and Clive Gamble, In Search of the Neanderthals, (New York: Thames and Hudson, 1993), p. 174
37. Paul Mellars, The Neanderthal Legacy, (Princeton: University Press, 1996), p. 152
38. Paul C. Mellars, The Neanderthal Legacy, (Princeton: University Press, 1996), p.250
39. Paul Mellars, The Neanderthal Legacy, (Princeton: University Press, 1996), p. 55
40. Paul Mellars, The Neanderthal Legacy, (Princeton: University Press, 1996), p. 149
41. Kathy D. Schick and Nicholas Toth, Making Silent Stones Speak, (New York: Simon and Schuster, 1993), p.128
42. K. P. Oakley, "Emergence of Higher Thought 3.0-0.2 Ma B.P.", Phil. Trans. R. Soc. Lond. B, 292, 205-211 (1981), p. 209-211
43. Christopher Wills, The Runaway Brain, (New York: Harper Collins, 1993), p.108-109
44. Wil Roebroeks, Jan Kolen and Eelco Rensink, "Planning Depth, anticipation and the organization of Middle Palaeolithic Technology: The 'Archaic Natives' Meet Eve's Descendants," Helinium XXVIII/1. 1988, pp 17-34, p. 30
45. Steven Kuhn notes, "The movement of artifacts over distances well in excess of 10-15 km demonstrates that Mousterian hominids regularly anticipated needs from one day to the next." Steven L. Kuhn, "On Planning and Curated Technologies in the Middle Paleolithic," Journal of Anthropological Research, 48:3(1992), pp 185-214, p. 193
46. Steven Mithen, The Prehistory of the Mind, (New York: Thames and Hudson, 1996), p. 90
47. D. Bruce Dickson, The Dawn of Belief, (Tuscon: The University of Arizona Press, 1990), p. 126
48. Charles C. Ryrie, Basic Theology, (Wheaton: Victor Books, 1986), p. 197
49. Charles C. Ryrie, Basic Theology, (Wheaton: Victor Books, 1986), p. 197
50. Richard G. Klein, The Human Career, (Chicago: The University of Chicago Press, 1989), p. 334
51. John Wiester, The Genesis Connection, (Nashville: Thomas Nelson Publishers, 1983), p. 181.
52. Alan Walker, M.R. Zimmerman, and R.E. F. Leakey, 1982. "A Possible Case of Hypervitaminosis A in Homo Erectus," Nature, 296(March 18, 1982), p. 248-250.
53. Alan Walker and Pat Shipman, Wisdom of the Bones, (New York: Alfred Knopf, 1996), p. 162
54. Alan Walker and Pat Shipman, Wisdom of the Bones, (New York: Alfred Knopf, 1996), p. 164
55. Alan Walker and Pat Shipman, Wisdom of the Bones, (New York: Alfred Knopf, 1996), p. 165

56. Alan Walker and Pat Shipman, Wisdom of the Bones, (New York: Alfred Knopf, 1996), p. 165
57. Lewis Binford, In Pursuit of the Past, (New York: Thames and Hudson, 1983), p. 68
58. Richard Leakey, "Recent Fossil finds From East Africa," in J.R. Durant ed. Human Origins, (Oxford: Clarendon Press, 1989) p 61
59. Jane Goodall, In the Shadow of Man (New York: Houghton Mifflin, 1971), p. 221-223
60. Steven Mithen, The Prehistory of the Mind, (New York: Thames & Hudson, 1996), p. 16
61. Steven Mithen, The Prehistory of the Mind, (New York: Thames and Hudson, 1996), p. 52
62. Steven Mithen, The Prehistory of the Mind, (New York: Thames and Hudson, 1996), p. 54
63. Steven Mithen, The Prehistory of the Mind, (New York: Thames & Hudson, 1996), p. 91-92
64. Steven Mithen, The Prehistory of the Mind, (New York: Thames & Hudson, 1996), p. 191
65. Steven Mithen, The Prehistory of the Mind, (New York: Thames & Hudson, 1996), p. 94
66. Steven Mithen, The Prehistory of the Mind, (New York: Thames & Hudson, 1996), p. 206
67. Steven Mithen, The Prehistory of the Mind, (New York: Thames & Hudson, 1996), p. 208
68. Steven Mithen, The Prehistory of the Mind, (New York: Thames & Hudson, 1996), p. 178
69. Kenneth E. Glander, "Selecting and Processing Food," in Steve Jones, et al. editors, Cambridge Encyclopedia of Human Evolution, (Cambridge: Cambridge University Press, 1992), p. 67; see also, "Jungle Medicine," Scientific American, (Dec. 1996), p. 20
70. Josephine Flood, Archaeology of the Dreamtime, (New Haven: Yale University Press, 1990), p. 173
71. See Figure 13.2 in Josephine Flood, Archaeology of the Dreamtime, (New Haven: Yale University Press, 1990), p. 177. Compare these with any Neanderthal tools such as in Paul Mellars, Neanderthal Legacy, (Princeton: Princeton University Press, 1996), p. 110-125. The technical abilities of Neanderthal were very high.
72. Josephine Flood, Archaeology of the Dreamtime, (New Haven: Yale University Press, 1990), p. 176
73. Steven Mithen, The Prehistory of the Mind, (New York: Thames and Hudson, 1996), p. 75 See also W. C. McGrew, "Tools to Get Food: the Subsistants of Tasmanian Aborigines and Tanzanian Chimpanzees Compared," Journal of Anthropological Research, 43(1987):247-258.
74. Rhys Jones, cited by Josephine Flood, The Archaeology of the Dreamtime, (New Haven: Yale University Press, 1990), p. 185
75. Josephine Flood, Archaeology of the Dreamtime, (New Haven: Yale University Press, 1990), p. 176
76. Rhys Jones, "Tasmanian Archaeology: Establishing the Sequences," Annu. Rev. Anthropol. 1995, 24:423-446, p. 425
77. Josephine Flood, Archaeology of the Dreamtime, (New Haven: Yale University Press, 1990), p. 181
78. Josephine Flood, Archaeology of the Dreamtime, (New Haven: Yale University Press, 1990), p. 185
79. Josephine Flood, Archaeology of the Dreamtime, (New Haven: Yale University Press, 1990), p. 182
80. Josephine Flood, Archaeology of the Dreamtime, (New Haven: Yale University Press, 1990), p. 182
81. Andre Leroi Gourhan, The Hunters of Prehistory, transl. Claire Jacobson, (New York: Atheneum, 1989), p. 57-58; Robert G. Bednarik, "Concept-mediated Marking in the Lower Palaeolithic," Current Anthropology, 36:4(1995), pp. 605-634, p. 607; F. C. Howell and L. G. Freeman, "Ivory Points From the Earlier Acheulean of the Spanish Meseta," in Martin Almagro Basch, Homenaje, (Madrid: Ministerio de Cultura, 1983), pp 41-61; Brian Hayden "The Cultural Capacities of Neandertals ", Journal of Human Evolution, 24(1993):113-146, p. 119-120.
82. Paul Mellars, The Neanderthal Legacy, (Princeton: University Press, 1996), p. 227; Hartmut Thieme, "Lower Palaeolithic Hunting Spears form Germany," Nature, 385(Feb. 27,1997), p. 807-810; Robin Dennell, "The World's Oldest Spears," Nature 385(Feb. 27, 1997), p. 767-768; Brian Hayden "The Cultural Capacities of Neandertals ", Journal of Human Evolution, 24(1993):113-146, p. 117; Kenneth P. Oakley et al, "A Reappraisal of the Clacton Spearpoint," Proceedings of the Prehistoric Society 43, 1977, pp 13-30; H. L. Movius, Jr., "A Wooden Spear of Third Interglacial Age from Lower Saxony," Southwestern Journal of Anthropology, 6, 1950, pp 139-142
83. Hartmut Thieme, "Lower Palaeolithic Hunting Spears form Germany," Nature, 385(Feb. 27,1997), p. 808
84. Simon Holdaway, "Tool Hafting with a Mastic" Nature 380, March 28, 1996, p. 288; Eric Boeda, et al, "Bitumen as a Hafting Material on Middle Palaeolithic Artefacts," Nature, 380, March 28, 1996, p. 337
85. Eileen M. O'Brian, "What was the Acheulean Hand Ax?" Natural History, July, 1984, p. 18
86. JoAnn Gutin, "Do Kenya tools Root Birth of Modern Thought in Africa?" Science 270 Nov. 17, 1995, p. 1118.
87. Wil Roebroeks, Jan Kolen and Eelco Rensink, "Planning Depth, anticipation and the organization of Middle Palaeolithic Technology: The 'Archaic Natives' Meet Eve's Descendants," Helinium XXVIII/1. 1988, pp 17-34, p. 30
88. From 900,000 years ago-Robert G. Bednarik, "Art Origins", Anthropos, 89(1994):169-180, p. 172; From 1.6

million years ago-K. P. Oakley, "Emergence of Higher Thought 3.0-0.2 Ma B.P.", <u>Phil. Trans. R. Soc. Lond.</u> B, 292, 205-211 (1981), p. 206-207;

89. See the cover of this Nature issue. Jean-Jacques Hublin, Fred Spoor, Marc Braun F.Zonnenveld and Silvana Condemi, "A Late Neanderthal Associated with Upper Palaeolithic Artefacts," <u>Nature</u>, 381: May 16, 1996, p. 224

90. Brian Hayden "The Cultural Capacities of Neandertals," <u>Journal of Human Evolution</u>, 24(1993):113-146, p. 120

91. Colin Groves, "The Neanderthals," in Goren Burenhult, ed. <u>The First Humans</u>, (San Francisco: HarperSanFrancisco, 1993), p. 72

92. B. Bower, "Child's bones found in Neandertal Burial," <u>Science News,</u> 148, October 21, 1995, p. 261; Takeru Akazawa et al, "Neanderthal Infant Burial," <u>Nature</u>, 378, Oct. 19, 1995, p. 586

93. Juan-Luis Arsuaga et al, "Three New Human Skulls from the Sima de los Huesos Middle Pleistocene Site in Sierra de Atapuerca, Spain," <u>Nature</u>, 362(1993):534-537; Paul G. Bahn, "Treasure of the Sierra Atapuerca," <u>Archaeology</u>, January/February 1996, p. 47

94. Alan Walker and Pat Shipman, <u>The Wisdom of the Bones</u>, (New York: Alfred Knopf, 1996), p. 240

95. Richard G. Klein, <u>The Human Career</u>, (Chicago: The University of Chicago Press, 1989), p. 206-207; Vadim A. Ranov, Eudald Carbonell, and Xose Pedro Rodriguez, "Kuldara: Earliest Human Occupation in Central Asia in Its Afro-Asian Context," <u>Current Anthropology</u>, 36:2, April 1995, p. 337-346, p. 342; R.W. Dennell, H. M. Rendell and E. Hailwood, "Late Pliocene Artefacts from Northern Pakistan," <u>Current Anthropology</u>, 29:3, June 1988, p. 498.

96. Jean-Jacques Hublin, "The First Europeans," <u>Archaeology</u>, January/February, 1996, pp 36-44, p. 39.

97. Roy Larick and Russell L. Ciochon, "The African Emergence and Early Asian Dispersals of the Genus *Homo*," <u>American Scientist</u>, 84(Nov/Dec, 1996), p. 547-548

98. Chris Stringer and Clive Gamble, <u>In Search of the Neanderthals</u>, (New York: Thames and Hudson, 1993), p. 64

99. Glenn R. Morton, <u>Foundation, Fall and Flood</u>, (Dallas: DMD Publishing Co., 16075 Longvista Dr. Dallas, TX 75248, 1995).

ADAM AND EVE WERE NOT MARRIED AND OTHER ODDITIES

The question which this book has been aiming to answer is "When was the time of Adam's existence?" We have looked at the behavioral evidence for Adam having existed prior to two million years ago. But there is one more piece of evidence that needs to be examined. This is the molecular data. Using population genetics and knowledge of various genetic systems, it is possible to calculate how long the human race has been in existence.

During each generation, there are new mutations that arise which occur at a known rate. As each generation is produced, and more mutations are accumulated along descendant lineages, two isolated populations will gradually drift apart genetically. The rate of mutation can be used to date how long it would take for two populations to accumulate the genetic differences. Consider the sequences below. They start identically and in each generation they are mutated at one location. This eventually produces two different sequences which appear to have little relationship to each other, yet they are both descendants of the same sequence. Similarly, with DNA, two populations can find themselves gradually diverging. The number of DNA mutations between the two populations can be used to tell how long ago the two populations split apart. This is accomplished by estimating the coalescence time between the various DNA sequences. In Table 5, if all you have are the sequences at generation 4, one can estimate how long ago the two sequences 'coalesced'. The coalescence time is the time that the last common progenitor of the DNA segment lived. All sequences at generation 4 were derived from the

Generation	Sequence 1	Sequence 2
1	4-6-9-2-8-5-1	4-6-9-2-8-5-1
2	4-3-9-2-8-5-1	4-6-9-2-8-1-1
3	4-3-9-7-8-5-1	3-6-9-2-8-1-1
4	4-3-2-7-8-5-1	3-6-0-2-8-1-1

Table 5 Randomly diverging sequences

one sequence at generation 1.

Mitochondria are small energy producing organelles that are found in each cell. Each mitochondrion has its own DNA that comes only from the mother. Since the sperm is little more than nuclear DNA with a tail, and contains very little cytoplasm, the only source of mitochondria for a fertilized egg comes from the egg's cytoplasm, which comes from the mother. Because of this, there is no mixing of mitochondrial DNA (mtDNA). Divergent sequences must be due to mutations in the matrilineal line. By measuring the coalescence time, it can be determined when the last common mother of the mitochondria of all human beings lived. This is mitochondrial Eve.

Molecular biologists Rebecca Cann, Mark Stoneking, and Allan Wilson studied the evolution of mtDNA (mitochondrial DNA). They took mitochondrial DNA (mtDNA) samples from 147 people with Asian, Australian, New Guinean, Caucasian and African origins, and established that there were 133 different sequences of mtDNA. The 133 different sequences could be used to determine how long ago there was only one sequence *if* one could calibrate the rate of mutation in mtDNA. Cann, Stoneking and Wilson used two points for calibration: 1) the colonization of Australia and New Guinea, which at the time was believed to have taken place about 40,000 years ago, and 2) the time of the human-chimpanzee split. They measured the amount of genetic divergence in Australian aboriginal mtDNA. Since all of the divergence was believed to have arisen after the colonization 40,000 years ago, a mutation rate could be calculated. Doing this gave a mutation rate of 2 to 4 per cent per million years. The 40,000-year date for the Australian colonization will become important in our discussion below. From this, using a program called PAUP, they attempted to construct a phylogeny of the human lineage. The assumptions were that the tree should require the fewest mutations. They calculated that the sequences converged to a single sequence, inherited from a single woman who lived somewhere between 142,500 and 285,000 years ago. They compromised on a 200,000-year figure.

The mechanics of losing mtDNA lineages can be explained by looking at the family names on Pitcairn Island. In 1790, six men mutinied on the H.M.S. Bounty. They fled from British justice with thirteen Polynesian women, landed on Pitcairn Island and founded a population with six family names. Today, six or seven generations later, the fifty people on the island have only four family names. Three of these are from the original founders and one is from a whaler who settled on Pitcairn Island much later. Thus, in less than 200 years, three of the six male lineages have disappeared.[1] Similarly, if there were a way to track the female mitochondria on Pitcairn Island, we would see an equivalent loss of mitochondrial lineages.

In 1987 Cann, Stoneking and Wilson published the results of their research in Nature.[2] The concept that all humans were descended from a single woman was quickly dubbed the Eve theory. Newsweek made it a cover story; Johnny Carson mentioned her in his monologue. Eve was an overnight sensation. Even Christians have attached some importance to this theory as a means of denying evolution and excluding fossil man from the human lineage. Davis and Kenyon write:

> "As we have said, it measures back to 100,000 to 200,000 years. Though startling to scientists, a 'molecular root' no longer than this seems to fit very nicely with the fossils of modern humans dating approximately 110,000 years ago.
>
> "If the theory turns out to be confirmed in some reasonable approximation of its current form, it would have three major implications in man's quest for his ancestry: 1. It would mean that humanity, as represented by its contemporary peoples, is dramatically younger than traditionally conceived by most scientists. 2. It would eliminate Neanderthal as a candidate for ancestry to European peoples. 3. It would eliminate the vast majority of *Homo erectus* populations across Europe and Asia as ancestral to man, leaving open only the possibility that some tiny *Homo erectus* population in Africa gave rise to modern *Homo sapiens*, if the two are related at all."[3]

Hugh Ross wrote:

> "Several years ago, I reported on a parallel investigation of women's genetic variation. Because the differences found were so slight, theorists concluded that women can trace their lineage only a couple hundred thousand years at the most to a common ancestor, whom the scientists called 'Eve' (I won't argue the name, but I would lean toward a more recent date of origin than 200,000 years).
>
> ...
>
> I have a different hope and expectation. It seems more likely that future research will continue to confirm only slight variations in the genetic material of humans. If this is the case, we should see biologists' date for 'Adam and Eve' drop from a maximum of about 200,000 years ago to a date within the biblical range of about 10,000 to 60,000 years ago."[4]

With men, there is the Y-chromosome that is passed on to sons, like the mitochondrial DNA is passed on to children by their mothers. The molecular divergence on the Y chromosome of men can be used, as was the mtDNA, to estimate the time for the Y-chromosome Adam. This is the person from whom all modern Y-chromosomes have descended. The Y-chromosome is not as variable as the mitochondrial DNA but it yielded a similar age of coalescence as the original mitochondrial Eve study. Michael F. Hammer says,

> "The age of the ancestral human Y chromosome is estimated to be 188,000 years, with a standard deviation of 94,000 years. The coalescence time has 95% confidence limits of 51,000 to 411,000 years. This age agrees with estimates of the ancestral human mtDNA molecule. An implication of these results is that the current distribution of Y chromosomes carrying the YAP element reflects migrations of male *H. sapiens* from Africa to Europe and Japan.
>
> "A coherent picture now emerges that places both our common ancestral Y chromosome and mitochondrial DNA molecule in the late middle Pleistocene only slightly before the hypothesized origin of anatomically modern humans based on fossil data."[5]

In an article that followed Hammer's article, Whitfield, Sulston and Goodfellow, reported another study of the Y chromosome which estimated the coalescence time of the Y chromosome as 90-120 thousand years.[6]

One of the problems with studies of coalescence concerns the occasional phenomenon called *crossing over*. Each cell has two of each chromosome, call them 1 and 1'. Chromosomes 1 and 1' are not identical to each other. During the process of forming the sperm and egg, each chromosome forms a duplicate of itself. Thus, temporarily each cell has two identical copies of 1 and two identical copies of 1', for a total complement of 1,1,1',1'. Occasionally, part of 1 and 1' will break and exchange the broken portions. This creates a new chromosome which is a combination of 1 and 1'. Each of the four chromosomes in the bottom frame of Figure 31 becomes the only representative of that chromosome in the sperm or egg. This process of crossing over makes the detection of a single parent for any chromosome difficult to detect.

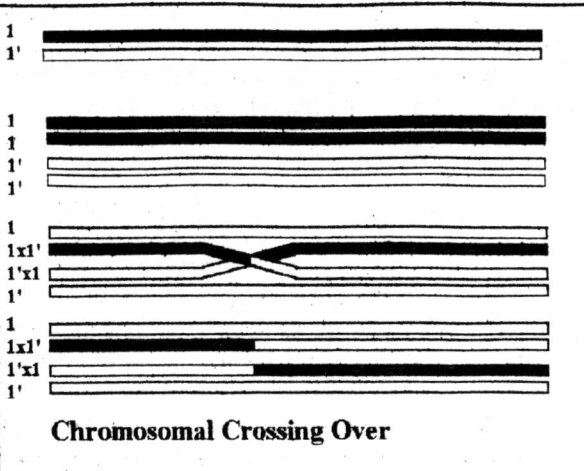

Figure 31 Chromosomal cross over

To avoid this problem, Dorit, Akashi and Gilbert studied a 729-base pair region of the Y chromosome.[7] This region is not involved with crossover mutations. They extracted this region from the Y chromosome of 38 men around the world. They found absolutely *no* variation in the 729 base pair region. From this they calculated a coalescence time which indicated that the most recent common ancestor for the male lineage lived no more than 270,000 years ago.[8] This, of course, is rather strange. As Jon Marks of Yale University pointed out, "[H]ow can there be a coalescence time for something that never diverged to begin with?"[9]

Does this data mean that two people 100-200,000 years ago were the parents of all of humankind? Has science shown the Bible correct here? Many Christians have reacted as if this were the case. Hugh Ross, reacting to the Dorit et al study, suggests that this study means that no evolution has occurred.[10] He also says that the study proved that all men were descended from a single individual who lived no more than 270,000 years ago. This is entirely wrong.

There are several problems with this response. First, the molecular data, far from showing that all people descended from two people who lived between 100 and 300 thousand years ago, shows that mankind is several times that old. It also shows that the Y-chromosome Adam and the mitochondrial Eve were not alone on the planet. In fact, it shows that they were not married and were not the common parents of us all (Figure 32). Newer data suggests that the mitochondrial Eve lived long before the Y chromosome Adam.

Remember that the rate of mutation was calculated from the amount of divergence among Australian Aborigines. The discovery in the late 1980s that Australia was inhabited at least by 60,000 years ago, would mean that the mutation rate was slower. Using this slower rate on the original data would imply that Eve lived between 214,000 and 423,000 years ago. In 1996, evidence was presented indicating that humans colonized Australia possibly around 140,000 years ago. Utilizing this even slower rate of mitochondrial divergence would place Eve between the years 500 thousand and 1 million years ago. With present day knowledge, Eve would have been a *Homo erectus*! Far from lowering the date of Eve, the changing calibration point pushes the date of mitochondrial DNA convergence, the date of Eve, back in time. This is long before the Y-chromosome Adam. Thus, these data show that Adam and Eve were not married, and in fact, did not even know each other.

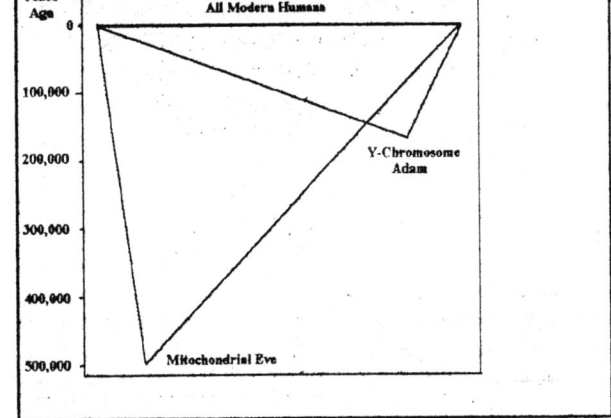

Figure 32 Mitochondrial Eve and Y Chromosome Adam were not married

But the real damage to the Eve theory came two years before Hugh Ross wrote of his hope that molecular genetics would yield a younger "Eve", and before the change in Australia's colonization date. Alan Templeton, a geneticist at Washington University, St. Louis, analyzed the data presented by Cann, Stoneking and Wilson.[11] He found that the PAUP program, which Cann et al, had used in their studies, would give different results depending upon the order of the input data. The reason was that there were 10^{250} different possible family trees for the mitochondrial data.[12] It is impossible to examine this many possible family trees to find the best fit to the data. It would take longer than the age of the universe for the fastest computer to look at all the possible phylogentic trees. Because of this difficulty, the time Eve lived and her location are unknowable. Trinkaus and Shipman note,

> "Basing his work on conservative assumptions, Templeton showed that the mtDNA divergence lay not in the relatively narrow band of time between 166,000 and 249,000 years ago, as had previously been estimated, but in a broad swath sometime between 191,000 and 772,000 years ago. This time interval embraces the period in which *Homo erectus* was spreading out of Africa and across Eurasia—meaning that the divergence in mtDNA might well have occurred long before the appearance of modern humans."[13]

When it comes to the Y chromosome Adam, there is a major criticism besides the calculation of a coalescence time when there has been no divergence. They very well may have measured the same Y chromosome over and over again. This is to be expected from the European conquests of the past four centuries. From time immemorial the practice of conquest has often meant the killing of the conquered males and the taking of the women for wives and mistresses. Judges 21:11 gives an example of this: all the men and non-virgin women were to be killed. The virgins were spared, presumably to become wives.

Shortly after the Dorit et al. study was published, paleoanthropologist Milford Wolpoff received a call from a fellow anthropologist, Clara Nii Ska, who was the wife of Wub-e-Ke-Niew, a member of the Ahnishinahbaeo'jibway Indian tribe. Nii Ska told Wolpoff that her husband had wanted to learn his genealogy. After inputing 60,000 names and relationships into the computer, he found that the large majority of the lineages were not traced back to American Indians, but were traced back to the European Americans. Ms. Nii Ska claims that 99% of all the Ahnishinahbaeo'jibway have European paternal lineages! In other parts of the world, consider the effects of the many visits to Pacific Islands of European sailors. All these men had not seen women in months and they eagerly left their Y chromosomes in abundance on these islands. Obviously, if this applies elsewhere in the world, their study does not mean what they or Hugh Ross thinks it does.

This is consistent with other genetic data which also indicates that there has not been a recent genetic bottleneck. A bottleneck is defined as a period of time in which few members of a species exist. Such a period limits the genetic variability of the species. The other genetic data clearly says that Adam and Eve must have lived *long before the mitochondrial Eve and the Y-chromosome Adam*. To explain this, a bit of genetics is required.

A gene is a bit of DNA that controls a trait or a series of traits. The gene is a sequence of DNA that gives instructions for the coding of certain proteins. Each gene can have several variant forms, called alleles. Each allele is a sequence of DNA that varies from all the other alleles by coding for at least one different amino acid in the protein produced by the gene. While genes control things like eye-color, the gene's alleles determine *which* color—blue, brown, green etc.

A person has two copies of each

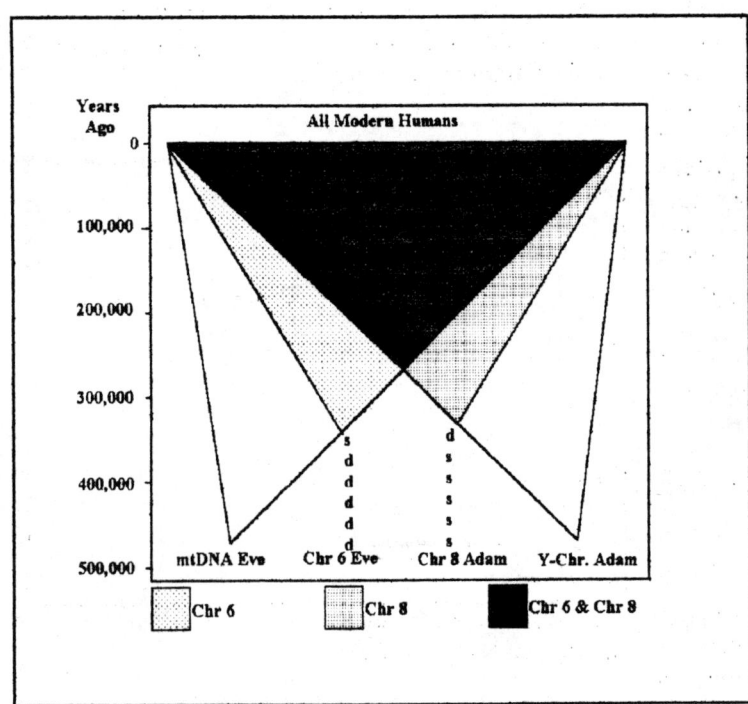

Figure 33 Other genetic Adams and Eves.

gene. Both copies may be the same allele, such as two copies of the blue-eyed allele. In that case they have blue eyes. If there is one copy of the blue-eyed allele and one of the brown, then they will have brown eyes. If there are two brown eyed alleles, they will also have brown eyes.

Because each person has only two copies of each gene, if Adam and Eve existed only 6,000 years ago, then there could only have been four copies of each gene, or a maximum of four different alleles on the earth 6,000 years ago. Thus those who believe that Adam lived six thousand years ago must believe that the observed genetic variation arose since that time. This requires a rate of mutation and a rate of evolution far faster than any evolutionist would be willing to admit. Observations in hospitals show that the mutation rate is not that rapid.[14] The young-earth creationists are wrong.

Similarly, old-earth creationists, who believe that humanity arose within the past 100,000 years, are also wrong in believing that all of the observed genetic variation could have arisen during that time. The new data for the age of the mitochondrial Eve place her more than 500,000 years ago.

Why the difference in age of the molecular Adam and Eve? Different parts of our molecular inheritance came from different people and mutated at slightly different rates. The different rates mean that different parts of our genetic heritage come from different times and from different people. Consider the circumstance outlined in Figure 33. There is a mitochondrial Eve. She gave rise to all the mitochondria used by all modern humans. Similarly, the Y-chromosome Adam fathered all the Y-chromosomes possessed by men today. But other people also contributed to our genetic heritage. Consider the Chromosome 6 Eve and the Chromosome 8 Adam. The Chromosome 6 Eve has a line of daughters (the d's) until finally her line produces no daughters but continues with a son who marries a daughter of mitochondrial Eve. The lack of daughters means that Chromosome 6's mitochondria have been lost to the world, but her chromosome 6 continues to this day having been passed on to the world through her male progeny. The lines radiating out from the marriage of Chromosome 6's son represents the replacement of all other chromosome 6's. Chromosome 8 Adam produces a line which finally fails to produce a son to transmit the Y-chromosome, but still produces a daughter who married a descendant of the Y-chromosome Adam. Chromosome 8 Adam does not pass on his Y-chromosome to humanity, but he passes on his copy of the 8th chromosome. The radiating lines from this point represent the replacement of all other human chromosome 8's. Similar reasoning would yield an Adam or Eve for each of the 23 pairs of human chromosomes. With an old population of humans, genetics is totally unable to prove descent from a single pair.

Thus, even if the Y-chromosome Adam and the mitochondrial Eve did live at the same time and were married to each other, one cannot rule out the existence of other people who also contributed to the modern genetic heritage. If fact, when you include the genetic process of crossing over, which we discussed above, there are literally billions of "Adams" and "Eves" for the various tiny pieces of genetic material they left us. Each chromosome we inherited from our parents was constructed from various parts of millions of other people's chromosomes. There actually could be more Adams than Eves or vice versa.

The biblical account of man's origin requires that two people be the genetic parents of all. There is one subsequent genetic bottleneck at the time of Noah's Flood where 5 people became the parents of all. These biblical bottlenecks require that all the observed genetic variation arise since that time. But writers who teach an Adamic creation within the past 150,000 years are flying in the face of modern genetic knowledge. Observed rates of mutation are too slow to have created the current genetic diversity within such a time frame.

The major histocompatibility complex controls much of the human immune response. Some of these genes, the DRB1 gene, have as many as 59 alleles.[15] Since an original pair of humans could only have 4 of these alleles, the other 55 alleles must have arisen since the first pair. The difficulty for the Christian is that some calculations of the coalescence time for the DRB1 gene would indicate that there has been no population bottleneck, no original pair, for the past 60 million years.[16] If this is true, then the entire biblical account of our origins is incorrect. Is there a solution? Only partially. The further back in time Adam and Eve lived, the less is the problem because there is more time for this much genetic diversity to have arisen. While at this moment there is not a solution for this problem, We cannot serve our case well by ignoring data like this and sweeping it under the rug. We cannot solve problems we do not discuss.

Were They Us Biologically?

The most important question which should answer our relationship with fossil man is that of our physical relationship. Is there evidence that we and they were biologically compatible? There is. The way to trace relationships is through unique similarities. While not absolutely conclusive, it is quite effective. One can recognize one of the Kennedy clan by the cut of his jaw and the shape of his nose. That clan has a set of characteristics that display the genetic relationship.

This same set of characteristics can be used to indicate relationships between different populations. There are three different views of human origins today: "replacement," "multiregional evolution," and "hybridize and replacement". The replacement view holds that anatomically modern man arose in Africa and spread throughout the world totally replacing the archaic forms of humanity: *Homo erectus*, archaic *Homo sapiens*, and Neanderthals. This view predicts that there would be few regional characteristics which are found in the earliest anatomically modern peoples. There is no genetic intermixing between the ancients and the moderns. Some anthropologists who are against this view call it the Pleistocene holocaust, since every non-modern individual on earth was believed to be wiped out. The multiregional view says that in every region of the world, the local population evolved into modern man. This view predicts that local, unique characters found in the ancients would continue into the local modern populations. It further predicts a relatively slow transition of traits from the ancient state to the modern state. The hybridize and replacement theory holds that anatomically modern men originated in Africa and as they spread out, they mated with the local populations. This would bring about individuals with intermediate morphology between the archaic forms and the modern forms, but the change would be rapid. Parents of different ethnic groups today, produce children with mixed their individual traits. Thus, each view has testable predictions, replacement, few transitional features; multiregional evolution, many, slowly developing transitional features; and hybridization, many, rapidly developed transitional features. What do the fossils say?

In the case of Neanderthals, the best argument for their legacy in the modern European gene pool is the

European	H-O Foramen %	Normal Foramen %
Neanderthal	53	47
African Eves	0	100
Skhul/Qafzeh	0	100
Early U. Paleolithic	18	82
Late U. Paleolithic	7	93
Mesolithic	2	98
Medieval Europeans	1	99

Table 6 Mandibular Foramen in Ancient and Modern Europeans

continuation of unique and non-adaptive traits across the Neanderthal/modern human transition.[17] A non-adaptive trait is one that has no survival value. Such a trait is that of the shape of the mandibular foramen. The mandibular foramen is a hole in the jaw bone through which the nerve for the teeth and lower jaw passes. This hole is on the inside and at the back of the lower jaw. It is the place that the dentist attempts to reach when he gives you a shot of zylocaine before he works on your teeth. There are two forms of the mandibular foramen. The normal form of the foramen has a long groove that contains the nerve. It also has a vertical groove that splits off from the main groove at the upper most end. In the other form, called the H-O form, the opening is oval in shape with the long axis of the oval lying horizontally on the inside of the jaw bone. The groove is covered by a bridge of bone. The form of the foramen makes no difference in function, but it makes a big difference in the discussion of whether or not Neanderthal is our ancestor.

The H-O form of the mandibular foramen is unique among European fossils. This trait is only rarely found in European fossils prior to the evolution of Neanderthals.[18] There is only one occurrence of this form of the foramen in any fossil outside of Europe. It is from Olduvai Gorge, (OH-22). Table 6 shows the percentage of this feature in the various European populations from Neanderthal to modern men. In general the peoples listed lower in the table are more recent. This trait is found in no other region of the world today. The fact that 18 percent of the Early Upper Paleolithic anatomically modern European's and 1 percent of today's Europeans possess this

Fossil Sample	Nasion Projection (mm)
Neanderthals	29.3
African Eve	17.8
Skhul/Qafzeh	12.4
Early Upper Paleolithic	21.9
Late Upper Paleolithic	19.3
Mesolithic	19.3
Medieval Hungarians	20.2

Table 7 Nasion projections in Fossil hominids.

feature would imply hybridization. The trait arises in modern humans rapidly, indeed almost instantaneously. The above would also imply that Neanderthal genes run through the blood of modern Europeans. One of the most certain conclusions is that it does not appear in the Skhul/Qafzeh remains. These peoples were the ancestors of the modern humans who entered Europe 35,000 years ago. Replacement must be wrong because the invaders didn't have this trait, yet their descendants do. Only hybridization or multiregional evolution can account for this.

Neanderthals had *big* noses. Their nose bone made nearly a 90 degree angle with their forehead. Anatomists use several indices to measure the strength of nasal projection. If Neanderthals were totally replaced by Africans who lacked this nasal projection, the earliest Upper Paleolithic peoples would not have strong nasal projection. But Table 7 shows the facts. The Skhul/Qafzeh peoples were those who lived in Israel and were the ancestors of the anatomically modern people who invaded Europe after 35,000 years ago. They had the smallest Nasion projection, their descendants, the first "replacements" for the Neanderthals, had the second largest nasion projections. This is inconsistent with replacement but is consistent with hybridization, since the trait appeared rapidly. The nasion projection is shown in Table 7.

Neanderthals had a large retromolar space which is a gap between the back tooth and the upright portion of the lower jaw. It can be seen on the cover drawing, which is the Amud Neanderthal skull. This trait is considered to be a Neanderthal trait, yet it is found in many early Upper Paleolithic fossil men. The retromolar space is found in the anatomically modern fossil men, Predmosti 3,4 and 21, Brno 2, and Stetten 1. The Stetten

Fossil Sample	Meric Index
European Neanderthals	79.6
Skhul/Qafzeh	83.1
Early Upper Paleolithic	77.6
Late Upper Paleolithic	78.0
Mesolithic	78.0
Medieval	80.4

Table 8 Meric index for Neanderthal and modern men

1 mandible also is one that has the Neanderthal trait of the H-O foramen.

Another non-adaptive trait is the small bump on the flange of Neanderthals. This flange shape is measured by the meric index shown in Table 8. Frayer writes,

"The low meric index in European Neanderthals compared to the Skhul/Qafzeh hominids which lack the lateral proximal flange, apparently relates to the marked development of this feature in European Neanderthals. All known early Upper Paleolithic specimens possess the proximal femoral flange, as do a number of later Upper Paleolithic and more recent specimens. Kidder, Smith and Jantz argue that this flange is likely a genetic trait since the presence of the feature is not correlated with a specific size or shape of the proximal femur. Also, according to them, the flange is found in infant femurs indicating 'a non-adaptive' etiology for the trait. The high frequency of this feature in European Neanderthals and in the subsequent Upper Paleolithic hominids, along with its apparent absence in the Skhul/Qafzeh femurs, provides evidence for evolutionary continuity in Europe and discontinuity between Skhul/Qafzeh and the Upper Paleolithic."[19]

Neanderthals possessed a curiously shaped skull which protruded in the rear. This feature is called the occipital bun. Few modern men possess this feature and those that do are most likely to be of European descent. Figure 3 in Frayer's article clearly shows that the back of the skulls of Early Upper Paleolithic are intermediate between Neanderthal and the African skulls. Frayer writes:

"Yet, the morphology of this region exhibits continuity between European Neanderthals and the early Upper Paleolithic."[20]

Sixty-five percent of all Neanderthals had a dorsal border type for their scapula (shoulder blade). No ancestor of the supposed replacements from Africa, the Skhul/Qafzeh peoples, had this type of border. Yet 16% of the Early Upper Paleolithic peoples of Europe still possessed the dorsal border type. In all of the features we have discussed, the replacement theory fails. To hold to the total replacement theory, one must assume that the invaders evolved in a fashion to become more Neanderthal-like. Replacement rules out interbreeding which can allow Neanderthal traits to continue in the population, even if it is at a lower frequency. The only skeletal feature which argues for replacement is the length of the arms and legs. The early Upper Paleolithic peoples clearly have longer arms and legs than the Neanderthals. But one fact in favor of a theory cannot overthrow much data against it.

There has been a recent, well-publicized article which claimed to prove that Neanderthal left no descendants on earth.[21] It is based upon the maternally inherited mitochondrial DNA. They determined part of the mtDNA sequence for a Neanderthal. The sequence showed evidence for the Neanderthal having 27 mutational differences from the human "standard" DNA sequence. However, the actual number of mutations could have been as low as 22 or as high as 36. The New York Times, quoting Christopher Stringer, reported:

"Describing it as "an incredible breakthrough in studies of human evolution," Stringer said the results showed Neanderthals "diverged away from our line quite early on, and this reinforces the idea that they are a separate species from modern humans." [22]

There are several problems with this idea. The work does show that the mother of *this* Neanderthal left no mtDNA on earth today. But it does not prove that she left no nuclear DNA. This is because it is easier for nuclear DNA to be transmitted than for mtDNA to be transmitted to future generations. No one has the mtDNA from their father's mother, but they have lots of nuclear DNA from her. My father's mother was the only red-head in the family. Most likely I inherited the gene for a red beard from her. Similarly, Krings et al, admit,

"These results do not rule out the possibility that Neandertals contributed other genes to modern humans."[23]

The real problem in the article concerns the amount of mtDNA variability among living humans and the implications that has for the author's conclusions. Modern humans have as many as 24 mutations compared to the estimated 27 Neanderthal mutations. Neanderthal has accumulated just about as many mutations as have modern human beings. Because of this, using the same mutation rate for both lineages, Neanderthal could only have split off from the modern human line, just slightly prior to the time that humans split apart from each other. Far from proving that Neanderthal was far removed from modern humans, the work really shows that his lineage and our lineage split apart at about the same time. It does prove, however, that there is no direct female lineage between them and us. But then again, there is none between my father's mother an me either.

One final note on this, this author posted a criticism of the statistics in the article on the Neanderthal Message board. This note received a mention in the E-Mail Anthropology News. It is included in the notes for anyone interested in the technical details.[24]

In the case of *Homo erectus*, there is a similar evidence of continuity of features between them and both modern Northern Asians and Australian aborigines. Zhoukoudian *Homo erectus* had shovel-shaped incisors, an internal thickening of the mandibular body, an extra bone in the cranium, known as an Inca bone or wormian bone, and a sagital keel. Modern Northern Asians exhibit these features also.[25] Early Chinese anatomically modern humans preserve "the overall characteristics of limb bones seen in Chinese *H. erectus* in contrast to the postcrania of early moderns from the Near Eastern sites of Qafzeh and Skhul ..."[26] Also the shape of the occipital bone and the archaic brain's blood supply pattern is retained even in many historical Chinese people.[27] The 35,000 year old Hetao, Mongolian skull's meningeal blood pattern found in the skull is very similar to that possessed by the Chinese *H. erectus*.[28] Other features possessed in common by modern and archaic Asian populations include laterally continuous supraorbitals and the angle of the torus.[29] Without going into more morphological details, Milford Wolpoff had to reconstruct the skull of Sangiran 17, of which Shreeve writes,

> "Instead of using rubber bands Wolpoff built a framework of toothpicks and in a couple of hours he had the face glued in place. He let the glue set for another half hour and only then picked up the specimen and turned it in profile. 'I nearly dropped dead. Instead of being just another *erectus*, here was this great big, hyper-robust Australian aborigine. I knew at that moment that Thorne was right, and I was wrong.
>
> "What astonished Wolpoff was the fossil's face, especially the way it projected out from the skull. Once he had completed the toothpick reconstruction, he could see that the jutting face was unlike anything he'd seen in *erectus* specimens from Africa or among the Peking Man casts in Beijing, where he had been just days before. Though some 700,000 years old, the face eerily resembled those of far younger, modern human fossils from Australia. The 'robust' Australian *sapiens* were as modern in brain size as any in the world, but they showed the same facial projection--big browridges, thick bones, sloping foreheads, and heavy molars--that Wolpoff saw in Sangiran 17. Many living Australian aborigines carry the same traits today."
>
> "Decades before, Weidenreich had suggested a connection between the *erectus* fossils of Java and modern Australian aborigines. But Weidenreich had only skullcaps to work from, like those of Solo Man from Ngandong. In the face of Sangiran, Wolpoff saw the missing anchor to Weidenreich's Australian lineage. Another researcher had suggested that the anatomy of Australian aborigines bore 'the mark of ancient Java.' For Wolpoff, Sangiran was stunning proof. In their arguments, Alan Thorne had been trying to convince Wolpoff that regional features would appear first at the remote edges of the hominid range, farthest away from the African birthplace of the earliest hominids. And here sat Sangiran 17, three quarters of a million years old and about as far from the African 'center' as one could get--but already full-fledged native Australasian."[30]

What all of this evidence seems to be saying is that yes, there was an outpouring of humanity from Africa around 100,000 years ago. But they did not fully replace every other hominid on earth. The traits indicate that there was some genetic continuity. But, the data does not support the view that each local population evolved into modern man independently. In fact, contrary to what most replacement theorists say, Wolpoff, today's leading multiregionalist, does believe in hybridization.[31]

Where Fossil Man Fits Biblically

If it is true that we, and the archaic forms of man, were able to interbreed, that is, that we are one species, then how do we fit fossil man into the biblical account? I have argued that the fossil and archeological evidence strongly supports the concept that mankind has been on this earth for many millions of years. They behaved as we would if we had the technology that they had. There is only one way to fit all of these facts into a biblical perspective.

As outlined in a previous book,[32] God created man about 5.5 million years ago. This is the time of the first occurrence of hominids on earth. Adam was both an evolutionary product of the apes *and* a special creation of God. There are three facts whose understanding are absolutely essential before one can unite Scripture and modern science. The first is that the African apes have 48 chromosomes and *Homo sapiens* have only 46. If mankind is a product of evolution, then there must have been a chromosomal fusion at some time during the past. The place I propose for this to occur was at the split between the apes and men. This is because none of the apes have 46 chromosomes, only humans do.

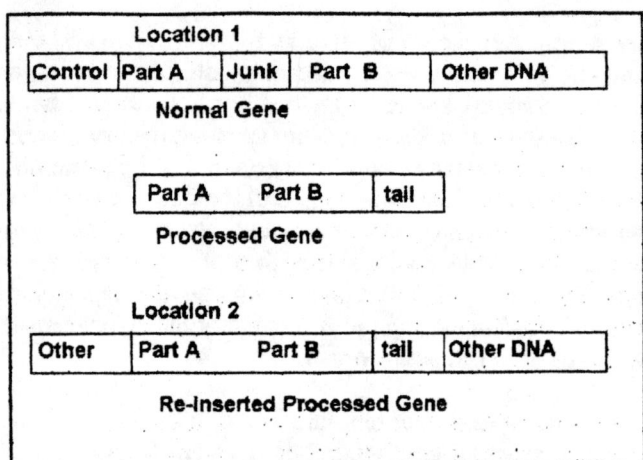

Figure 34 The construction of a pseudogene.

The second fact is that the genes of humans and chimpanzees are 98% identical. This amazing similarity means that only modest changes in the genetics of a chimpanzee would produce a human. Creationists have often used this similarity as evidence for similar design. According to this view, God used similar genetics to construct similar animals. But this view is disproven by the third fact.

The third fact is the existence of pseudogenes which strongly support the genetic relationship between man and ape. A pseudogene is a broken gene; it is a gene with no functionality. A gene which codes for a protein is constructed as is shown in Figure 34. In the nuclear DNA (location 1 in the figure) the working gene consists of a control section, followed by some DNA that codes for part of the protein, called an exon. Then there is a region called an intron which is also known as junk DNA. This region does not provide any information for the manufacture of the protein. Following this, there is another exon, (part B) that contains the information for the rest of the protein. When the nuclear DNA is copied prior to the manufacture of the protein, parts A and B are united, the control portion is removed, and a tail is added to the united parts A and B. This processed gene is then used in the ribosome to manufacture a copy of the protein.

Occasionally, a mistake is made and the processed gene is reinserted back into the nuclear DNA at a second location. But the processed gene lacks the control portion and lacks the intron. In this form, it cannot manufacture a protein any longer. It is absolutely useless. This lack of utility is what demonstrates that evolution has occurred *and* that man is related genetically to the great apes.

Recently, Edward Max has brought forth some information which contradicts the common design argument of the creationists.[33] Max cites two court cases, Colonial Book Co, Inc. v. Amsco School Publications, Inc., District Court, S.D. New York, Sept 9, 1941 and SubContractors Register, Inc. v McGovern's Contractors & Builders Manual, Inc. et al. District Court, S.D. New York, August 2, 1946. In these cases, the plaintiff claimed that their books had been plagiarized. But since in both of these cases, the two products were designed to serve a similar function and to convey similar information, it was difficult to say that the correct information had been acquired in an illicit fashion. After all, anyone can get a business address and if it is correct, there is no way to know if they got it out of a copyrighted book or from the tax office. But, if the first published book contains an error that is not in the records of the tax office, then the existence of the identical error, in the identical place in the second book would be proof of copyright infringement. This is exactly what the courts decided. The two defendants had transmitted errors made in the plaintiff's books. This transmission proved to the jury that the source of the information was illicitly acquired.

A processed pseudogene plays exactly the same function in the biological court of law. A common pseudogene has been found at the identical location in four species: humans, chimpanzees, gorilla, orangutan and Old World monkeys.[34] Common design simply does not explain this data.

Consider the following situation. You ask five of your friends to copy the Encyclopedia Britannica, all thirty volumes. You ask them to go to five different cities, with instructions that at some randomly chosen point in the typing of the encyclopedia, they are to stop, randomly select a paragraph, and insert that paragraph into the Britannica. Then they are to continue the rest of the copying job. When they come back together and show you their copies, you find that all five people had copied the identical paragraph into the identical spot of the Britannica. You would think that they had conspired to achieve this effect. There is no way that you would think that this was simply the result of chance. Similarly, to find the same error in the genes of five different species, but not in all species, is evidence of common descent.

There are other examples of useless DNA embedded in the DNA of humans and chimpanzees. Occasionally, retroviruses are able to insert their DNA code into the nuclear DNA of an animal. There are cases where identical retroviral DNA is found at the same location in both chimpanzees and humans. This means that some common ancestor caught a retroviral disease, the retrovirus inserted itself into the genes of that individual and it was passed on to all the descendants. These nonuseful genes prove that we arose from the apes.

Because of these three facts, there is only one explanation which allows for the incorporation of the data into a biblical perspective. Five and a half million years ago, an ape-like creature gave birth to a child that had a grievous genetic defect. This child had 46 chromosomes rather than the normal 48 of her species. The child was born dead. God took pity on this creature, fixed the defect and breathed life into it. This was Adam.

Adam did not know how to talk. His biological mother didn't either. Thus, God brought all the animals to Adam and let him name them. Whatever Adam named the creatures was what they were called. But among all the animals, no mate was found for him. He had 46 chromosomes and all of his relatives had 48. Because of this, he was a lone creature on this earth with no ability to reproduce. While a chromosomal fusion is not necessarily a barrier to reproduction, it often is. Following Adam's recognition that there was no mate for him, God caused a deep sleep to come on Adam and removed a rib. From this rib, God fashioned a woman for Adam. God told them not to eat from the fruit of a particular tree, which they did. God cursed them and ejected them from the place where they were living. Humanity descended from this primeval pair who were both the product of evolution and special creation.

Was Adam an australopithecine? There is no answer to this question. Ernst Mayr, the great taxonomist, preferred to classify them in the genus *Homo*. But there are so many differences between them and us, it seems unlikely. What I would prefer is that Adam's descendants were able to live for millions of years in an isolated pocket, maybe in a forest where bones are unlikely to be preserved. Acidic soils found in tropical forests destroy bones in less than a year. Eventually, they reproduced and became widespread enough that they left fossil evidence of themselves around the world.

If Adam were a *Homo erectus*, then one can probably place the origin of man long prior to 2.0 million years. *Homo erectus* was so widespread around the world when he first appears on earth that a long time is required for the population to spread out that far. While anthropologists have not been looking for early *Homo erectus* in places other than Africa, it is likely that he had been on earth for millions of years prior to his first fossil example. In this fashion, all of the data of anthropology can be placed in a biblical framework, leaving no contradictions.

Conclusion

We have seen the evidence anthropology has gathered over the past century. It, along with the genetic data, shows that humanity is quite old. As we have noted several times throughout this book, the usual Christian response to fossil man has been entirely unsatisfactory to anyone who seriously examines the details of the data. To claim, as almost all do, that Adam was a recent creation, flies in the face of the data we have just seen. Christians of all stripes tend to ignore the data. This is true regardless of whether one is a young-earth creationist who believes that Adam was created 6,000 years ago or an old-earth creationist who believes that Adam was created either 35,000, 60,000 or 100,000 years ago. Mankind is much, much older than any of these dates. This does not violate the truth of the Scripture. As Bernard Ramm wrote,

> "The Bible itself offers no dates for the creation of man. We mean by this that there is no such statement in the text of the Bible at any place. We may feel that 4000 B.C. or 15,000 B.C. is more consonant with the Bible than a date of 500,000 B.C. But we must admit that any date of the antiquity of man is an inference from Scripture, not a plain declaration of Scripture.
>
> "If the anthropologists are generally correct in their dating of man (and we believe they are), and if the Bible contains no specific data as to the origin of man, we are then free to try to work out a theory of the relationship between the two, respecting both the inspiration of Scripture and the facts of science."[35]

It is time for Christian apologists to cease the constant rejection of scientific data in order to hold to a comfortable theological position, no matter whether the position is a young-earth or old-earth position. Young-earth creationists reject all scientific data that indicates an old universe, but accept the humanity of fossil man. Old-earth creationists, while looking down their noses at the young-earth creationists, accept all scientific data *except* that indicating the humanity of fossil man. But both camps are selective in their use of data. This must cease. We must live in *this* world, not the world of our imagination. We must know and deal with the observational data of *this* world. Anything less buries our heads in the sand.

The fact that fossil men were human in every theological sense of the word leaves us with a theological problem. Our apologetical systems do not account for this. These systems must be altered. The events of early Genesis must have taken place millions of years ago. Does this reduce our position in the universe? No. It merely means that God has been active in human affairs much longer than most of us have been willing to admit.

References

1. James R. Shreeve, The Neandertal Enigma, (New York: William Morrow and Co., 1995), p. 77
2. R. L. Cann, M. Stoneking, and A. C. Wilson, "Mitochondrial DNA and Human Evolution," Nature, 325(1987):31-36
3. Percival Davis and Dean H. Kenyon, Of Pandas and People, (Dallas: Haughton Publishing Co. 1993), p. 112
4. Hugh Ross, "Chromosome Study Stuns Evolutionists," Facts & Faith, 9(1995):3, p. 3.
5. Michael F. Hammer, "A Recent Common Ancestry for Human Y Chromosomes," Nature, 378(1995):376-378, p. 377-378
6. L. Simon Whitfield, John E. Sulston, and Peter N. Goodfellow, "Sequence Variation of the Human Y Chromosome," Nature 378(1995):379-380.
7. Robert L. Dorit, Hiroshi Akashi, Walter Gilbert, "Absence of Polymorphism at the ZFY locus on the Human Y Chromosome," Science, May 26, 1995, p. 1183
8. Robert L. Dorit, Hiroshi Akashi, Walter Gilbert, "Absence of Polymorphism at the ZFY locus on the Human Y Chromosome," Science, May 26, 1995, p. 1184
9. Quoted by Milford Wolpoff and Rachel Caspari, Race and Evolution, (New York: Simon & Schuster, 1997), p. 362
10. Hugh Ross, "Chromosome Study Stuns Evolutionists," Facts & Faith, 9(1995):3, p. 3.
11. Alan R. Templeton, "The 'Eve' Hypothesis: A Genetic Critique and Reanalysis," American Anthropologist 95(1): 51-72. p. 52
12. Erik Trinkaus and Pat Shipman, The Neandertals, (New York: Alfred Knopf, 1993), p. 394-396
13. Erik Trinkaus and Pat Shipman, The Neandertals, (New York: Alfred Knopf, 1993), p. 394-396
14. "The total number of nucleotides in a gamete is very large (3 billion), and the mutation rate per generation may be of the order of 1 in 200 million nucleotides." L.Luca Cavalli-Sforza, Paoli Menozzi and Alberto Piazzi, The History and Geography of Human Genes, (Princeton: Princeton University Press, 1994); p. 6
15. Francis J. Ayala, "The Myth of Eve: Molecular Biology and Human Origins," Science, 270(1995):1930-1936.
16. Francis J. Ayala, "The Myth of Eve: Molecular Biology and Human Origins," Science, 270(1995):1930-1936, p. 1932
17. The data for the following tables are taken from David W. Frayer, "Evolution at the European Edge: Neanderthal and Upper Paleolithic Relationships," Prehistoire Europeenne, 2:9-69.
18. David W. Frayer, "Evolution at the European Edge: Neanderthal and Upper Paleolithic Relationships," Prehistoire Europeenne, 2:9-69, p. 29
19. David W. Frayer, "Evolution at the European Edge: Neanderthal and Upper Paleolithic Relationships," Prehistoire Europeenne, 2:9-69, p. 36
20. David W. Frayer, "Evolution at the European Edge: Neanderthal and Upper Paleolithic Relationships," Prehistoire Europeenne, 2:9-69, p. 21
21. Matthias Krings, et al, 1997. "Neandertal DNA Sequences and the Origin of Modern Humans," Cell, 90:19-30
22. http://www.nytimes.com/yr/mo/day/news/national/sci-neanderthal-dna.html
23. Matthias Krings, et al., "Neandertal DNA Sequences and the Origin of Modern Humans," Cell, 90:19-30, p. 27
24. http://www.dealsonline.com/origins/message/neanderthals/messages/2074.html

I just read the Cell article by Krings et al. I see a couple of areas for concern, statistically. I don't know if I am being redundant but here goes. I checked my concerns with a Statistics prof friend of mine.

They write:

"Whereas these modern human sequences differ among themselves by an average of 8.0+/- 4.1 (range 1-24) substitutions, the difference between the humans and the Neandertal sequence is 27.2+/-2.2 (range 22-36) substitutions. Thus the largest difference observed between any two human sequences was two substitutions larger than the smallest difference between a human and the Neandertal. In total, 0.002% of the pairwise comparisons between human mtDNA sequences were larger than the smallest difference between the Neandertal and a human.

"The Neandertal sequence, when compared to the mitochondrial lineages from different continents, differs by 28.2 +/- 1.9 substitutions from the European lineages, 27.1 +/-2.2 substitutions from the African lineages, 27.7+/- 2.1 substitutions from the American lineages and 28.3+/- 3.7 substitutions from the Australian /Oceanic lineages. Thuse, whereas the Neandertals inhabited the same geographic region as contemporary Europeans, the observed differences between the Neandertal sequence and
modern Europeans do not indicate that it is more closely related to modern Europeans than to any other population of contemporary humans.

"When the comparison was extended to 16 common chimpanzee lineages,. the number of positions in common among the human and chimpanzee sequences was reduced to 333. This reduced the number of human lineages to 986. The average number of differences among humans is 8.0+/- 3.0 (range 1-24), that between humans and the Neandertal, 25.6 +/- 2.2 (range 20-34), and that between humans and chimpanzees, 55.0 +/- 3.0 (range 46-67). Thus, the average number of mtDNA sequence differences between modern humans and the neandertal is about three times that among Humans,
but about half of that between modern humans and modern chimpanzees."Krings, Matthias, et al, 1997. "Neandertal DNA Sequences and the Origin of Modern Humans," Cell, 90:19-30, p. 24-25

First, it seems to me that the comparison of a range of experimental uncertainty (Neanderthal range 22-36 substitutions-an experimental uncertainty) with mtDNA divergence across the human race is an
invalid comparison. It is an equivocation error in logic. They are also comparing an average (8 in modern humans) with the single measurement from one individual of Neandertal--an equally inappropriate mathematical measure.

Secondly, the last paragraph above seems to be an invalid comparison. Since humans have up to 24 substitutionary differences, and Neandertal is about that distance from the reference sequence, how can their conclusion that there is a 3 times difference be meaningful? I mean a divergence of 24 in human populations, is almost the same divergence as the Neanderthal. Human divergence is not 3 times less than Neandertal, it is almost the same. They further compound this when they write:

"To estimate the time when the most recent ancestral sequence common to the Neandertal and modern human mtDNA sequences existed, we used an estimated divergence date between humans and chimpanzees of 4-5 million years ag and corrected the observed sequence differences for multiple substitutions at the same nucleotide site. This yielded a date of 550,000 to 690,000 years before present for the divergence of the Neandertal mtDNA and contemporary human mtDNAs. When the age of the modern human mtDNA ancestor is estimated using the same procedure, a date of 120,000 to 150,000 years is obtained, in agreement with previous estimates.(Cann et al., 1987 Vigilant et al. 1991)." Krings, Matthias, et al, 1997. "Neandertal DNA Sequences and the Origin
of Modern Humans," Cell,90:19-30, p. 25

The time for humans to accumulate 24 substitutions should be almost the same as the time for Neandertal to accumulate 27 substitutions. While there may have been isolation between Neandertal women and modern people the time of divergence should be the same unless they can show that human mitochondrial substitutions occur at a more rapid rate.

glenn

This report was cited in the August 1997 Anthropological E-mail News.

25. Milford Wolpoff and Rachael Caspari, Race and Human Evolution, (New York: Simon and Schuster, 1997), p. 194-195

26. Dennis A. Etler, "The Fossil Evidence for Human Evolution in Asia," Ann. Rev. Anthropol. 1996, 25:275-301, p. 290

27. Dennis A. Etler, "The Fossil Evidence for Human Evolution in Asia," Ann. Rev. Anthropol. 1996, 25:275-301, p. 290

28. Dennis A. Etler, "The Fossil Evidence for Human Evolution in Asia," Ann. Rev. Anthropol. 1996, 25:275-301, p. 290-291

29. Dennis A. Etler, "The Fossil Evidence for Human Evolution in Asia," Ann. Rev. Anthropol. 1996, 25:275-301, p. 294

30. James R. Shreeve, The Neandertal Enigma, (New York: William Morrow and Co., 1995), p. 102-103

31. Milford Wolpoff and Rachael Caspari, Race and Human Evolution, (New York: Simon and Schuster, 1997), p. 202

32. Glenn R. Morton, Foundation, Fall and Flood, (Dallas: DMD Publishing, 1995). ISBN 9648227-1-7, 16075 Longvista Dr., Dallas, TX 75248

33. This can be found on the Internet at http://earth.ics.uci.edu:8080/faqs/molecular%2Dgenetics.html

34. see http://earth.ics.uci.edu:8080/faqs/molecular%2Dgenetics.html

35. Bernard Ramm, The Christian View of Science and Scripture, (Grand Rapids: Eerdmans Publishing Co., 1954), p. 220

Appendix A
Chronology of Human Technology

Age BP	Oldest example of...	Place	Species
2.0 kyr	continuous industrial worksite	Bigo, Uganda	Homo sapiens[1]
2.0 kyr	grammar text	Greece	Homo sapiens[2]
3.0 kyr	census	Israel	Homo sapiens[3]
3.0 kyr	iron mine	Velem St. Vid, Hungary	Homo sapiens[4]
4.0 kyr	oldest song	Ur, Iraq	Homo sapiens[5]
4.5 kyr	Maya farming	Belize	Homo sapiens[6]
4.8 kyr	fixed calendrical date	Egypt	Homo sapiens[7]
5.0 kyr	extrabibilcal person's name	Egypt	Homo sapiens[8]
6.0 kyr	iron object	Egypt	Homo sapiens[9]
6.0 kyr	existant ship	England	Homo sapiens[10]
6.0 kyr	horse domesticated	Ukraine	Homo sapiens[11]
6.5 kyr	gold jewelry	Varna, Bulgaria	Homo sapiens[12]
7.0 kyr	wine	Zagros Mtns, Iran	Homo sapiens[13]
7.5 kyr	two-stage domed kiln	Yarim Tepe, Iraq	Homo sapiens[14]
8.4 kyr	dugout canoe	Pesse, Netherlands	Homo sapiens[15]
9.0 kyr	fragment of cloth	Turkey	Homo sapiens[16]
9.0 kyr	pigs domesticated	Greece	Homo sapiens[17]
9.0 kyr	continuously existing religion	Queensland, Australia	Homo sapiens[18]
9.4 kyr	boat paddle	Star Carr, England	Homo sapiens[19]
9.5 kyr	sheep domesticated	Turkey	Homo sapiens[20]
10 kyr	goats domesticated	Asiab, Iran	Homo sapiens[21]
10 kyr	wooden Australian boomerang	Wyrie Swamp, Australia	Homo sapiens[22]
10 kyr	depiction of warfare	Australia	Homo sapiens[23]
11 kyr	confirmed basket	Danger Cave, Utah	Homo sapiens[24]
11 kyr	wooden arrow	Hamburg, Germany	Homo sapiens[25]
12 kyr	city	Mureybet Syria	Homo sapiens[26]
12 kyr	arrowhead	San Teodoro, Sicily	Homo sapiens[27]
13 kyr	flint sickle	Middle East	Homo sapiens[28]
13 kyr	first pottery	Japan	Homo sapiens[29]
13 kyr	copper working	various	Homo sapiens[30]
13-14 kyr	toothpick	South Africa	Homo sapiens[31]
13-14 kyr	extant wooden hut remains	Monte Verde, Chile	Homo sapiens[32]
14 kyr	fish hooks	Europe	Homo sapiens[33]
14 kyr	homo sapiens religious sanctuary	El Juyo Cave, Spain	Homo sapiens[34]
19.3 kyr	twisted fiber cord	Ohalo, Israel	Homo sapiens[35]
19. kyr	fishing net	Ohalo II, Israel	Homo sapiens[36]
19.6 kyr	basket or mat	Meadowcroft, PA	Homo sapiens[37]
20 kyr	ivory boomerang	Oblazowa, Poland	Homo sapiens[38]
23 kyr	ground stone tools	Malangangerr, Australia	Homo sapiens[39]
25 kyr	net fishing	Willendra Lake, Australia	Homo sapiens[40]
25 kyr	puppet	Brno, Moravia	Homo sapiens[41]
26 kyr	bone sewing needle	central Europe	Homo sapiens[42]
27 kyr	impression of woven cloth	Eastern Europe	Homo sapiens[43]
27 kyr	ceramics	Dolni Vestonice	Homo sapiens[44]
30 kyr	human cremation	Lake Mungo Australia	Homo sapiens[45]
32 kyr	coal mine	Landek, Czechoslovakia	Homo sapiens[46]
35 kyr	fossil collection	Arcy-sur-Cure	Homo sapiens neanderthalensis[47]
36 kyr	glue	Umm el Tlel, Syria	?[48]

Age	Item	Location	Species
39 kyr	cave painting	Carpenter's Gap, Australia	archaic Homo sapiens[49]
43-67 kyr	7 note diatonic musical scale	Divje Babe I, Slovenia	Homo sapiens neanderthalensis[50]
45 kyr	Neanderthal flute	Divje Babe I, Slovenia	Homo sapiens neanderthalensis[51]
47 kyr	successful surgical amputation	Shanidar, Iraq	Homo sapiens neanderthalensis[52]
50 kyr	shaman's cape	Hortus, France	Homo sapiens neanderthalensis[53]
60 kyr	Neanderthal hut/tent	Molodova, Russia	Homo sapiens neanderthalensis[54]
68 kyr	Murder by spear	Shanidar, Iraq	Homo sapiens neanderthalensis[55]
70-80 ky	Neanderthal art, pseudoVenus	Wildenmannlisloch, Switz.	Homo sapiens neanderthalensis[56]
70-80 kyr	musical instrument-flute	Haua Fteah, Libya	Homo sapiens neanderthalensis[57]
73 kyr	use of coal for fire	Les Canalettes	Homo sapiens neanderthalensis[58]
75-116 kyr	rock engraving	Jinmium, Australia	archaic Homo sapiens[59]
80 kyr	Homo sapiens wooden spear	Mt. Carmel, Israel	Homo sapiens[60]
80 kyr	Neanderthal bone tool	Regourdou, France	Homo sapiens neanderthalensis[61]
80 kyr	religious sanctuary	Drachenloch, Switzerland	Homo sapiens neanderthalensis[62]
90-100 kyr	whistles	Prolom II, Crimea	archaic Homo sapiens[63]
110 kyr	underground mining	Lion Cave, Swaziland	?[64]
110-130 kyr	Neanderthal spear	Lehringen, Germany	Homo sapiens neanderthalensis[65]
130 kyr	burial	Krapina, Croatia	Homo sapiens neanderthalensis[66]
130 kyr	food processors	various	archaic Homo sapiens[67]
140 kyr	over-horizon sailing	Australia	archaic Homo sapiens[68]
140 kyr	Shellfish exploitation	Italy	Homo sapiens neanderthalensis[69]
160 kyr	European blade tools	England Belgium	Homo sapiens neanderthalensis[70]
200 kyr	post hole	Lunel-Viel	archaic Homo sapiens[71]
200 kyr	Neanderthal bedding	Lazaret, France	Homo sapiens neanderthalensis[72]
200 kyr	Neanderthal living floor	Grotte d'Aldene	Homo sapiens neanderthalensis[73]
230 kyr	European human dung	Terra Amata, France	Homo erectus?[74]
240 kyr	upper Paleolithic blade tools	Kenya	?[75]
240-700 kyr	woodworking	Gesher Benot Ya'aqov	Homo erectus[76]
250 kyr	Invention of Mousterian tools	Vaufry Cave, France	?[77]
300 kyr	geometric engraving	Pech de l'Aze	Homo erectus[78]
300 kyr	Siberia inhabited-clothing needed	Diring Yuriakh	?[79]
300 kyr	jewelry	various	Homo erectus[80]
>300 kyr	ritual corpse disposal	Atapuerca, Spain	archaic H.s.[81]
330 kyr	Depiction of human form	Berekhat Ram, Israel	Homo erectus/archaic H.s[82]
>350 kyr	stone wall	Bhimbetka, India	archaic H.s/Homo erectus[83]
3-400 kyr	European huts	Bilzingsleben, Germany	Homo erectus[84]
3-400 kyr	paved social area	Bilzingsleben, Germany	Homo erectus[85]
3-400 kyr	oldest evidence of counting	Bilzingsleben, Germany	Homo erectus[86]
3-400 kyr	oldest religious altar	Bilzingsleben, Germany	Homo erectus[87]
400 kyr	wooden spears	Schoningen, Germany	Homo erectus[88]
400 kyr	3 component composite tools	Schoningen, Germany	Homo erectus[89]
400 kyr	tools made by other tools	Schoningen, Germany	Homo erectus[90]
400 kyr	wooden boomerang	Schoningen, Germany	Homo erectus[91]
500 kyr	mineral collection	Zhoukoudian, China	Homo erectus[92]
500 kyr	man and canine assoc.	Zhoukoudian, China	Homo erectus[93]
500 kyr	Asian fire	Zhoukoudian, China	Homo erectus[94]
600 kyr	scalping	Bodo, Ethiopia	Homo erectus[95]
700 kyr	ocean travel	Flores, Indonesia	Homo erectus[96]
750 kyr	European fire	Escale Cave, France	Homo erectus[97]
800-900 kyr	Homo erectus bedding	Wonderwork Cave, S. A.	Homo erectus[98]
970 kyr	European structure	Soleihac Cave	Homo erectus?[99]
1.0 MYR	tanning hides	Swartkrans, South Africa	Homo erectus[100]
1-1.5 MYR	toolmaking by *Australopithecus*	Swartkrans, South Africa	A. robustus[101]
1.5 MYR	evidence of fire	Swartkrans South Africa	H. erectus A. robustus[102]

1.5 MYR	woodworking	Koobi Fora, Kenya	Homo erectus[103]
1.6 MYR	Man-made representational art	Olduvai Gorge	Homo erectus?[104]
1.6 MYR	working with animal hides	Swartkrans	A. robustus[105]
1.6 MYR	bone tool	Swartkrans	A. robustus[106]
1.7 MYR	Human compassion	East Africa	Homo erectus[107]
1.9 MYR	Right-handedness	Koobi Fora	Homo erectus[108]
1.9 MYR	Stones Thrown as Weapons	Olduvai Gorge, Tanzania	Homo habilis[109]
1.95 MYR	Larynx capable of Speech	Africa	Homo erectus[110]
2.0 MYR	Brain structure for Language	Lake Rudolf	Homo habilis[111]
2.0 MYR	Windbreak structure	Olduvai	Homo erectus[112]
2.0 MYR	Oldest Toothpick use	Ethiopia	Homo erectus[113]
2.5 MYR	Genus Homo	Lake Rudolf	Homo rudolfensis[114]
2.6 MYR	Stone tools	Gona, Ethiopia	?[115]
3.0 MYR	recognized art	Makapansgat, South Afr.	A. africanus[116]
3.4 MYR	TMJ Disease	Laetoli, Tanzania	A. africanus[117]
4.0 MYR	Bipedalism	Kanapoi, Kenya	A. anamensis[118]
5.5 MYR	earliest hominid fossil	Lothagam, Kenya	*Ardipithecus* ?[119]

REFERENCES

1. Christopher Wills, The Runaway Brain, (New York: Harper Collins, 1993), p. 101
2. Wayne M. Senner, "Theories and Myths on the Origins of Writing: A Historical Overview," in Wayne M. Senner, editor, The Origins of Writing, (Lincoln: University of Nebraska Press, 1989), p. 4
3. Bible Exodus 30:12
4. R. Shepard, Prehistoric Mining and Allied Industries, (New York: Academic Press, 1980), p. 210
5. Anne Draffkorn Kilmer, Richard L. Crocker and Robert R. Brown, Sounds from Silence (Berkeley, California: Bit Enki Publications, 1976) and Bob Fink, "The Oldest Song in the World," in Archaeologia Musicalis (Study Group on Music Archaeology, Feb., 1988), pp. 98-100.
6. B. Bower "Maya Beginnings Extend Back at Belize Site," Science News, April 30, 1994, p. 279
7. Colin Renfrew, Before Civilization, (New York: Penguin Books, 1976), p. 29-30
8. Henry George Fisher, "The Origin of Egyptian Hieroglyphs," in Wayne M. Senner, Ed. The Origins of Writing, (Lincoln: University of Nebraska press, 1989), p. 61
9. Microsoft Encarta Multimedia Encyclopedia, 1994. "Iron"
10. Dallas Morning News, Thursday, 5-8-97, p. 16A
11. Victor Barnouw, An Introduction to Anthropology: Physical Anthropology and Archaeology, 1, (Homewood, Ill: The Dorsey Press, 1982), p.193
12. A. Gopher, et al, "Earliest god Artifacts in the Levant," Current Anthropology, 31:4(1991), pp 436-443, p. 441
13. B. Bower, "Wine Making's Roots Age in Stained jar", Science News June 8, 1996, p. 359
14. Joan Oates, "The Emergence of Cities in the Near East," in Andrew Sherratt, editor, The Cambridge Encyclopedia of Archaeology, (New York: Cambridge University Press, 1980), p. 112
15. Grahame Clark and Stuart Piggott, Prehistoric Societies, (New York: Alfred A. Knopf, 1965), p. 106
16. Science News, Vol. 144, p. 54
17. Victor Barnouw, An Introduction to Anthropology: Physical Anthropology and Archaeology, 1, (Homewood, Ill: The Dorsey Press, 1982), p.193
18. Josephine Flood, The Archeology of the Dreamtime, (New Haven: Yale University Press, 1989), p. 143
19. Victor Barnouw, An Introduction to Anthropology: Physical Anthropology and Archaeology, 1, (Homewood, Ill: The Dorsey Press, 1982), p.185
20. O. Bar-Yosef, "The Role of Western Asia in Modern Human Origins," Phil. Trans. R. Soc. Lond. B, 1992, p. 193-200 esp. p. 199
21. Victor Barnouw, An Introduction to Anthropology: Physical Anthropology and Archaeology, 1, (Homewood, Ill: The Dorsey Press, 1982), p.193
22. Josephine Flood, "The Archeology of the Dreamtime, (New Haven: Yale University Press, 1989), p. 154
23. Bruce Bower, "Seeds of Warfare Precede Agriculture," Science News, Jan. 7, 1995, p. 4.
24. D. Nadel, et al. "19,000-Year-Old Twisted Fibers from Ohalo II," Current Anthropology, 35:4(1994), pp. 451-457, p. 456-457; and Jacquetta Hawkes, Prehistory, (New York: Mentor Books, 1963), p. 153
25. Richard Klein, "Later Pleistocene Hunters," in Andrew Sherratt, editor, The Cambridge Encyclopedia of Archaeology, (New York: Cambridge University Press, 1980), p. 90
26. Jean Guilaine, "The First Farmers of the Old World," in Jean Guilaine, editor, Prehistory: The World of Early Man, (New York: Facts on File, 1986), p. 80-81
27. L. Bachechi, P.F. Fabbri and F. Mallegni, "An Arrow-Caused Lesion in a Late Upper Paleolithic Human Pelvis," Current Anthropology, 38:1(Feb. 1, 1997):135-140, p. 139-140
28. The Software Toolworks Multimedia Encyclopedia, Grolier, 1992, Timeline.
29. L. Luca Cavalli-Sforza, Paoli Menozzi and Alberto Piazzi, The History and Geography of Human Genes, (Princeton: Princeton University Press, 1994), p. 202
30. Microsoft Encarta Multimedia Encyclopedia, 1994. "Copper"
31. Richard Klein, "Later Pleistocene Hunters," in Andrew Sherratt, editor, The Cambridge Encyclopedia of Archaeology, (New York: Cambridge University Press, 1980), p. 90
32. E. James Dixon, Quest for the Origins of the First Americans,(Albuquerque: University of New Mexico Press, 1993), p. 96
33. Richard Klein, "Later Pleistocene Hunters," in Andrew Sherratt, editor, The Cambridge Encyclopedia of Archaeology, (New York: Cambridge University Press, 1980), p. 90

34. L. G. Freeman and J. G. Echegaray, "El Juyo: A 14,000-year-old Sanctuary From Northern Spain," History of Religion, Aug. 1981, p. 15-16.
35. D. Nadel, et al. "19,000-Year-Old Twisted Fibers from Ohalo II," Current Anthropology, 35:4(1994), pp. 451-457
36. D. Nadel, et al. "19,000-Year-Old Twisted Fibers from Ohalo II," Current Anthropology, 35:4(1994), pp. 451-457
37. Donald Johanson and Blake Edgar, From Lucy to Language, (New York: Simon and Schuster, 1997), p. 47.
38. Adrian Lister and Paul Bahn, Mammoths, (London: Boxtree, 1995), p. 112
39. Josephine Flood, "The Archeology of the Dreamtime, (New Haven: Yale University Press, 1989), p. 85
40. Josephine Flood, "The Archeology of the Dreamtime, (New Haven: Yale University Press, 1989), p. 50-51
41. Adrian Lister and Paul Bahn, Mammoths, (London: Boxtree, 1995), p. 115
42. Donald Johanson and Blake Edgar, From Lucy to Language, (New York: Simon and Schuster, 1997), p. 99
43. B. Bowers, "Stone Age Fabric Leaves Swatch Marks," Science News, 147:276, May 6, 1995, p. 276
44. Chris Stringer and Clive Gamble, In Search of the Neanderthals, (New York: Thames and Hudson, 1993), p. 206
45. James R. Shreeve, The Neandertal Enigma, (New York: William Morrow and Co., 1995), p. 103
46. R. Shepard, Prehistoric Mining and Allied Industries, (New York: Academic Press, 1980), p.231-232, 236
47. Andre Leroi Gourhan, The Hunters of Prehistory, transl. Claire Jacobson, (New York: Atheneum, 1989), p. 93
48. Eric Boeda, et al, "Bitumen as a Hafting Material on Middle Palaeolithic Artefacts," Nature, 380, March 28, 1996, p. 336-337
49. Paul G. Bahn, "Further Back Down Under," Nature, Oct 17, 1996, p. 577-578, p. 578
50. Bob Fink, http://www.webster.sk.ca/greenwich/fl-compl.htm also se
51. Ivan Turk, Janez Dirjec and Boris Kavur, 'Ali so v Sloveniji Nasli Najstarejse glasbilo v Evropi?' Razprave, IV, razreda SAZU, XXVI(1995), pp 288-293.
52. J.B. Birdsell, Human Evolution, (Chicago: Rand McNally & Co., 1972), p. 285
53. James R. Shreeve, The Neandertal Enigma, (New York: William Morrow and Co., 1995), p. 52
54. Chris Stringer and Clive Gamble, In Search of the Neanderthals, (New York: Thames and Hudson, 1993), p.157
55. Rose L. Solecki, "More on Hafted Projectile Points in the Mousterian," Journal of Field Archaeology, 19(1992):207-212, p. 211
56. Ivar Lissner, Man, God and Magic, (New York: G. P. Putnam's Sons, 1961), p. 189-191; Dr. Emil Bachler, Das Alpine Palaeolithikum der Schweiz, Monographien Zur Ur-Und Fruhgeschichte der Schweiz (Basel: Verlag Birkhauser & Cie, 1940), Bd II, figure 111.
57. C.B.M. McBurney, Haua Fteah (Cyrenaica),(Cambridge: Cambridge University Press, 1967), p. 90
58. I. Théry, J. Gril, J. L. Vernet, L. Meignen, J. Maury,"Coal used for Fuel at Two Prehistoric Sites in Southern France: Les Canalettes (Mousterian) and Les Usclades (Mesolithic)," Journal of Archaeological Science, v 23, n 4, July 1996, p. 509-512
59. R.K.Fullagar, D.M. Price & L.M. Head, "Early Human Occupation of Northern Australia: Archaeology and Thermoluminescence Dating of Jinmium Rock-Shelter, Northern Territory," Antiquity 70(1996):751-773, p. 771
60. Grahame Clark and Stuart Piggott, Prehistoric Societies, (New York: Alfred A. Knopf, 1965), p. 60
61. Brian Hayden "The Cultural Capacities of Neandertals ", Journal of Human Evolution, 24(1993):113-146, p. 119-120
62. Emil Bachler, Das Drachenloch (St. Gallen: Druck der Buchdruckerei Zollikofer & Cie., 1921).
63. Vadim N. Stpanchuk, "Prolom II, A Middle Palaeolithic Cave Site in the Eastern Crimea with Non-Utilitarian Bone Artefacts," Proceedings of the Prehistoric Society 59, 1993, pp 17-37, p. 33-34.
64. P.M. Vermeersch and E. Paulissen, "The Oldest Quarries Known: Stone Age Miners in Egypt," Episodes, 12:1(March 1989), p. 35-36, p. 36
65. Paul Mellars, The Neanderthal Legacy, (Princeton: University Press, 1996), p. 227
66. L. A. Schepartz, "Language and Modern Human Origins," Yearbook of Physical Anthropology, 36:91-126(1993), p. 113
67. Brian M. Fagan, The Journey From Eden, (London: Thames and Hudson, 1990), p. 61
68. Josephine Flood, "The Archeology of the Dreamtime, (New Haven: Yale University Press, 1989), p. 102;R.K.Fullagar, D.M. Price & L.M. Head, "Early Human Occupation of Northern Australia: Archaeology and Thermoluminescence Dating of Jinmium Rock-Shelter, Northern Territory," Antiquity 70(1996):751-773.
69. Lewis R. Binford, "Subsistence-a Key to the Past," in Steve Jones et al, editors, The Cambridge Encyclopedia of Human Evolution, (Cambridge: Cambridge University Press, 1992), p. 367

70. J. A. J. Gowlett, "Tools--The Palaeolithic Record," in Steve Jones et al, editors, The Cambridge Encyclopedia of Human Evolution, (Cambridge: Cambridge University Press, 1992), p. 353
71. Brian Hayden "The Cultural Capacities of Neandertals ", Journal of Human Evolution, 24(1993):113-146, p. 132
72. Paul C. Mellars, The Neanderthal Legacy, (Princeton: University Press, 1996), p. 285
73. Brian Hayden "The Cultural Capacities of Neandertals ", Journal of Human Evolution, 24(1993):113-146, p. 132-133
74. Clive Gamble, Timewalkers, (Cambridge: Harvard University Press, 1994), p. 138
75. JoAnn Gutin,"Do Kenya Tools Root Birth of Modern Thought in Africa?" Science 270 Nov. 17, 1995, p. 1118.
76. S. Belitszky et al, "A Middle Pleistocene Wooden Plank with Man-made Polish," Journal of Human Evolution, 20(1991):349-353.
77. James R. Shreeve, The Neandertal Enigma, (New York: William Morrow and Co., 1995), p. 139
78. Robert G. Bednarik, "Art Origins", Anthropos, 89(1994):169-180, p. 170
79. Michael R. Waters, Steven L. Forman, and James M. Pierson,"Diring Yuriakh: A Lower Paleolithic Site in Central Siberia,", Science, 275(Feb. 28, 1997):1281-1283
80. Robert G. Bednarik, "Concept-mediated Marking in the Lower Palaeolithic," Current Anthropology, 36:4(1995), pp. 605-634, p. 606
81. Paul G. Bahn, "Treasure of the Sierra Atapuerca," Archaeology, January/February 1996, p 47.
82. Desmond Morris, The Human Animal, (New York: Crown Publishing, 1994), p. 186-188.
83. Robert G. Bednarik, "Stone Age Stone Walls," The Artefact, 16(1993), p. 60
84. D. Mania and U. Mania and E. Vlcek, "Latest Finds of Skull Remains of Homo erectus from Bilzingsleben (Thruingia)", Naturwissenschaften, 81(1994), p. 123-127, p. 124
85. D. Mania and U. Mania and E. Vlcek, "Latest Finds of Skull Remains of Homo erectus from Bilzingsleben (Thruingia)", Naturwissenschaften, 81(1994), p. 123-127, p. 124
86. Compare Alexander Marshack, The Roots of Civilization,(New York: McGraw-Hill Book Co., 1972), p. 139 with Robert G. Bednarik, "On Lower Paleolithic Cognitive Development," 23rd Chacmool Conference Calgary 1990, pp 427-435, p. 432
87. Rick Gore, "The First Europeans," National Geographic, July, 1997, p. 110
88. Robin Dennell, "The World's Oldest Spears," Nature 385(Feb. 27, 1997), p. 767; Hartmut Thieme, "Lower Palaeolithic Hunting Spears form Germany," Nature, 385(Feb. 27,1997), p. 810
89. Hartmut Thieme, "Lower Palaeolithic Hunting Spears form Germany," Nature, 385(Feb. 27,1997), p. 810
90. Hartmut Thieme, "Lower Palaeolithic Hunting Spears form Germany," Nature, 385(Feb. 27,1997), p. 810
91. Hartmut Thieme, "Lower Palaeolithic Hunting Spears form Germany," Nature, 385(Feb. 27,1997), p. 810
92. W.C. Pei, "Notice of the Discovery of Quartz and Other Stone artifacts in the Lower Pleistocene Hominid-Bearing Sediments of the Choukoutien Cave Deposit," Bulletin of the Geological Society of China, 11:2:1931:109-146, p.120
93. Juliet Clutton-Brock, "Origins of the Dog: Domestication and Early History," in James Serpell, ed. The Domestic Dog, (Cambridge: Cambridge University Press, 1995), p. 8-10
94. Victor Barnouw, An Introduction to Anthropology: Physical Anthropology and Archaeology, Vol. 1, (Homewood, Illinois: The Dorsey Press, 1982) p. 141
95. Ian Tattersall, The Fossil Trail (New York: Oxford University Press, 1995), p. 244; new 1994 dating shows Bodo to be 600 kyr. Donald Johanson and Blake Edgar, From Lucy to Language, (New York: Simon and Schuster 1997), p. 194
96. P.Y. Sondaar, et al., "Middle Pleistocene faunal turnover and Colonization of Flores(Indonesia) by Homo erectus," Comptes Rendus de l'Academie des Sciences. Paris 319:1255-1262, p. 1260
97. Victor Barnouw, An Introduction to Anthropology: Physical Anthropology and Archaeology, Vol. 1, (Homewood, Illinois: The Dorsey Press, 1982) p. 143
98. Robert G. Bednarik, "Wonders of Wonderwork Cave," The Artefact, 16(1993), p. 61
99. Richard G. Klein, The Human Career, (Chicago: The University of Chicago Press, 1989), p. 212-213
100. Richard G. Klein, The Human Career, (Chicago: The University of Chicago Press, 1989), p. 164
101. Ian Tattersall, The Fossil Trail (New York: Oxford University Press, 1995), p. 202-203
102. J. A. J. Gowlett, J. W. K. Harris, D. Walton and B. A. Wood, "Early archaeological sites, Hominid Remains and Traces of Fire from Chesowanja, Kenya," Nature, 294, Nov. 12, 1981, p. 128; C. K. Brain and A. Sillen, "Evidence from the Swartkrans cave for the earliest use of fire," Nature, 336, Dece. 1, 1988, p. 464-465

103. Kathy D. Schick and Nicholas Toth, <u>Making Silent Stones Speak</u>, (New York: Simon and Schuster, 1993), p.160

104. M.D. Leakey, <u>Olduvai Gorge</u> 3 <u>Excavations in Beds I and II, 1960-1693</u>, (Cambridge: Cambridge University Press, 1971), p. 269

105. Donald C. Johanson, Lenora Johanson, and Blake Edgar, <u>Ancestors,</u> (New York: Villard Books, 1994), p. 163-165

106. Donald C. Johanson, Lenora Johanson, and Blake Edgar, <u>Ancestors,</u> (New York: Villard Books, 1994), p. 163-165

107. Alan Walker and Pat Shipman, <u>The Wisdom of the Bones</u>, (New York: Alfred Knopf, 1996), p. 168; A. Walker, M.R. Zimmerman, and R.E. F. Leakey, "A Possible Case of Hypervitaminosis A in *Homo Erectus,"* <u>Nature</u>, 296, March 18, 1982, p. 248-250, p. 249-250

108. Richard G. Klein, <u>The Human Career</u>, (Chicago: The University of Chicago Press, 1989), p. 169

109. Barbara Isaac, "Throwing," in Steve Jones et al, editors, <u>The Cambridge Encyclopedia of Human Evolution</u>, (Cambridge: Cambridge University Press, 1992), p. 358

110. Even critics of erectus abilities agree they had speech Alan Walker and Pat Shipman, <u>The Wisdom of the Bones</u>, (New York: Alfred Knopf, 1996), p. 281

111. Dean Falk, <u>Braindance</u>,(New York: Henry Holt and Co., 1992), p. 50

112. Victor Barnouw, <u>An Introduction to Anthropology: Physical Anthropology and Archaeology</u>, Vol. 1, (Homewood, Illinois: The Dorsey Press, 1982), p. 126

113. Juan Luis Arsuaga Ferreras, "Faces *from the* Past," <u>Archaeology</u>, May/June 1997, p. 31-33, p. 32

114. "New Relation. Upper Jaw from Hadar, Ethiopia, Helps Fill in a Crucial Period" <u>Human Evolution</u>. 274, Number 5291, Issue of 22 November 1996, p. 1298

115. Bernard Wood, "The Oldest Whodunnit in the World," <u>Nature</u> , 385(Jan.23, 1997), p. 292; S. Semaw, et al, "2.5-Million-year-old stone Tools from Gona Ethiopia," <u>Nature</u>, 385(January 23, 1997), p. 333-336, p. 335

116. R.A. Dart, "The Waterworn Australopithecine Pebble of Many Faces from Makapansgat," <u>South African Journal of Science</u>, 70(June 1974), pp 167-169

117. Tim D. White, et al, "New Discoveries of *Australopithecus* at Maka in Ethiopia", <u>Nature</u>, Nov. 18, 1993, p. 263 TMJ is Temporomandibular Joint disease.

118. M.G. Leakey et al, "New Four-million-year-old Hominid Species from Kanapoi and Allia Bay, Kenya", Nature, 376, August 17, 1995, p. 567-568

119. J. W. K. Harris, "Early Man," in Andrew Sherratt, editor, <u>The Cambridge Encyclopedia of Archaeology</u>, (New York: Cambridge University Press, 1980), p. 64

Index

A

Abel . . . 62, 75
aborigines . . . 44, 62, 100, 103, 135, 136, 138, 144, 146, 148, 173, 177, 183
Abri Blanchard . . . 111
Abri Lartet, France . . . 142
Abri Mege, France . . . 81
Abri Vaufrey, France . . . 146
Achenheim, Germany . . . 145
Acheulean . . . 10, 46, 57, 58, 63, 65, 80, 84, 87, 103, 104, 106, 114, 115
Acheulean . . . 122, 143, 145, 146, 155, 168, 170, 173
Adam . . . 2, 4, 13, 15, 27, 29, 37-39, 41, 42, 49, 55, 56, 60, 82, 85, 88, 90, 92, 94
. . . 97-99, 103, 111, 114, 115, 125, 128, 130, 132, 135, 136, 143
. . . 156, 158, 171, 175-180, 184, 185
Ahlberg, Per . . . 48
Ahnishinahbaeo'jibway . . . 178
Aiello, L. C. . . . 94
Ainu . . . 68
Akashi, Hiroshi . . . 177, 186
Akazawa, Takeru . . . 78, 79, 86, 174
Albrecht, G. . . . 149
Algeria . . . 136
Algonkins . . . 67
Almli, C. R. . . . 171
Alps . . . 30, 71, 112
Altaics . . . 81
altar . . . 69, 77, 190
altricial . . . 93
altriciality . . . 93
amber . . . 45, 48
Ambrona, Spain . . . 82, 131, 146
amputation . . . 190
Amud Cave, Israel . . . 22, 148, 181
Anderson-Gerfaud, Patricia . . . 132, 151
Andersson, Johan Gunar . . . 16
Andes . . . 30
Angeloni, Elvio . . . 95, 150
animal hides . . . 125, 126, 191
Anthropopithecus . . . 15
antler . . . 58, 81, 105, 106, 123, 132, 146, 161, 169, 170
Anton, S. C. . . . 26, 27, 152
Apaches . . . 63
ape . . . 5, 6, 8, 9, 11, 13-17, 20, 21, 29-36, 41, 49-51
. . . 54-56, 59, 60, 72, 90, 120, 127, 152, 159, 184, 185
apocrine glands . . . 89-91
Arago, France . . . 19
archaic *Homo sapiens* . . . 1, 4, 18, 19, 25, 37, 66, 82, 114, 120, 123, 129-131,
. . . 135, 136, 138, 139, 143, 159, 163, 169, 180
Arcy-sur-Cure . . . 105, 107, 121, 122, 189
Arensburg, B. . . . 60
arrowheads . . . 44, 129, 131, 189
arrows . . . 44, 47, 68, 129, 189, 192
Arsuaga, Juan Luis . . . 79, 144, 154, 174, 195
art . . . 38, 39, 41, 42, 43, 46, 47, 48, 58, 60, 61, 63, 65, 72, 75, 81, 83,
. . . 97-104, 106-108, 111, 112, 114-118
. . . 120, 130, 135, 140, 143, 144, 151, 154, 171, 174, 190, 191, 194
artefact . . . 18, 26, 31, 43-48, 57, 58, 63, 68, 69, 77, 83, 84, 85, 86, 98, 99
. . . 103, 106, 108, 109, 114, 116-118, 120, 121, 123, 127-129, 131,132, 135
. . . 137, 142, 143, 145, 146, 149, 152-154, 168, 169, 170-174, 192-194
Artemis . . . 68
artist . . . 98, 99, 101, 108
Asian . . . 19, 26, 41, 76, 95, 96, 117, 133, 136, 152
. . . 174, 175, 183, 190
Atapuerca, Spain . . . 19, 79, 86, 174, 190, 194
Atapuercans . . . 80
Aterian Industry . . . 129
Auditorium Cave . . . 102-104
Augustus, Caesar . . . 137
Aurignacian . . . 64, 65, 103, 104, 107, 110, 113, 116
Australia . . . 5, 23, 39, 44, 62, 82, 100, 101, 103, 116
. . . 124, 128, 133-135, 137, 143, 152, 154
. . . 165, 169, 175, 177, 183, 189, 190, 193
Australian . . . 5, 52, 62, 103, 112, 134, 135, 141
. . . 144, 148, 152, 175, 177, 183, 187, 189
australopithecine . . . 4, 8-10, 12, 25, 31-34, 57, 93, 108, 117
. . . 127, 131, 158, 185, 195
Australopithecus afarensis . . . 4, 8, 9, 31, 145
Australopithecus anamensis . . . 4, 9
awl . . . 106, 127, 146, 170

axes . . . 10, 25, 57, 58, 104, 128, 129, 131, 133, 139, 146, 161, 168-170
Ayala, Francis J. . . . 186

B

baboons . . . 6, 11, 12, 32, 108, 165, 168
Babylonian . . . 40
Bachechi, L. . . . 192
Bachler, Emil . . . 65, 71-75, 85, 121, 149, 193
Bacho Kiro . . . 103, 107
Bahn, Paul G. 42, 48, 83, 86, 101, 111, 116, 117,136, 144, 149, 151-154, 174, 193, 194
Bale, Switzerland . . . 139
Bali . . . 136
bamboo . . . 109, 115
Bar-Yosef, Ofer . . . 121, 123, 148, 192
Bara Bahau, France . . . 140
Barbaza, Michael . . . 69, 84
Barbe Cormio, France . . . 139
Baringo, Kenya . . . 33
Barlow, G. W. . . . 6, 8
Barnouw, Victor . . . 83, 84, 150, 153, 192, 194, 195
Bartstra, Gert-Jan . . . 152
basalt . . . 65
Basch, Martin Almagro . . . 155, 173
baskets . . . 46, 128, 132, 133, 140, 141, 189
Batavia, Indonesia . . . 15
Bates, Elizabeth . . . 54
Baume Bonne, France . . . 125
Baume des Peyrards, France . . . 120
bear cult . . . 67-69, 72, 74
bears . . . 20, 22, 65, 67-72, 74, 75, 78, 79, 81, 84, 85, 104, 110, 122, 124, 128, 164
Beaumont, P. B. . . . 123, 140, 153
Becov, Czechoslovakia . . . 82
bedding . . . 122-124, 190
Bednarik, Robert G. . . . 26, 42, 45, 48, 61, 72, 83, 85, 102, 103, 107, 112, 116, 123
. . . 128, 138, 140, 142, 143, 149, 150, 153-155, 171, 173, 174, 194
Begouen, Count . . . 69
Beldibi Cave, Turkey . . . 141
Belfer-Cohen, Anna . . . 77
Belitszky, S. . . . 83, 151, 194
Berekhat Ram . . . 18, 65, 66, 84,100, 104, 115, 190
Bickerton, Derek C. . . . 53, 59, 165
Bigo, Uganda . . . 189
Bilzingsleben, Germany . . . 39,58,61, 67,123, 125, 132, 143, 146, 149-151, 170, 190, 194
Binford, Lewis R. . . . 72, 85, 124, 129-131, 148, 151, 154, 166, 173, 193
bipedal . . . 9, 18, 23, 38, 39, 56, 60, 81, 88, 93, 112, 115
bipedality . . . 95
Birdsell, J. B. . . . 76, 85, 137, 153, 193
Biship, W. W. . . . 118
Bishop, Edmund . . . 29
bison . . . 67, 81, 163
bitumen . . . 129, 151, 172, 173, 193
Black, Davidson . . . 7, 16
blade . . . 25, 104-106, 111, 114, 132, 170, 182, 190
Blanchard Cave, France . . . 142
Blanford, Mr. . . . 148
boards . . . 16, 46, 99, 129, 182
boats . . . 128, 133, 134, 136, 137, 152, 189
Boaz, Noel . . . 17, 25, 26, 95
Bocksteinschmiede, Germany . . . 107
bodkins . . . 58, 123
Bodo, Ethiopia . . . 80, 190, 194
body painting . . . 62, 63, 82, 107, 114, 140
Boeda, Eric . . . 151, 172, 173, 193
Boesch, C. . . . 162, 163 172
Boesch, H. . . . 162, 163 172
Bohlin, Birger . . . 17
Boker Tachtit . . . 24, 111
bone tools . . . 24, 58, 104, 105, 111, 123, 127, 139, 140, 146, 169, 170, 190, 191
Bonifay, Eugene . . . 69, 84
Boswell, Percy . . . 11
bottleneck . . . 178-180
Boule, Marcelin . . . 35, 36, 71
Bowden, Malcom . . . 41
bower birds . . . 76, 127, 128
Bower, Bruce . . . 60,111, 112, 117,150, 151, 152,174,192-193
Bowler, J. M. . . . 152
Brace, C. Loring . . . 25
bracelets . . . 138

Brady, James E. ... 80, 86
brain ... 5, 6, 8, 9, 12, 13, 15, 18, 22, 24-26, 30, 32-34, 41, 51,
... 53-56, 59, 60, 88-95, 116, 127, 131, 135, 149, 151, 152, 154
... 156-159, 166-168, 171, 172, 183, 191, 192, 194
Brain, C. K. ... 154, 194
Brauer, Gunter ... 27, 152
Braun, Marc ... 174
Breuil, Abbe ... 69
Brno, Czechoslovakia ... 110, 181, 189
Broca's area ... 13, 18, 22, 32, 33, 53-57, 59, 138, 168
Bronze Age ... 116, 128
Brooks, Allison S. ... 105
Broom, Robert ... 7-9
Brown, Robert R. ... 192
Brownjohn, J. Maxwell ... 84-86, 115, 149, 150
Bruniquel Cave, France ... 66, 67, 69, 122
Budapest-Farkasret, Hungary ... 139
Buffon, Georges L. ... 147
Buka, South Pacific ... 133
bull-roarers ... 111
Burenhult, Goren ... 85, 86, 116, 117, 150, 151, 174
burial ... 10, 11, 45-47, 62, 63, 71, 74-80, 85, 86, 98, 107
burial ... 117, 127, 139, 140, 148, 150, 169, 170, 174, 190
burin ... 103-106
bushmen ... 100
butchery ... 130, 131, 142
Butzer, K. W. ... 82, 87

C

Caesar ... 137
Cain ... 62, 158, 159
Cairo ... 5
calendar ... 142
Calvin, William H. ... 145, 154, 155
camel ... 89, 90, 91, 95
Campbell, Bernard ... 26, 60, 90, 95
canids ... 147, 148
Cann, Rebecca ... 175-178, 186, 187
cannibalism ... 63, 72, 79, 80
Carbonell, Eudald ... 26, 86, 95, 174
Carpathian Mountains ... 30
carpenter ... 103, 114, 127, 128
Carpenter's Gap ... 103, 190
Caspari, Rachel ... 27, 186-188
Casteret, Norbert ... 69
Castro, Jose de ... 19
cats ... 31, 32, 49, 51, 59, 63, 66, 93, 136
Caucasus, Mountains ... 30, 143
Cavalli-Sforza, L. Luca ... 27, 87, 150-152, 186, 192
cave bear ... 20, 65, 69, 71, 72, 74, 75, 79, 84, 85, 110
Ceprano Man ... 136, 137
ceramics ... 45, 120, 189
Chad ... 9, 25
Channel Islands, England ... 136
Chapman, Roy ... 16
charcoal ... 101, 112, 121, 122, 134, 135
Chauvet Cave, France ... 69, 103
Chenggang, China ... 48
chest ... 10, 71, 73, 74, 78, 79
Chichkino, Siberia ... 101
childbirth ... 88, 92-94
chimpanzee ... 6, 8-10, 12, 15, 49-51, 56, 60, 76, 90, 91, 94, 120, 122, 123, 127
... 128, 156, 159-163, 165-169, 172, 173, 184, 187
chin ... 19, 22, 23, 35, 55, 66
China ... 16-19, 23, 26, 31, 46-48, 63, 83, 91, 92, 107, 109, 117,
... 119, 129, 133, 135, 140, 147, 190, 194
Chinese wolf ... 147, 155
Chippewas ... 66
Chomsky, Noam ... 53
chopping ... 58, 123, 139
Choukoutien Cave, China ... 17, 26, 83, 117, 147, 155, 194
Chowa, Zambia ... 140
chromosomes ... 42, 60, 176-179, 184-186
Churchill, S. E. ... 27
Clochon, Russell, L. ... 26, 41, 95, 96, 152, 174
cities ... 40, 43, 46, 84, 101, 120, 128, 140, 141, 184, 189, 192
Clacton-on-Sea, England ... 46, 130
Clara Nii Ska ... 178
Clark, G. A. ... 102, 116, 151
Clark, Grahame ... 151, 152, 192, 193
Clark, Robin ... 152
Clark, William ... 142, 162

clothing ... 2, 32, 43, 44, 46, 88, 91, 92, 94, 107, 114, 125-127, 150, 168, 190
clubs ... 46, 128, 130
Clutton-Brock, Juliet ... 155, 194
coal ... 138, 139, 143, 153, 189, 190, 193
coffin ... 79, 80
cognition ... 132, 154, 155
cognitive fluidity ... 167-170
Cohn, B. A. ... 91, 95
Cold Spring Harbor ... 5, 17
Coles, J. M. ... 71, 85, 109, 110, 117, 148
Combe Grenel, France ... 130
Combe Sauniere ... 107
compassion ... 21, 57, 163, 164, 165, 167, 191
composite tools ... 131, 190
Conderni, Silvana ... 174
Cook, Capt. James ... 133
cooking ... 69, 120, 143, 144
Coon, Carelton ... 85
corpse ... 76, 190
corpus callosum ... 157
cortex ... 5, 6, 13, 53, 54, 59, 139, 157, 158
cotton ... 43, 125, 160
Cougnac ... 111
counting ... 2, 140, 141, 143, 190
Cova Beneito, Spain ... 142
coyote ... 146
cranial capacity ... 9, 12, 18, 19, 24, 25, 28, 30, 34, 71, 94
creativity ... 24, 97, 98, 111
creator ... 10, 29, 54, 57, 66
creole languages ... 53, 59
Critchfield, Howard J. ... 95, 150
Cro-magnons ... 22, 23, 30, 36, 38, 83, 105, 124
Crocker, Richard L. ... 192
cross over ... 177
crystals ... 17, 63, 64, 83, 107, 124, 140
Cuddie Springs, Australia ... 143, 154
Cueva Morin, Spain ... 120
culture ... 18, 22, 23, 39, 43, 44, 47, 52, 58, 63, 67, 68, 75, 76, 77, 80, 81
culture ... 97-100, 109, 113, 129, 135, 146, 148, 161, 168, 171
Cunningham, Daniel ... 16
Cuozzo, John W. ... 56, 60
Curtis, G. H. ... 26, 27, 152
Curwen, E. Cecil ... 154
Custance, Arthur C. ... 89, 95
Cyrenaica ... 113, 118, 193
cytoplasm ... 175
Czechoslovakia ... 45, 82, 110, 138, 189

D

Dagg, Anne Innis ... 95
Dams, Lya ... 117
Danger Cave, Utah ... 189
Dart, Raymond ... 5-10, 12, 16, 17, 108, 117, 140, 153, 195
Darwin, Charles ... 14, 17, 20, 29, 41, 42, 52
Davidson, Iain ... 59, 172
Davis, Percival ... 33, 34, 36, 39-42, 57, 60, 75-77, 85, 97, 115, 127, 150, 171, 176, 186
Dawson, J. William ... 30, 40
Day, Michael H. ... 25, 27
de Bruyn, M. ... 6
de Lumley, Henri ... 82, 87, 122, 123
de Puydt, Marcel ... 21
De Beaune, Sophie A. ... 116
Deacon, Terrence W. ... 13, 26, 54, 55, 58-61, 124
Dederiyeh Cave ... 79
deer ... 77, 78, 81, 82, 139, 141
Dennell, Robin ... 131, 151, 161, 172-174, 194
diatonic scale ... 113
Dibble, Harold L. ... 153
Dickson, D. Bruce ... 71-73, 81, 84-87, 107, 117, 153, 172
Dillehay, Thomas ... 129
Diring Yuriakh, Siberia ... 91, 150, 190, 194
Dirjec, Janez ... 118, 193
Divje Babe, Slovenia ... 112, 190
Dixon, E. James ... 150, 192
Dmanisi mandible, (Georgia) ... 18, 91, 126, 171
Dobzhansky, Theodosius ... 4, 25
dogs ... 12, 123, 146-148, 155, 164, 165, 168, 194
Dolni Vestonice, Czechoslovakia ... 45, 110, 120, 122, 189
domestication ... 129, 146, 148, 155, 194
Domingo Garcia, Spain ... 101
Donoghue, M. J. ... 48
Dorit, Robert L. ... 177, 178, 186

Douglas-Hamilton, Iain 76
Douglas-Hamilton, Oria 76
Drachen Cave 110
Drachenloch Cave, Switzerland 65, 70, 71, 73, 72, 74, 73-75, 85, 149, 190, 193
Dragon Cave, Austria 74, 75
drawings 8, 73, 74, 81, 97, 99, 101, 115
drums 109, 111
Dubois, Eugene 5, 14-17, 30, 34
Duckworth, W. L. H. 6
dwelling 58, 105, 120, 123, 132
Dzhuruchula Cave, Molodova 143

E

East Rudolf, Kenya 164
eccrine gland 89-91
Echegaray, J. G. 66, 84, 193
Edey, Maitland 25
Edgar, Blake 25, 26, 60, 86, 150, 151, 155, 171, 193-195
Edwards, Diane 48
egg 175, 177
Egypt 19, 43, 46, 137, 139, 143, 153, 154, 189, 193
Ehringsdorf, Germany 158
Einstein, Albert 2, 119
Eitzman, W. I. 108
El Juyo Cave, Spain 66, 84, 189, 193
eland 89, 90, 94, 95
Eldredge, Niles 25
elephant 11, 58, 72, 76, 82, 123, 127, 130, 131, 136, 163
emissary veins 90, 92, 94
endocast 6, 8, 13, 33, 54, 56
energy 88, 89, 91, 94, 175
Engihoul, Belgium 20
Engis Cave, Belgium 19
epilepsy 157
es-Skhul Cave, Mt. Carmel, Israel 130
Escale Cave, France 190
eskimos 67, 98
Etler, Dennis A. 155, 187, 188
Eve 37-39, 57, 80, 88, 94, 97, 158, 159, 172, 173, 175-180, 186
evulsion 82
exon 184

F

Fabbri, P. F. 192
fabric 41, 60, 97, 115, 116, 150, 164, 193
Fagan, Brian M. 19, 26, 48, 82, 87, 92, 96, 105, 106, 116, 125, 149, 150, 152, 193
Falk, Dean 13, 25, 26, 41, 59, 60, 86, 89, 90, 95, 141, 154, 172, 195
Fall 89, 102, 118, 125, 127, 158
Fang, Zhongjie 48
Feraud, G. 84
Ferreras, Juan Luis Arsuaga 144, 154, 195
figurines 18, 45, 64-66, 84, 98, 100, 101, 104, 115
Finger, S. 171
Fink, Bob 113, 116, 192, 193
Finns 67, 68
fire 7, 18, 43, 63, 74, 92, 94, 98, 101, 112, 113, 124, 128, 135, 138, 143, 144, 152, 154, 190, 194
fireplace 74, 77
Fisher, Henry George 192
flexed burial 10, 77-79
flexing 77
flint 10, 11, 20, 44, 46, 58, 71, 77-79, 104, 107, 109, 121, 123, 130, 131, 137-140, 142, 160-162, 170, 189
FLK ravine 12, 108
flood 2, 46, 94, 142
Flood, Josephine 155, 169, 173, 192, 193
Flood, Noahic 2, 29, 36, 39, 40, 114, 130, 171, 174, 179, 188
Flores, Indonesia 136, 137, 152, 190, 194
Florisbad, South Africa 130, 150, 151
flute 2, 74, 97, 109-115, 118, 120, 190
Fogarty, John H. 86
footprints 9, 34
foramen magnum 6, 9
Forman, Steven L. 150, 194
Fortes, Meyer 98, 99, 115
Foster, Mary Lecron 155
fox 107, 122, 142
Frair, Wayne 34, 36, 39-42, 127, 150
Franchthi Cave, Greece 133
Franciscus, Bob 93
Frayer, David W. 22, 181, 182, 186
Freeman, Leslie G. 66, 82, 84, 87, 146, 148, 155, 173, 193

Fuegians 126
Fuhlrott, Johan Karl 20, 21
Fullagar, R. L. K. 116, 135, 152, 154, 193
fur 78, 90, 91, 93, 94, 122, 123, 125, 126
Furbringer, Max 14, 16
Furby, J. H. 154

G

Gabori-Csank, Veronika 153
Gaffney, Eugene S. 48
Galgenberg figurine 64
Galilee 46
Gamble, Clive 48, 52, 59, 85-87, 116, 122, 130, 131, 148-150, 152, 171, 172, 174, 193, 194
Gange, Robert 39, 40, 42
gaps 9, 44, 45, 54, 152, 181, 190
Garcia-Pozuelo-Ramos, Celedonio 32, 41
Gargett, Robert H. 78, 86
garments 69, 81, 88, 125
Garrod, Dorothy 69
Gauthier-Pilters, Hilde 95
Gay, John 115
gazelle 89
Geissenklosterle 102
genealogies 37, 97, 114
Georgia 18, 91, 92, 94, 126, 171
Germany 1, 10, 14, 16, 19-21, 39, 46, 58, 61, 67, 72, 75, 91, 102, 107, 123, 125, 126, 129-132, 143, 145, 146, 149, 151, 158, 160, 161, 170, 173, 189, 190, 194
Gesher Benot Ya'aqov 63, 107, 132, 190
gestation 93
Ghana 98
gibbon 12, 14, 15, 30, 34
Gibraltar, Strait of 20, 21, 136, 137
Gibson, K. R. 154, 155
Gijselings, G. 153
Gilbert, Walter 177, 186
Gill, A. M. 152
Gilyaks 67-69, 100
Gimbutas, Marija 83
Gish, Duane 31, 32, 35, 40, 41, 59, 60
glaciation 24, 85, 86, 117, 132, 148, 149, 152
Glander, Kenneth E. 173
Gobi Desert 16
God 1, 2, 13, 27, 29, 30, 38, 39, 49, 54, 60, 62, 67, 68, 82, 84-86, 88, 90, 94 97-99, 105, 115, 125, 128, 130, 150, 156, 158, 159, 163, 165, 184-186, 192, 193
Godin, Henri 69
Golan Venus 65, 66, 100, 142
Gona, Ethiopia 140, 154, 191, 195
Goodall, Jane 50, 148, 166, 173
Goodfellow, Peter N. 177, 186
Gopher, A. 192
Gore, Rick 67, 84, 153, 194
Goren-Inbar, N. 65, 66, 84
Gowlett, J. A. J. 145, 154, 155, 159, 172, 194
grammar 7, 51-54, 188, 189
Gran Dolina, Spain 80, 137
graves 43, 71, 73, 76-78, 81, 98, 120, 139, 142
Gravettian 110, 113
graveyard 43, 79
Greek 1, 64
Greeks 2, 125
Gril, J. 153, 193
Grimaldi, D. 48
Grotta dell'Addaura, Sicily 163
Grotte d'Aldene 125, 190
Grotte de Goyet 103
Grotte de Rigabe 121
Grotte du Prince 121
Grotte du Renne 121, 146
Groves, Colin 151, 174
Grun, Ranier 150, 151
Gudenus Cave, Austria 63
Guilaine, Jean 84, 148, 152, 192
Gutin, JoAnn 173, 194

H

habilines 54-57
habitations 52, 91, 119-121, 123, 124, 126, 130, 135, 161, 162
Haeckel, Ernst 14, 15
hafted tools 129, 160, 170, 172, 193

hafting	129, 139, 151, 161, 169, 172, 173, 193
Hailwood, E.	174
hair	91-95, 108, 165, 166
Halifax, Joan	88
Hamburg, Germany	129, 189
Hamm, Jim	44, 47
hammer	112, 113
Hammer, Michael F.	176, 177, 186
hand axes	10, 25, 58
Harding, Anthony	116
Harris, J. W. K.	60, 154, 194, 195
Haua Fteah, Libya	113, 118, 190, 193
Hawkes, Jacquetta	171, 192
Hayden, Brian	56, 60, 69, 84, 86, 100, 102, 115-117, 121, 123, 124, 129, 146, 148, 149, 151, 155, 172-174, 193, 194
hearths	58, 63, 74, 110, 112, 120-123, 125, 126, 130, 138, 139
height	9, 18, 71, 73, 92, 94, 133, 152
hematite	128, 140, 142
hemidecortication	158, 171
hemispherectomy	157, 158
Henry, Donald O.	149
Hertwig, Richard	30
Hibbett, D. S.	48
hides	5, 18, 46, 81, 125-127, 130, 169, 190, 191
Higgs, E. S.	85, 109, 110, 117, 148
Hill, Andrew	9
Hillman, Gordon C.	154
Himalayas	30
Hogup Cave, Utah	46
Hohler Fels	102
Holdaway, Simon	172, 173
Holloway, Ralph L.	13, 25, 171
Homo erectus	1, 2, 4, 5, 7, 13-19, 24-27, 29-35, 38, 39, 41, 45, 46, 55-58, 60, 61, 63, 64, 66, 67, 72, 75, 77, 82, 83, 91, 92, 94, 97, 103, 104, 105, 107, 108, 114, 120, 123-126, 130-132, 135-137, 140, 141, 143, 146-149, 151, 152, 156, 161-172, 176-178, 180, 182, 183, 185, 190, 191, 194, 195
Homo habilis	4, 10, 12, 13, 19, 28, 31-33, 38, 55-57, 92, 93, 108, 145, 156
Homo heidelbergensis	4, 19, 80
Homo rudolfensis	4
Homo sapiens	1, 2, 4, 5, 13, 14, 17-19, 23-27, 30, 33, 35-39, 48, 55, 56, 66, 75, 77, 79, 80, 82, 83, 100, 107, 108, 111, 113, 114, 120, 122, 123, 125, 129-131, 135, 136, 138, 139, 141, 143, 152, 158, 162, 163, 169, 176, 177, 180, 183, 184, 189, 190
Homo transvaalensis	5, 17
Horodyski, Robert J.	47
horse	93, 101, 142, 168, 189
Hortus, France	81, 87, 126, 190
housing	119, 124
Hovers, Erella	77
Howell, F. Clark	71, 85, 146, 155, 173
Hublin, Jean-Jacques	174
human sacrifice	67
hunter-gatherers	46, 78, 116, 122, 124, 143
hunting	61, 68, 72, 85, 97, 105, 125, 126, 130, 131, 144, 149, 151, 170, 173, 194
Huse, Scott M.	41
huts	18, 39, 67, 66, 100, 110, 114, 119-124, 126, 129, 130, 132, 189, 190
Huxley, Thomas	29, 30, 40
hydroencephalics	156, 157
hypervitaminosis	60, 164, 172, 195

I

ibex	101
idols	62-64
Ildefonso	49
image	1, 2, 23, 37, 38, 62, 67, 82, 97, 99, 104, 105
image of god	2, 38, 62, 82, 99, 105
images	45, 59, 63-65, 98, 116, 117, 134, 141, 149
Immune response	180
Inca bone	183
Incas	63
Indonesia	135-137, 152, 190, 194
Ingold, T.	154, 155
instinct	51, 53, 55, 59, 83, 87
intelligence	33, 49, 51, 59, 74, 97, 99, 100, 105, 119, 129, 130, 144, 146, 148, 156-158, 167, 168
intron	184
inventions	98, 102-104, 106, 119, 125, 140, 141, 143, 168, 190
inventiveness	144, 169
IQ	156-158
Iroquois	63

Isaac, Glynn	113, 118, 131, 145, 154, 155, 195
isard	101
Istallosko	110
Isturitz	109, 110, 117
Italy	64, 103, 136, 144, 190
ivory	45, 77, 104, 107, 110, 146, 155, 159, 169, 170, 173, 189
ivory tools	169, 170,
Ivory Coast	162

J

Jabal	119
jackals	165
James, E. O.	83
Janssen, J.	153
Jantz, Mr.	181
Jantzen, R.	152
Japan	23, 46, 48, 68, 119, 129, 151, 177, 189
Java	5, 14, 17, 18, 26, 27, 29-31, 34, 35, 41, 92, 135, 152, 171, 183
Java man	5, 17, 29-31, 34, 35, 41
javelins	58, 62, 131, 161, 170
Jerison, H. J.	59
Jerison, I.	59
jewelry	64, 100, 106, 107, 189, 190
Jews	98
Jinmium, Australia	103, 116, 134, 135, 143, 152, 154, 190, 193
Johanson, Donald C.	8, 10, 25, 26, 34, 60, 86, 87, 94, 127, 149-151, 155, 171, 193-195
Johanson, Lenora	127, 150-151, 155, 171,
Johnson, D. A.	171
Johnson, Phillip E.	37, 42, 57, 60
Jones, Rhys	173
Jones, Steven	94, 148, 151, 154, 155, 172, 173, 193-195
Jordan	92, 122, 149
Jubal	97

K

Kalahari Desert	98
Kalambo Falls, Zambia	46, 130
Kanam, Kenya	11
Kanapoi, Kenya	149, 191, 195
Kanjera, Kenya	11
Karlich-Seeufer, Germany	132, 151
Kattwinkel, Mr.	10
Kavur, Boris	118, 193
Kebara Cave, Israel	78, 121, 123, 124, 148, 161, 172
Keeley, Lawrence	46, 132
Keith, Arthur	6-8, 10, 16, 31, 40
Kenya	8, 10, 11, 13, 18, 33, 56, 123, 132, 144, 145, 154, 164, 173, 190, 191, 194, 195
Kenyon, Dean H.	33, 39-42, 57, 60, 75-77, 85, 97, 115, 171, 176, 186
Kerlich, Germany	129
Kershaw, A. Peter	152
Kesslerloch	147
Keys, David	112, 117
Kiev, Ukraine	111
Kilmer, Anne Draffkorn	109, 192
Kimeu, Kamoya	92, 164
Kitching, James W.	48
Klasies River Mouth Cave, S.A.	23, 27, 39, 40, 130, 135, 158
Klein, K.	140, 148-151, 153
Klein, Richard G.	27, 32, 41, 42, 116, 140, 148-151, 153, 163, 172, 174, 192, 194, 195
Klotz, John	32, 33, 36, 41
Knauth, L. Paul	47
Knecht, H.	116, 117
knives	58, 109, 123, 126
KNM-ER 1470	13, 56, 57, 60
KNM-ER 1808	164-166
Kolb, Brian	171
Kolen, Jan	172, 173
Konigsaue, Germany	129
Koonalda Cave, Australia	103, 137
Korea	129
Koryaks	67
Kostenki, Russia	107, 120, 122
Krapina Cave, Yugoslavia	79, 190
Krings, Matthias	182, 186, 187
Kromdraai, South Africa	8
Kubik, Stefan	90, 94, 95
!Kung	98
Kurten, Bjorn	69, 72-75, 84, 85

L

L'Ermitage, France	143
La Chapelle-aux-Saints Cave, France	71, 152, 163

La Ferrassie Cave, France 78, 107, 124, 142, 143
La Madeleine 110
La Quina, France 107, 142, 145
Laban 97
Laetoli, Tanzania 9, 34, 191
Lagrave, France 139
Lake George, Australia 134, 135, 152
Lake Mungo, Australia 82, 135, 189
lamp 102
Lan-tien, China 147
lances (see spears) 131
Landek, Czechoslovakia 189
language 13, 19, 25-27, 30, 33, 38, 47, 48, 49-60, 72, 83, 84, 86, 87
93, 97, 116, 131, 132, 138, 150, 152, 154, 155, 158, 167, 168, 171, 172, 191, 193, 194
Lapita 133
Lapps 67
Larick, Roy 26, 41, 95, 96, 152, 174
Lascabannes, France 139
Lascaux Cave, France 81, 116, 117, 140
lateralization, brain 18, 54
Lazaret Cave, France 122, 123, 190
Le Gros Clark, Wilfred 8
Le Moustier Cave, France 73, 77, 78, 143, 170
Le Placard Cave, France 109, 113
Leahy, Michael 52
Leakey, Frida 11, 12
Leakey, Jonathon 12
Leakey, Louis 8-12
Leakey, Mary D. 11, 12, 34, 108, 117, 124, 149
Leakey, Meave G 9, 195
Leakey, Richard 31, 33, 34, 54, 56, 60, 92, 117, 164, 166, 172, 173, 195
Lehringen, Germany 72, 130, 131, 160, 190
leopard 66, 81, 94, 126, 164
Leroi-Gourhan, Andre 121, 126, 149
Leroi-Gourhan, Arlette 78, 86
Les Canalettes, France 138, 143, 153, 190, 193
Les Fieux, France 111
Les Trois Freres Cave, France 68, 69, 109
Levallois method 129, 139
Levalloisian 160, 170
Lewin, Roger 117, 171
Lewis, Meriwether 162
Lewontin, Richard 34
Li, Xingxue 48
Lieberman, Philip 55, 56, 60
lighting 108, 112, 122, 124, 137, 138
Lion Cavern, Swaziland 140
Lissner, Ivar 64, 67, 68, 71, 74, 75, 81, 84-86, 115, 128, 149, 150, 193
Lister, Adrian 136, 153, 193
lithophones 111, 117
Liujiang, China 23
Lloyd, Barbara 115
Lohest, Marie Joseph Maximin 21
Lokve, Hungary 110
Lombok, Indonesia 136
Lorber, John 156, 157, 171
Lothagam, Kenya 191
Lovas, Hungary 140, 146
Lubenow, Marvin L. 1-3, 26, 32, 35, 36, 41, 56, 60, 156, 158, 171
Lucy 8, 25, 26, 32, 34, 60, 86, 87, 94, 149, 150, 193, 194
Lyon, Daniel 28, 33, 156

M

madonna 64
Magdalenian 64, 66, 69, 81, 104, 109
Makapansgat, South Africa 108, 117, 191, 195
Malagasy 63
Malangangerr, Australia 189
Mallaha, Palestine 120
Mallegni, F. 192
Malta 64, 66, 136
mammoths 77, 111, 120, 121, 136, 146, 153, 193
mandibular foramen 179-181
manganese 140
Mania, Dietrich 58, 61, 67, 123, 132, 149, 151, 194
Mania, Ursula 58, 61, 67, 123, 132, 149, 151, 194
Maoris 63
Marks, Anthony 154
Marks, Jon 177
Marshack, Alexander 65, 66, 78, 82, 84, 86, 87, 103, 115-117, 141, 142, 154, 155, 194
Martin, Henri 142
Martin, Jobe 35, 36, 41

Martin, R. D. 48
Martyn, John 33
Mas de Azil 68
Massat, France 69
Matenkupkum Cave, New Guinea 133
mathematics 140, 156
Mauer mandible 19
Maury, J. 153, 193
Mawson, Douglas 164, 165
Max, Edward 184
Maya 80, 189, 192
Mayer, A. F. 21
Mayr, Ernst 5, 17, 25, 108, 171, 185
McBurney, C. B. M. 112-114, 118, 193
McCrone, John 133, 152
McGregor (chimpanzee) 166, 167
McGrew, W. C. 173
McIntyre, A. 152
McKie, Robin 84, 86
meander 103, 104
Mediterranean Sea 24, 64, 68, 125, 133, 136, 137
Megali Vrisi, Thessaly 64
Megaw, J. V. S. 85, 110, 117
Meignen, L. 153, 193
Mellars, Paul C. 24, 27, 48, 86, 87, 118, 122, 130
148, 149, 151-153, 155, 160, 172, 173, 193, 194
Menez-Dregen, France 144
Menozzi, Paoli 27, 87, 150-152, 186, 192
mental modules 167, 168
meric index 181
Mertens, Steven 149
Mesolithic 78, 142, 153, 179-181, 193
Mesopotamia 40, 68, 82, 109
metabolism 88
Mezaros, Mr. 140
Mezin, Russia 111
microscopic 44, 46, 65, 66, 125, 126, 130, 132
middle paleolithic 24, 25, 58, 60, 66, 77, 79, 84-86, 102, 111, 117, 118
131, 132, 140, 148, 149, 151, 153, 154, 162, 172, 173, 193
Miller, J. 118
Milner, Andrew R. 48
Mindel Glaciation 125, 149
mines 114, 137
mining 2, 5, 16, 26, 114, 119, 124, 137-140, 146, 153, 155, 190, 192, 193
missiles 144, 145
Mithen, Steven 51, 59, 62, 77-79, 81, 83, 84, 86
88, 94, 132, 143, 146, 151, 154, 155, 163, 167-173
mitochondria 175, 176, 179
mitochondrial DNA 146, 175-177, 182, 186
Mivart, Sir George . . . .
Mixnitz Cave, Austria 29, 30, 40, 74, 75, 110
moccasins 125
modules 51
molluscs 122
Molodova, Ukraine 121, 122, 143, 190
mongol 81
Monnier, Jean-Laurent 144
Montagu, Ashley 61
Monte Verde, Chile 129, 189
Montespan Cave, France 69
Montet-White, A. 153
Moravia 110, 120, 189
Morris, Desmond 26, 194
Morris, Henry M. 29-34, 36, 40-41, 56, 60, 114, 118
Morris, John 39, 42
mortars 129, 143
Morton, Glenn R. 94, 118, 154, 174, 186
mortuary 77, 79, 80
Mount Carmel, Israel 89, 90, 91, 95, 130
Movius, H. L. 85, 151, 173
Mukuri, Helsen 124
Muller-Beck, H. 149
multitasking 159, 161, 162
Mumford, Lewis 107
Mungo, Australia 82, 135, 189
mushroom 45
music 1, 37, 57, 97, 109-115, 117, 118, 192
musical instrument 24, 39, 57, 109-115, 117, 190
musical scale 109, 113, 118, 190
mutation 55, 175, 177, 179, 182, 186
mutation rate 175, 177, 179, 182, 186

N

Nadel, D. . . . 48, 192, 193
Nahr Ibrahim, Lebanon . . . 82
Napier, John . . . 10, 12
Napoleon . . . 21
Nariokotome, Kenya . . . 18, 55, 92-94
nasion . . . 180, 181
Natron, Tanzania . . . 57, 58
natural history intelligence . . . 51, 146, 168
Nazlet Khater, Egypt . . . 139, 140
Neanderthals . . . 1, 2, 4, 13-16, 18-27, 29, 30, 34-36, 38, 39, 40, 42
. . . 47, 48, 55-57, 59, 60, 64-67, 69, 71-79, 81-87
. . . 137-144, 146-153, 155, 156, 158-164, 169-172, 174
. . . 97-99, 103, 104, 106, 107, 111-118, 120-126, 129-133
. . . 176, 179-182, 186-188, 190, 193, 194
Nebraska . . . 63
Nebraska Man . . . 34
necklaces . . . 106, 107, 170
needles . . . 106, 119, 125, 126, 189
Nelson, Byron C. . . . 31, 40
nephesh . . . 38, 60
Nerja, Spain . . . 111
nets . . . 46, 56, 134, 189
Netti, Bruno . . . 109, 114, 117
Neumann, Joachim . . . 1
New Britain . . . 133
New Guinea . . . 23, 52, 129, 133, 154, 175
New Guinean . . . 52, 175
New Ireland . . . 133
Ngandong, Indonesia . . . 18, 135, 183
Niah Cave, Indonesia . . . 135
Nile River . . . 26, 43, 139, 153
Nim Chimpsky . . . 51
Nishiyagi, Japan . . . 46
Noah . . . 130, 131, 171, 179
Noble, William . . . 59, 172
numerals . . . 140, 141

O

O'Brian, Eileen M. . . . 155, 170, 173
Oakley, Kenneth . . . 117, 172-174
Oates, Joan . . . 66, 84, 192
Oberkassel, Germany . . . 146
obsidian . . . 133
occipital bun . . . 20, 22, 182
ochre . . . 66, 77, 78, 82, 97, 107, 108, 114, 124, 135, 139, 140, 170
OH 7 . . . 12
Ohalo, Israel . . . 46, 48, 189, 192, 193
Okinawa . . . 133
Oldowan . . . 108, 146, 162
Oldoway man . . . 10, 11
Olduvai . . . 10-12, 18, 82, 124, 131, 140, 145
Olduvai Gorge, Tanzania . . . 10, 11, 82, 108, 117, 124, 145, 181, 191, 195
Olorgesailie, Kenya . . . 123, 145, 163
Olsen, John W. . . . 147, 155
Olsen, Stanley J. . . . 147, 155
Olson, E. C. . . . 85
Onondaga . . . 142
Opdyke, N. D. . . . 152
open-air sites . . . 47, 79, 101, 102, 121, 132, 139, 151
ornamentation . . . 82, 107
ornaments . . . 97, 107, 116, 117, 170
Orochi . . . 67, 68
oryx . . . 89, 90, 94, 95
Osborn, Henry Fairfield . . . 16
Ostrava Petrkovice, Czechoslovakia . . . 138
Ostyaks . . . 67
Ouyang, Shu . . . 48

P

paddles . . . 128, 133, 189
Pakistan . . . 14, 92, 171, 174
parkas . . . 125
Parker, Gary E. . . . 32, 34, 36, 41
Parsees . . . 76
Passemard . . . 117
Patterson, F. P. . . . 50
Paulissen, E. . . . 153, 193
Paunescu, Mr. . . . 121
pavements . . . 114, 124, 125
Pavlov, Czechoslovakia . . . 110
Pavlov, Moravia . . . 120, 126

Pech de l'Aze, France . . . 104, 121, 143, 190
Pech-Merle, France . . . 111
Pei, W. C. . . . 17, 26, 83, 117, 194
Pekarna Cave . . . 110
Peking man . . . 5, 7, 27, 31, 34-36, 63, 147, 183
Pelcin, Andrew . . . 65, 66, 84
Peleg . . . 39
Peltz, Sergio . . . 65, 66
Perigord Region France . . . 162
personal ornaments . . . 107, 170
Pesse, Netherlands . . . 133, 189
Petershohle Cave, Germany . . . 75
Peterson, Dennis R. . . . 36, 56
Petitto, Laura . . . 50, 51, 59
phalanges . . . 110
Piazzi, Alberto . . . 27, 87, 150-152, 186, 192
pictographs . . . 141
Pierson, James M. . . . 150, 194
Piggott, Stuart . . . 151, 152, 192, 193
pigments . . . 82, 91, 101, 114, 140, 142
Pinker, Steven . . . 49-53, 59, 83, 87
Pitcairn Island . . . 175, 176
Pithecanthropus erectus . . . 4, 5, 14-18, 29, 34
plank, wooden . . . 46, 48, 63, 83, 128, 129, 132, 133, 151, 194
planning . . . 53, 62, 104, 138, 139, 158-163, 172, 173
Plato . . . 2
Pleistocene . . . 24, 26, 83, 86, 102, 112, 113, 117, 118, 124, 131,
. . . 137, 145, 147, 149-154, 169, 174, 177, 180, 192, 194
Pliocene . . . 24
Plouhinec, France . . . 144
pole star . . . 68
polio . . . 166
Polkey, Charles e. . . . 171
pollen . . . 78, 135, 141
Polynesians . . . 63, 133
pontifex . . . 128
population . . . 17, 36, 47, 54, 79, 91, 100, 119, 140, 147
. . . 169, 175, 176, 179, 180, 182, 183, 185, 187
Portugal . . . 101, 111
post holes . . . 110, 120, 123, 130, 190
pottery . . . 43, 97 126, 133, 189
pounders . . . 126
precocial species . . . 93
Predmost . . . 110
Premack, David . . . 49
Prentice, Adam . . . 15
Price, D. M. . . . 152, 154, 193
Price, George McCready . . . 30, 31, 40
Proconsul . . . 11
Prolom, Crimea . . . 114, 118, 142, 143, 154, 190, 193
Pronyatin, Ukraine . . . 143
pseudogene . . . 183, 184
Ptolemy . . . 119
pyramid . . . 77, 137
pyrite . . . 107

Q

Qafzeh, Israel . . . 23, 39, 161, 179-183
Qena, Egypt . . . 139, 140, 153
querns . . . 143, 154
Quneitra, Israel . . . 66, 142, 154

R

Rachel . . . 97
radiocarbon . . . 101, 110, 120, 122, 129, 140, 153
Rainbow Serpent Religion . . . 103
Rak, Yoel . . . 56, 78, 86
Ramm, Bernard . . . 31, 37, 38, 40, 42, 111, 112, 114, 115, 117, 118, 185, 188
Ranov, Vadim A. . . . 26, 95, 174
rattles . . . 111
Reck, Hans . . . 10, 11
Regourdou Cave, France . . . 69, 71, 74, 75, 78, 146, 190
reindeer . . . 81, 106, 110
Relethford, John . . . 28, 83, 171
relics . . . 23, 38, 75, 127
religion 58, 62-64, 66, 67, 69, 72, 73, 76, 80, 82-85, 103, 109, 114, 126, 170, 189, 193
Renfrew, Colin . . . 46, 48, 152, 153, 192
Rensberger, Boyce . . . 95, 126, 150
Rensink, Eelco . . . 172, 173
retia mirabilia . . . 89
rhesus . . . 157
Rimmer, Harry . . . 31

Rink, W. J. 26, 27, 152
Ripiceni-Izvor, Romania 121
Riss glaciation 125, 152
Ritter, Malcom 48
ritual 62, 63, 65, 67-69, 71, 75, 77-82 109, 114, 171, 190
Robinson, George 169
Robinson, John 9
robot 119
Rocky Mountains 30, 140
Rodriguez, Xose Pedro 26, 95, 174
Roe, Derek 47, 61
Roebroeks, Wil 172, 173
Roman 2, 18, 64, 94
Romanelli, Italy 103
Romans 128, 158
Rosas, Antonio 19
Rose, F. D. 169, 171, 172
Ross, Hugh 23, 27, 37-42, 56, 57, 60, 75, 78, 85, 86, 97, 100
           104-106, 111-113, 115-118, 123, 127, 150, 176-178, 186
Roucadour, France 111
Royal Society 7, 11, 124
Ruddman, W. F. 152
Rudwick, Martin S. J. 40
Ryrie, Charles 172

S

Sahara Desert 6, 101, 106
Sahul 133, 154
sails 133
Salmons, Josephine 6
Salzhofen, Austria 110
San Teodoro, Sicily 189
sanctuaries 63, 66, 67, 84, 189, 190, 193
Sangiran, Indonesia 135, 183
sanpan 133
Sansom, Ivan J. 48
Sauer, Carl O. 150
scaffolding 138, 140
scarification 82, 107
Schaaffhausen, Herman 20, 21
Schaller, Susan 49
Schepartz, L. A. 24, 27, 77, 79, 84, 86, 193
Schepers, Gerrit 6
Schick, Kathy D. 46, 48, 59, 122, 124, 127, 149-151, 162, 172, 195
Schmandt-Besserat, Denise 141, 154
Schmerling, Charles 19, 20
Schmid, E. 139, 153
Schoetensack, Otto 18, 19
Schoningen, Germany 131, 132, 161, 170, 190
Schwalbe, Gustav 16
Schwarcz, H. P. 26, 27, 150-152
Schwartz, Jeffrey H. 23, 27
scrapers 58, 123, 125, 126, 129, 132
scythe 44
Selden, Paul A. 48
self-conscious 57
Semaw, S. 154, 195
Semenov 107
Senner, Wayne 154, 192
Sereno, Paul C. 48
sewing 97, 98, 106, 119, 125, 169, 189
shaman 63, 80, 81, 86, 190
shamanism 80
Shanidar Cave, Iraq 20, 72, 78, 163, 164, 170, 190
shark 45, 48
shellfish 144, 190
Shepard, R. 153, 155, 192, 193
Sherratt, Andrew 47, 60, 61, 84, 148-150, 192, 195
Shipman, Pat 12, 15, 16, 25, 26, 40, 41, 55, 59, 60
                         85, 92-96, 165, 172-174, 178, 186, 195
shovel 22, 183
Shreeve, James R. 10, 13, 22, 25-27, 42, 47, 59, 62, 69, 81, 83, 84, 86, 87
                         94, 115,135, 149, 150, 152, 155, 183, 186, 188, 193, 194
shrine 66, 69, 98
Siberia 39, 64, 68, 81, 91, 100, 125, 126, 150, 190, 194
siberians 67, 76, 81, 98, 101
Sicily 136, 163, 189
sickle 44, 189
sign language 49-51
Sillen, A. 154, 194
Sima de los Huesos 79, 80, 144, 170, 174
Singh, G. 152

Singi Talav, India 63
Sioux 142
Siwalik, Pakistan 15
Skhul Cave, Israel 161, 179-183
Smirnov, Yuri 69, 77, 85, 86, 117, 148
Smith, Aaron 158, 171
Smith, Arthur 5, 11
Smith, Fred H. 161, 172
Smith, G. Elliot 5-7, 16, 17
Smith, M. A. 154
Smith, M. M. 48
Smith, M. P. 48
Smuts, Jan 7
social intelligence 51, 167, 168
social module 167
Soffer, Olga 86, 150
Solecki, Ralph 78
Solecki, Rose L. 78, 172, 193
Solo, Indonesia 15, 135, 183
Solomon Islands 133
Solutre 110
Solutrean 110
Somerset, England 46
Sondaar, P. Y. 136, 152, 194
soul 1, 10, 57, 67, 78, 80, 113, 148, 156, 163, 167, 168, 171
Spain 131, 137, 142, 144, 146, 174, 189, 190, 193
Spain 19, 66, 69, 72, 79, 80, 82, 84, 86, 87, 101, 111
spears 2, 46, 48, 58, 61, 62, 66, 69, 72, 77, 85, 127-132, 146, 149, 151,
                  160, 161,168, 170, 172, 173, 190, 194
speech 7, 13, 14, 18, 22, 32, 33 49, 51, 53-56, 59, 93, 191, 195
sperm 175, 177
spirit 23, 29, 38, 39, 57, 63, 68, 74, 76, 77, 80, 100, 117, 127, 148, 156
Spiro Mound 81
Spoor, Fred 174
Spy d'Orneau Cave, Belgium 21
Star Carr, England 128, 133, 189
stegodon 136
Stellenbosch, South Africa 124
Stellenbosch University 7
Sterkfontein, South Africa 8, 9
Stetten mandible 181
Stoneking, Mark 175-177, 186
Stpanchuk, Vadim N. 118, 193
Stranska Skala 143
Strauss, L. G. 142
Stringer, Christopher 22, 23, 26, 27, 48, 84-86, 116, 122
Stringer, Christopher 148, 149, 152, 171, 172, 174, 182, 193
Sulston, John E. 177, 186
Sumerians 2, 109
Sungir, Russia 77, 86, 107, 125, 149, 150
Suprijo, A. 26, 27, 152
swan 107
Swartkrans, South Africa 127, 131, 144, 154, 190, 191, 194
Swaziland 140, 190
sweat 88-92, 94, 95
sweating 89-94
Swisher, C. C. 26, 27, 152
Switzerland 139, 147, 190
symbol 77, 86, 87, 94, 103, 104, 155
syntax 50, 52, 53
Syria 120, 129, 189

T

taboo 11, 98
Tahitian 133
Talgua Cave, Honduras 80
Talikamatan Desert 46
Tallensi 98-100, 115
tally sticks 141-143
Tanum Bohuslan, Sweden 101
Tanzania 9, 10, 58, 131, 145, 191
taphonomy 72, 102
Tarim Basin, China 48
tarsiers 44
Tasmania 169, 170
Tasmanians 146, 169, 170, 173
Tassili n'Ajjer, Sahara 101
Tattersall, Ian 23-27, 41, 80, 83, 86, 111, 112, 117, 149, 194
Taubach, Germany 75
Taung, South Africa 8-9, 17
taurodont teeth 22
Taylor, C.R. 94, 95

Tchernitcheff, General . . . 21
tear ducts . . . 23
technical intelligence . . . 51, 168
temples . . . 1, 43, 64, 75, 127
Templeton, Alan R. . . . 60, 177, 178, 186
tents . . . 119, 124, 127, 190
Terblanche, Gert . . . 8
termites . . . 43, 159
Terra Amata, France . . . 82, 190
Terrace, H. . . . 50, 51, 58
Teshik Tash, Russia . . . 72, 78
Thailand . . . 44
Theime . . . 131
Thery, I. . . . 153, 193
Thieme, Hartmut . . . 61, 131, 149, 151, 161, 173, 194
thinking . . . 10, 19, 49, 57, 58, 72, 123, 134, 165, 169
Thorne, Alan . . . 135, 183
throwing . . . 62, 80, 130, 144-146, 154, 155, 195
thumbs . . . 54
Tierra del Fuegians . . . 126
Timonovka Russia . . . 147
Timor, Indonesia . . . 134, 152
Tobias, Philip . . . 10, 12, 23
Tocharian language . . . 46
tools . . . 10-12, 15, 16, 21, 23-25, 32, 35, 37, 38, 44, 48, 54, 58
. . . 62, 63, 65, 71, 75-77, 80, 83, 97, 98, 100, 103-108, 110-112
. . . 123, 125-133, 136-140, 143, 144, 146, 151, 154, 155
. . . 159-163, 166, 169-173, 189-191, 194, 195
toothpicks . . . 144, 183, 189, 191
Tor Faraj, Jordan . . . 122, 149
Toralba, Spain . . . 82
Toth, Nicholas . . . 46, 48, 59, 122, 124, 127, 149-151, 162, 172, 195
Trinil, Indonesia . . . 15
Trinkaus, Erik . . . 40, 25-27, 59, 60, 93, 178, 186
Trois Freres Cave, France . . . 68, 69, 81, 109
Truk Islander . . . 134
Tuc d'Audoubert Cave . . . 163
Tungus . . . 67, 81
Turing Test . . . 2, 3
Turk, Ivan . . . 118, 193
Turkana . . . 92

U

upper paleolithic . . . 24, 25, 45, 56, 63-66, 69, 81-83, 97, 100-104
. . . 107, 109-112, 114, 117, 124, 128, 129, 133, 138
. . . 141-143, 146, 154, 161, 170, 174, 180-182, 186, 190, 192

V

Vallois, H. . . . 142, 154
Vaquer, Jean . . . 152
Vargha-Khadem, Faraneh . . . 171
Vaufrey, R. . . . 142, 146, 154
Veddahs . . . 156
Velem St. Vid, Hungary . . . 189
Venus figurines . . . 64-66, 100
Vermeersch, P. M. . . . 26, 153, 193
Vernet, J. L. . . . 153, 193
Vertes, Mr. . . . 140
Vertut, Jean . . . 101, 111, 116, 117, 149
Vila, Carles . . . 146, 147, 155
village . . . 39, 49, 58, 67, 120, 123, 132, 139, 148
Villerest . . . 121
Vinca, Yugoslavia . . . 64
Vindija Cave . . . 103
Virchow, Rudolf . . . 14, 16, 21, 30
Vlcek, Emanuel . . . 58, 61, 123, 149, 194
vocabulary . . . 49-51
vocalizations . . . 54
Voguls . . . 67
Volman, Thomas P. . . . 151, 153
von Meyer, H. . . . 20
von Rudolf Feustel . . . 149
vultures . . . 76, 109, 110, 166

W

Wadi Kubbaniya, Egypt . . . 143, 154
Walker, Alan . . . 12, 15, 16, 25, 26, 41, 55, 59, 60, 92-96, 164, 165, 172-174, 195
walking . . . 9, 15, 32, 35, 69, 88, 93, 100, 158, 162
wall . . . 64, 74, 75, 101-104, 120-123, 135, 190
Walton, D. . . . 154, 194
Wang, Yi . . . 48
Washburn, Sherwood . . . 17
Waters, Michael R. . . . 36, 128, 150, 194

Weber, Max . . . 15
wedges . . . 58, 105, 123, 138, 141
Weidenreich, Franz . . . 4, 36, 135, 183
Wernicke's Area . . . 32, 13, 18, 53, 54, 168
West Tofts . . . 107
Wheeler, P. E. . . . 89, 95
whistles . . . 85, 109-111, 113-115, 117, 190
White, Andrew D. . . . 29, 40
White, Randall . . . 99, 107, 116, 117
White, Tim D. . . . 195
Whitfield, Simon . . . 177, 186
Whiting, John R. . . . 152
Whitney, Eli . . . 160
Widiasmoro . . . 26, 27, 152
Wiester, John . . . 38, 40, 42, 56, 57, 60, 62, 164, 172
Wilberforce, Samuel . . . 29
Wilcox, David L. . . . 23, 27, 37, 38, 40-42, 56-58, 60, 75, 77, 85, 97-99, 115, 158, 171
Wildenmannlisloch Cave, Switzerland . . . 64, 85, 190
Wildkirchli Cave, Switzerland . . . 121, 147
Willendra, Australia . . . 189
Wills, Christopher . . . 9, 25, 26, 41, 116, 149, 151, 152, 172, 192
Wilson, A. C. . . . 175-177, 186
Wilson, Daryl . . . 94
Wilson, E. O. . . . 49, 50, 59
Witswatersrand, South Africa . . . 5
wolf . . . 72, 107, 122, 146-148, 155, 168
Wolpoff, Milford . . . 27, 135, 178, 183, 186-188
wolves . . . 146-148
Wonderwork Cave, South Africa . . . 123, 124, 149, 190, 194
wood . . . 18, 43, 44, 48, 58, 62, 74, 81, 98-100, 104, 109, 114,
. . . 119, 120, 123, 127-130, 132, 133, 144, 154, 159-161, 170
Wood, Bernard A. . . . 154, 194, 195
wooden . . . 44, 48, 58, 62, 63, 66, 69, 72, 81, 83, 85, 100, 106, 109, 114,
. . . 127-133, 138, 140, 151, 160, 161, 170, 173, 189, 190, 194
wool . . . 125
wormian bone . . . 183
woven . . . 125, 126, 164
woven cloth . . . 169
Wub-e-Ke-Niew . . . 178
Wurm glaciation . . . 114, 121, 132
Wynn, Thomas . . . 61
Wysong, Randy . . . 41

X

xylophones . . . 111

Y

y chromosome . . . 176-179, 186
Yarim Tepe, Iraq . . . 189
Yuanmou, China . . . 147
Yukaghirs . . . 76
Yurak-Samoyedes . . . 67

Z

Zaire . . . 105
Zambia . . . 46, 128, 130, 140
Zdansky, Otto . . . 16
zebra . . . 168
Zenker, Wolfgang . . . 90, 94, 95
Zeuner, Frederick E. . . . 147, 155
Zhoukoudian Cave, China . . . 7, 16, 17, 18, 31, 63, 107, 140, 147, 183, 190
zigzag motif . . . 103
Zihlman, A. L. . . . 91, 95
Zimmerman, M. R. . . . 60, 164, 172, 195
Zimmerman, Paul A. . . . 41
Zinjanthropus . . . 4, 12
Zitkov, B. . . . 68
Zonnenveld, F. . . . 174
Zoroastrians . . . 76
Zulus . . . 63